A CULTURAL HISTORY OF PLANTS

VOLUME 6

A Cultural History of Plants
General Editors: Annette Giesecke and David J. Mabberley

Volume 1
A Cultural History of Plants in Antiquity
Edited by Annette Giesecke

Volume 2
A Cultural History of Plants in the Post-Classical Era
Edited by Alain Touwaide

Volume 3
A Cultural History of Plants in the Early Modern Era
Edited by Andrew Dalby and Annette Giesecke

Volume 4
A Cultural History of Plants in the Seventeenth and Eighteenth Centuries
Edited by Jennifer Milam

Volume 5
A Cultural History of Plants in the Nineteenth Century
Edited by David J. Mabberley

Volume 6
A Cultural History of Plants in the Modern Era
Edited by Stephen Forbes

A CULTURAL HISTORY OF PLANTS

IN THE MODERN ERA

VOLUME 6

Edited by Stephen Forbes

BLOOMSBURY ACADEMIC
LONDON • NEW YORK • OXFORD • NEW DELHI • SYDNEY

BLOOMSBURY ACADEMIC
Bloomsbury Publishing Plc
50 Bedford Square, London, WC1B 3DP, UK
1385 Broadway, New York, NY 10018, USA
29 Earlsfort Terrace, Dublin 2, Ireland

BLOOMSBURY, BLOOMSBURY ACADEMIC and the Diana logo
are trademarks of Bloomsbury Publishing Plc

First published in Great Britain 2022
Paperback edition published in 2025

Series design: Raven Design
Cover image © VCG Wilson/Corbis/Getty Images

A catalogue record for this book is available from the British Library.

Library of Congress Control Number: 2021932844

ISBN: PB: 978-1-3505-5063-6
Set: 978-1-3505-5064-3
ePUB: 978-1-3502-5942-3
ePDF: 978-1-3502-5941-6

Series: The Cultural History Series

Typeset by Integra Software Services Pvt. Ltd.
Printed and bound in Great Britain

To find out more about our authors and books visit www.bloomsbury.com
and sign up for our newsletters.

CONTENTS

ILLUSTRATIONS

TABLES

SERIES PREFACE

The connectedness of humans to plants is the most fundamental of human relationships. Plants are, and historically have been, sources of food, shelter, bedding, tools, medicine, and, most importantly, the very air we breathe. Plants have inspired awe, a sense of well-being, religious fervor, and acquisitiveness alike. They have been collected, propagated, and mutated, as well as endangered or driven into extinction by human impacts such as global warming, deforestation, fire suppression, and over-grazing. *A Cultural History of Plants* traces the global dependence of human life and civilization on plants from antiquity to the twenty-first century and comprises contributions by experts and scholars in a wide range of fields, including anthropology, archaeology, art history, botany, classics, garden history, history, literature, and environmental studies more broadly. The series consists of six illustrated volumes, each devoted to an examination of plants as grounded in, and shaping, the cultural experiences of a particular historical period. Each of the six volumes, in turn, is structured in the same way, beginning with an introductory chapter that offers a sweeping view of the cultural history of plants in the period in question, followed by chapters on plants as staple foods, plants as luxury foods, trade and exploration, plant technology and science, plants and medicine, plants in (popular) culture, plants as natural ornaments, and the representation of plants. This cohesive structure offers readers the opportunity both to explore a meaningful cross-section of humans' uses of plants in a given period and to trace a particular use—as in medicine, for example—through time from volume to volume. The six volumes comprising *A Cultural History of Plants* are as follows:

Volume 1: *A Cultural History of Plants in Antiquity* (*c.* 10,000 BCE–500 CE)
Volume 2: *A Cultural History of Plants in the Post-Classical Era* (500–1400)
Volume 3: *A Cultural History of Plants in the Early Modern Era* (1400–1650)
Volume 4: *A Cultural History of Plants in the Seventeenth and Eighteenth Centuries* (1650–1800)
Volume 5: *A Cultural History of Plants in the Nineteenth Century* (1800–1920)
Volume 6: *A Cultural History of Plants in the Modern Era* (1920–present).

By way of guidance to our readers, it should be noted that the plant names used in these volumes accord with those in the fourth edition of *Mabberley's Plant-book* (Cambridge University Press, 2017). When they are discussed, individual plants are identified using their common names and, at their first mention in each chapter, with their scientific names: e.g. bay laurel (*Laurus nobilis*). As is recommended for general works such as this, the authorities to whom the scientific names are attributed (e.g. *Laurus nobilis* L., where L. identifies Linnaeus as the identifying authority) have been omitted.

Annette Giesecke and David J. Mabberley,
General Editors

Introduction

Plants and Culture

STEPHEN FORBES

Our hearts beat with the sunlight captured by plants as chemical energy through the miracle of photosynthesis. Plants are the basis of our past, present, and future. The astonishing capacity of plants for innovation in biochemistry has provided our food and medicine; shelter and construction materials; fuel, oils, and waxes; fibers and dyes; resins and solvents; and fragrances and psychoactive drugs as well as fodder for domestic stock. Our intimate relationship with plants has shaped our environment and us throughout history. However, in the modern era, the continuing evolution of industrialization and globalization together with commodification, consumerism, and the digital revolution have profoundly changed this relationship.

There is a story to tell about every plant species, and there are 391,000 known plant species besides perhaps 2.2–3.8 million species of fungi (Antonelli et al. 2019). The relationship between people, plants, and place—the story of "us"—is a complicated, continuously unfolding story. The scope and scale of our relationship with plants limits any encyclopaedic endeavor in this arena. An anecdote here is illustrative: in an interview of growers on raising the perfect melons grown for the Japanese luxury gift market, one farmer, grinning, observes, "I think only about melons—I'm a melon fool" (Bosker 2017). Yet, each strand of that story is manageable in and of itself, and painstaking curation of such individual portraits, carefully layered, may serve as the best introduction to a larger narrative (e.g. Murdock 2019).

Our relationship with plants can, and should, be viewed from the innumerable perspectives that shape us, and us them. There is no single narrative or Big Theory to scaffold this relationship. However, the relationship between plants and people may be categorized to illuminate particular fields and disciplines. Separating plants in nature from those in cultivation is a common starting point. This separation has ancient roots—the so-called Father of Botany, Theophrastus (*c.* 371–*c.* 285 BCE), observed "the fulfillment of [a plant's] nature also comes about when whatever it happens to lack is added through art (*dia technēs*), such as appropriate and abundant nourishment, and the removal of things that would hinder and harm it" (Theophrastus 1976–90 1.16.11: 141; Secord 2015 (trans)). Classicist Benedict Einerson saw Theophrastus separating two "great divisions" for the plant kingdom: spontaneous phenomena belonging either to the plant or to the country, and "phenomena initiated by human art, which either helps the nature of the plant achieve its goal or goes beyond it" (Einarson 1976: xi). However, the line between these two "great divisions" might be less clear-cut—even from Theophrastus' perspective.

Historian Jared Secord suggested that "Theophrastus considered that the efforts of people to cultivate and improve plants qualified as natural: 'cultivation (therapeia)' aimed for the best results in plants, just as nature did" (Secord 2015: 215).

A later refinement of this division came from Roman orator and philosopher Cicero (106–43 BCE), who wrote of what he called a "second nature": "We sow corn, we plant trees, we fertilize the soil by irrigation, we dam the rivers and direct them where we want. In short, by means of our hands we try to create as it were a second nature within the natural world" (Cicero 45 BCE: LX, 271). John Dixon Hunt equated these cultural landscapes, including agriculture, as an alternative to the landscapes of "the natural world" or "first nature" (2000). He then pointed to Jacopo Bonfadio's (1508–50) notion of the garden as a "third nature" where art and nature are melded (Hunt 2000: 32; Beck 2002). A "fourth kind of nature" has been proposed for "[w]oodlands and other succession stages in abandoned urban-industrial sites" lacking horticultural planning or design (Kowarik 2005: 15, 22). In his acclaimed works on urbanism and suburban landscapes, Alan Berger referred to these dystopic new wildernesses illustrating post-industrial abandonment as "drosscapes" (Berger 2006; Hunt 2015: 211–28). Geographer Steve Carver presented an alternative view in co-opting "fourth nature" to represent "re-wilding" as a transition returning to "first nature" within a cycle of nature-culture (Carver 2013: 384).

The nature of our language in our association with plants provides another lens through which to view a cultural history of plants. Languages applied to plants range through "common" and commercial names for plants and plant products, linguistically encoded environmental knowledge (LEEK), botanical Latin as a formal scientific language, to the symbolic role of plants explored in the discipline of semiotics. While all of these languages are fluid, linguist Kevin Harrison has highlighted the significance of environmental linguistics, and the rapid erosion of LEEK and of the plant knowledge it represents during the modern era (2019). Ironically this is occurring at the same time as the project for an encyclopedic scientific catalog of life continues to expand the scientific naming of plants (Harrison 2007).

As already noted, the recruitment of plants as symbols in our lives reflects another form of language and continues as an enduring cultural phenomenon (e.g. De Carvalho 2011). The field of semiotics explores the meaningful use of these symbols as signs and of their relationships, although the field is often viewed as abstruse and arcane (e.g. Johnson 2011). One of the most successful advertising slogans of the modern era underscores the connections between plants and peoples' memories and emotions: "Say it with flowers" (*Creative Review* 2018). A few examplars are illustrative here: the 1921 formal enlistment of poppies for remembrance following the First World War (1914–18) (Imperial War Museum 2014), the sacred lotus as an enduring symbol of purity in Hinduism, Buddhism and other traditions (e.g. Ward 1952), and the public grief expressed with floral tributes on the death of the British Princess Diana in 1997 (Monger and Chandler 1997).

Finally, the philosophy of plants might be considered as a framework. Philosophy professor Michael Marder observed that philosophers have been unwilling to address plants directly:

> ... after the hard-won independence from theologico-philosophical dogma in early modernity, philosophers, for the most part, refrained from problematizing vegetal life, which they entrusted to the care of botanists and, later on, geneticists, ecologists, and microbiologists. The being of plants was no longer question-worthy; it did not present itself as a problem to those who took the time to contemplate it, let alone to those who made immediate us of the fruit or the flower, the root or parts of the tree trunk.
> (Marder 2013:2)

Marder's own work has made a significant contribution in this arena. Nevertheless, the field is a challenging one, and even Marder's introductions to plant philosophy and phytosemiotics can be difficult reading (Marder 2013; 2018). Meanwhile, liberal arts research professor Jeffrey Nealon has explored the progress of the philosophical study of animals and biopower in the twentieth century—a period where plants remained neglected in the humanities (2015: x). Nonetheless, the twenty-first century has seen a greater interest in plants where, he wrote, "... the winding road of plant theory also leads back to more practical, movement- and rights-based treatments of the earth, the environment, and the battle over genetically modified plants or the politics of food production" (xiii).

Any of these approaches can make valuable contributions to exploring the cultural history of plants. However, from a global perspective in the modern era, the focus must be on the obscuration of plants.

THE MODERN ERA: PULLING UP ROOTS

The modern era has seen a substantial decline in the direct connections between people and plants, reflecting accelerating trends of urbanization, industrialization, and consumerism—the latter shifting in meaning during the twentieth century from an emphasis on consumer rights to include economic policies relying on ever-increasing consumption to generate prosperity and "excessive materialism" (Crocker 2016: 3; Swagler 1994: 354–5).

A cultural history of the modern era then has to begin by addressing this disconnection between people and plants. The modern era's population has grown from less than two billion people in 1920 to almost eight billion in 2020 (Roser et al. 2020). Advances in science and technology during this period have, remarkably, allowed the world's agricultural production to accommodate this growth at the same time as improving broader health outcomes. However, while the number of people facing hunger and malnutrition continued to decline in recent decades, this decline has now been reversed (FAO 2019).

The industrialization that has underpinned feeding the world's population has also been accompanied by global political, economic, environmental, social, and cultural change. The ascendancy of neoliberalism and globalization has seen an acceleration of the global exchange of goods and services as well as an exchange of values across the world. Consumerism, branding, and religion have adopted one another's strategies (during a period now, somewhat paradoxically, viewed as one of global desecularization) (Berger 1999; Gauthier et al. 2013: 9; 2018; Pew Research Center 2015). Burgeoning demand has also accelerated CO_2 emissions and driven global warming to the point that it now represents a clear and present danger for humanity (Ripple et al. 2019).

The majority of the world's population was still rural in 1920, although the global move to urban areas was already established. Most people retained a day-to-day connection with plants in the places where they lived. However, in 1920 the population of United States cities exceeded the population of rural areas for the first time. By 2007 half of the world's population lived in cities, and the UN predicts that by 2050 68 percent of the world's population will be urbanized (Ritchie and Roser 2020; UN DESA/Population Division 2018). This move to cities, and the remoteness of urban communities from intensive crop production and natural or semi-natural areas, has profoundly changed peoples' day-to-day connections with living plants.

The impact of industrialization extends to changes to our connection with the harvest from plants. The substitution of plants by coal and oil as resources for the chemical and pharmaceutical industries, together with the rise of ultra-processed foods have further reduced direct engagement between plants and people in the modern era. The first commercial production of a synthetic dye derived from coal tar dates only from 1856. Yet by 1930 only 30 percent of industrial organic chemicals was derived from plants, and by 1960 this proportion had been reduced to less than 1 percent as the petrochemical industry developed cheaper or improved alternatives (Ohlrogge 1999).

While food-processing has ancient origins, the development of ultra-processed foods is essentially a modern-era phenomenon. Carlos Monteiro (who coined the phrase) defined ultra-processed foods as "formulations of ingredients, mostly of exclusive industrial use, that result from a series of industrial processes (hence 'ultra-processed')" (Monteiro et al. 2013: 22). Processes enabling the manufacture of ultra-processed foods involve several steps and different industries. It starts with the fractioning of whole foods into substances that include sugars, oils and fats, proteins, starches, and fiber. These substances are often obtained from a few high-yield plant foods (corn, wheat, soya, cane, or beet) and from puréeing or grinding animal carcasses, usually from intensive livestock farming (with little or no whole foods) (Monteiro et al. 2019: 937). The complexity of this definition of ultra-processed foods illustrates a distinction between ancient traditions of food processing based on whole foods to the denaturing of foods into constituents through complex industrial processing.

Two inflection points for the development of ultra-processed foods serve to illustrate technical and cultural drivers that have influenced the industrialization of food supply and food-processing. From a technical perspective Yoshiyuki Takasaki's 1967 development of a cost-effective industrial process to convert glucose in cornstarch to fructose was a signal in shaping a market in High Fructose Corn Syrup (HFCS). HFCS is now commonly utilized as a substitute for cane sugar and is cheaper in many markets, while, as a liquid, may have certain manufacturing advantages (Warner 2006). The economics of this pricing advantage are complex—in addition to refinement of Takasaki's technology, the rise of HFCS in the US reflects production quotas for domestic sugar, import tariffs on foreign sugar, and subsidies for corn (Dilk & Savaiano 2017). Writer Michael Pollan wryly observed the insinuation of HFCS into Western diets in the last half-century. He warned, "Don't get your fuel from the same place your car does" and observed that "[g]as stations have become processed corn-stations: ethanol outside for your car and high-fructose corn-syrup inside for you" (2008: 192).

The release of Astrofood in 1971 illustrates the integration of technical advances in food processing within cultural settings authorizing the value of ultra-processed foods. Astrofood was engineered as a breakfast cake and claimed to be the nutritional equivalent of "four ounces of orange juice, two strips of bacon, an egg, and a piece of buttered toast." Robert Cotton, ITT's Vice President at Continental Bakery, boasted: "We achieve what nature could not do" (quoted in Logan and Prescott 2017: 1). Astrofood was delivered to certain disadvantaged schools through a partnership between International Telephone and Telegraph Corporation (ITT) and its subsidiary, Continental Bakery, with the USDA. Astrofood signals the development of commercially marketed, collaborative government-industry programming (Logan and Prescott 2017).

In the modern era consumption of ultra-processed foods has accelerated. In Canada their consumption, expressed as a percentage of total purchased calories, has increased from 24.4 percent in 1938 to 54.9 percent in 2001 (Moodie et al. 2013). In Europe, by 2017, the lowest figure was 10.2 percent in Portugal and the highest was 50.7 percent in the United Kingdom (Monteiro et al. 2017). The rise of ultra-processed foods is significant yet is only one element of changes to our relationship with food generally, and plants in particular, in urban areas. Cultural and social historian Adam Shprintzen (2018) chronicled changes in food culture in American cities in the twentieth century, observing the steady decline in the importance of regional geography for food supply, and, in the second half of the century, a decline in the importance of even global geography—these changes reflecting advances in food distribution and storage as well as the impacts of globalization. The linkages between urbanization and food systems have been articulated by researchers Karen Seto and Navin Ramankutty, who endeavored to identify and unravel these, noting that, "... as cities grow larger, the food retail market becomes increasingly dominated by large supermarket chains, and this has secondary effects on traditional retailers, small-scale producers, traditional food brokers, and the entire supply chain" (2016). Such changes further reinforce urban populations' disconnection from plants in their daily lives.

Michael Pollan (b. 1955) criticized the contemporary food industry from ethical, environmental, and health perspectives in his 2008 book *In Defense of Food*. His maxims on our relationship to plants here are memorable: "Eat mostly plants, especially leaves," and "[c]ook, and if you can, plant a garden" (2008: 167, 197). Pollan continues to explore the relationship between people and plants, and remains an effective critic, communicator, and activist.

The result of urbanization and industrialization in the modern era is that people have reduced direct contact with living plants in their immediate environment, and fewer identifiable plant products in their food and medicine. This dissociation from plants (as well as in the constructions humans inhabit, the clothes that they wear, and the utensils they use) may be the root cause of "plant blindness" in the modern era; plants have certainly become less visible to our daily lives. Environmental psychology researchers Mung Balding and Kathryn Williams have explored the notion of "plant blindness" and observed that, "[i]f immersed in a plant-affiliated culture, the individual will experience language and practices that enhance capacity to detect, recall, and value plants" (2016: 1195).

This disconnection from plants is reinforced in other cultural studies. For example, Selin and Pelin Kesebir (2017) have analyzed cultural products as an indicator of people's connection with nature and observed a steady decrease in references to nature in books of fiction, song lyrics, and film storylines since the 1950s. They suggested this decrease is linked to the growth in indoors and virtual recreation options in this period rather than to urbanization per se (see Pergams and Zaradic 2006). The decline in ability to identify British wildlife (plants and animals) by children reinforces this argument. Children in Britain proved better able to recognize sub-sets from a hundred common Pokemon than sub-sets from a hundred common British wildlife species (Balmford et al. 2002).

The connection between plants and people is also demonstrated by people's fluency in identifying plants and their values. Language identifying and describing plants in developed countries has shifted from a vernacular folk taxonomy developed through direct connections with plant life to one largely determined by marketing and branding, by property rights associated with plant genetic resources, and by a restricted authorized scientific taxonomy (e.g. Nursery & Garden Industry of Australia 2013; Turland et al. 2018; World Intellectual Property Organization 2016). Such regulated classifications are required to regulate global trade and supplant folk taxonomies that categorize plants according to "artificial" systems based on cultural traditions. Sarah Pilgrim and colleagues elegantly demonstrated the inverse correlation between plant knowledge and income levels in, and among, people in India, Indonesia, and the United Kingdom (Pilgrim et al. 2008: 1005). The measures of ecological knowledge were based on the identification of local plant species, the number of uses of them, and the difference in plant knowledge between the most and least knowledgeable members of a community (knowledge variance). Language then might both provide an indicator for the disconnection between people and plants, and the potential for communities to interpret the foundations that plant taxa contribute to livelihoods, well-being, and sustainability.

This disconnection between people and plants underlies the limited effectiveness of communities, institutions, and governments to protect plant life and, ultimately, the global ecosystem services that sustain food production, regulate climate and disease, support water and nutrient cycles as well as delivering cultural and recreational values. The evidence for this destruction is overwhelming, and while the direct causes are apparent, the root causes are complex. Failure to invest in conservation may reflect a paucity in ecological understanding and limited local community support, even in "wealthy nations [with] the financial resources to invest in conservation efforts" (Pilgrim et al. 2008: 1008). Even with local community support, decision-making for resource management remains complex. Arun Agrawal and Clark Gibson, political scientists and environmental scholars, have explored the disillusionment of conservationists with both the state and the market. They noted the valorization of a "mythic community" as an effective agent for change: such communities—described by small size, territoriality, group homogeneity, and shared understandings and identities—are rarely the reality. Indeed, there remain challenges for even defining "community" in the context of conservation, and, even where this is possible, communities have limited capacity to "withstand external threats (even from other community groups competing for access to the same resources), or manage resources that have a wide geographical spread." In such settings they suggested a focus on institutions, rather than on communities, in furthering conservation (Agrawal and Gibson 1999: 640).

ENVIRONMENTALISM: GROWING NEW ROOTS

While a general decline in societies' connections with plants can be traced through urbanization, industrialization, and language, there is another side to our relationship with plants in the modern era. Research relating to plants, and the practical application of this in plant technology, cultivation, and production, as well as in developing capabilities for ecosystem reconstruction have—at the same time as the disconnection between people and plants more generally has accelerated—been the subject of remarkable advances. Specialists have explored new frontiers in crop selection and applied gene technology to increase yields and, at the same time, to protect crops from environmental stress and from competition. Such advances have been critical to reducing hunger globally.

Science provides a powerful lens for understanding plants, as well as influencing our relationship with them. However, our relationship with plants extends well beyond science to intersect with every element of our lives. Unsurprisingly there is no accepted overarching framework for this relationship. Biologist Edward O. Wilson's (b. 1929) exposition of the relationship between humans and other forms of life published in his 1984 book *Biophilia* has been influential in this arena. Wilson suggested "that the urge to affiliate with other forms of life is to some degree innate, [and] hence deserves to be called biophilia" (1984: 85). His thesis that "to explore and affiliate with life is a deep and complicated process in mental development" has catalyzed further research in this field

(1). In a recent review environmental scholars Daniel Baxter and Luc Pelletier explored Wilson's biophilia hypothesis and assessed whether available literature supports his view that nature-relatedness meets fundamental criteria for a human psychological need. Their conclusion "supports the claim that human beings have a basic psychological need to feel a secure and pleasant experiential connection to nature in a cognitive, emotional, and physical sense" (2019: 30). Psychologist and socio-culturalist Peter Kahn Jr. concurred, arguing that the biophilia hypothesis does indeed provide "an overarching framework by which new scientific ground across many disciplines can be charted that bear on understanding the human relationship with nature" (1997: 2).

Wilson's biophilia underpins the notion of "Nature-Deficit Disorder®," introduced by writer Richard Louv (b. 1949) in *Last Child in the Woods: Saving our Children from Nature Deficit Disorder* and further developed in *The Nature Principle: Reconnecting with Life in a Virtual Age* (Louv 2005, 2012). While a powerful trope that has been influential in public discourse and environmental education, this hypothesis is confounded. Eco-culturalist Elizabeth Dickinson "identified three ways Louv and educators similarly present human-nature relationships and their underlying cultural assumptions. In both, nature is constructed: (a) as something from which children recently have fallen and need to return; (b) through a natural science lens; and (c) through naming practices." She argued, "incorporating techniques that involve communicating connectedness are essential yet they are largely absent in contemporary environmental discourses" (Dickinson 2013: 6, 15). Environmental anthropologist Robert Fletcher has agreed that framing the separation of culture from nature (the "Raw and the Cooked" of Claude Levi-Strauss) in Louv's rhetoric in fact undermines environmental education. He argued for "developing forms of environmental education that problematize rather than reinforce both the constructed opposition between nature and human culture and the (neoliberal capitalist) political-economic structures exacerbating the environmental degradation that these perspectives seek to combat" (Fletcher 2017: 228–32). Both essentially viewed so-called Nature-Deficit Disorder as a symptom rather than the underlying pathology of the problem.

There is good evidence for the value of connecting with nature; forest walkers may indeed "fulfil a psychological need to feel a secure and pleasant experiential connection to nature in a cognitive, emotional, and physical sense." At the same time, as a result of inhaling the forest atmosphere, such walkers also accumulate volatile organic compounds from the forest in their blood. Scientist Kazuhiro Sumitomo and his colleagues have demonstrated the accumulation of monoterpenes (organic compounds) in the blood of walkers in a Japanese conifer forest, and posited that one of these (α-pinene) demonstrates anticholinesterase activity and may provide a benefit against the progression of Alzheimer's disease (Sumitomo et al. 2015). Such research underscores the complexity of our relationship to plants in terms of our physiological, emotional, and spiritual well-being. A broader review of the connections between nature and human health and well-being concluded that "More robust analyses of relationships between human health, nature, and biodiversity remain as key gaps in ecological and medical research, especially regarding mechanisms of causation. Sufficient observational and correlational evidence now exists to support the basic premise of a wide range of health benefits, but for the most part how these benefits are mediated remains unknown" (Sandifer et al. 2015: 10).

The origins of environmentalism as a movement are variously attributed. Enlightenment science, supported by Romantic philosophies, provided the foundation for the nature conservation and environmental movements of the nineteenth and twentieth centuries

(Grove 1995). The environmental degradation resulting from the colonial empires' rampant exploitation of economic resources elicited a response from "a coterie of men—and some women—predisposed to rigorous analytic thinking about the processes of ecological change and the need for land control" (Grove 1992: 43). Individuals credited with influencing community opinion for the value of nature and natural areas in the nineteenth century include John Ruskin (1819–1900), William Morris (1834–96), and Octavia Hill (1838–1912) in Britain; and Henry Thoreau (1817–62) and John Muir (1838–1914) in the United States.

In Britain broader considerations shaped an environmentalism that endeavored to address the threats from industrialization to natural landscapes, open spaces, wildlife, and antiquities. Philip Lowe cataloged the progenitors of modern environmental conservation movements in Britain as outlined here (1989: 114):

1863 Society for Checking the Abuses of Public Advertising
1865 Commons Preservation Society (now the Open Spaces Society)
1877 Society for the Protection of Ancient Buildings
1884 National Footpaths Preservation Society
1885 Selbourne Society for the Protection of Birds, Plants and Pleasant Places
1889 Society for the Protection of Birds (now the Royal Society for the Protection of Birds)
1894 National Trust for Places of Historic Interest and Natural Beauty
1903 Society for the Preservation of the Wild Fauna of the Empire
1905 British Empire Naturalists' Association (now the British Naturalists' Association)
1912 Society for the Promotion of Nature Reserves (now the Royal Society for Nature Conservation)

This abridged catalog illustrates a broader context of concerns for the impacts of industrialization on people and on their culture, society, and environment. This is especially well illustrated in the establishment of the Kyrle Society, also known as the Society for the Diffusion of Beauty, by Miranda and Octavia Hill in 1875. The Society's 1910 annual report noted the importance for

> ... less fortunate members of the community whose lives of hard toil and little recreation give them but few opportunities of enjoying the better influences of life (including) Natural beauty, the beauty of trees and plants and flowers of distant view, of lake and river—these above all things help towards that *mens sana in corpore sano* which is the ambition of everyone for himself and others.
>
> (quoted in Whelan 2016: 97)

Concerns for the Commons and heritage were paralleled in the United States by concerns for the future of America's landscape heritage. The influence of Thoreau's and Muir's philosophies, and the grandeur of America's landscapes, resulted in early conservation milestones. In 1864 President Lincoln signed legislation protecting the Mariposa Grove and Yosemite Valley for "public use, resort, and recreation." Yellowstone National Park was dedicated by President Ulysses S. Grant in 1872 in the light of Ferdinand Vandeveer Hayden's (1829–87) exploratory report. Early conservation societies include the Audubon Society founded in 1886 (Library of Congress 2002).

The modern era that followed the establishment of these movements is characterized by major political, economic, social, and environmental catastrophes including two world wars (1914–18 and 1939–45), the Great Depression (1929–39) and, in the United

States, the Dust Bowl (1931–8). These events had implications for the environmental movement. However, landscape scale conservation advocates, including, most famously, Aldo Leopold (1887–1948), were able to establish the Wilderness Society in 1935 as an "organization of spirited people who will fight for the freedom and preservation of the wilderness" (Wilderness Society 2019).

Aldo Leopold's *A Sand County Almanac*, published posthumously in 1949, argued eloquently for a "land ethic" and has been influential in establishing a defining philosophy for conservation: "The land ethic simply enlarges the boundaries of the community to include soils, waters, plants, and animals, or collectively: the land" (Leopold 1949: 204; Callicott 1993). After the Second World War environmentalism was able again to gain traction, the Ecologists' Union, established by American ecologists in 1946, led to the incorporation of The Nature Conservancy (TNC) in 1951 (Nature Conservancy 2019). TNC is now the world's largest nature conservation charity with over a million members worldwide and an annual income over $1 billion (Barrett 2018).

Rachel Carson's (1907–64) *Silent Spring*, published in 1962, provided a landmark for environmentalism in the modern era and is often regarded as the most important environmental book ever written; renowned writer Margaret Atwood (born 1939) even canonized Carson in her 2009 novel *The Year of the Flood* (Atwood 2012). Scholar and rhetorician Mollie Murphy quoted Carson's biographer, Linda Lear: "Carson ... attacked the integrity of the scientific establishment, its moral leadership, and its direction of society" and saw *Silent Spring* as "... not merely a warning against the effects of pesticides; it was a direct threat to a fast-growing American ideology" (2019: 194). Carson's attack on the indiscriminate use of pesticides such as DDT to control insects is most commonly remembered, although her attack on the indiscriminate use of herbicides and advocacy for plants is also clear in chapter 6, "Earth's Green Mantle." She began this chapter with a comment on our relationship with plants,

> Although modern man seldom remembers the fact, he could not exist without the plants that harness the sun's energy and manufacture the basic foodstuffs he depends upon for life. Our attitude toward plants is a singularly narrow one. If we see any immediate utility in a plant we foster it. If for any reason we find its presence undesirable or merely a matter of indifference, we may condemn it to destruction forthwith. Besides the various plants that are poisonous to man or his livestock, or crowd out food plants, many are marked for destruction merely because, according to our narrow view, they happen to be in the wrong place at the wrong time. Many others are destroyed merely because they happen to be associates of the unwanted plants.
>
> (Carson, Darling, and Darling 1962: 63)

Carson's influence is considered the foundation for the establishment of the US Government's Environmental Protection Agency in 1970 (Lewis 1985). The Natural Resources Defense Council, a charity set up to ensure the rights of all people to clean air, clean water, and healthy communities, was established in the US in the same year.

While a holistic view of the environment is critical for conservation, specific challenges in biodiversity conservation saw increasing specialization (or perhaps atomization) across environmentalism. The establishment of the International Union for the Protection of Nature in 1948 (renamed as the International Union for the Conservation of Nature, IUCN, in 1956) represented a milestone for global biodiversity conservation efforts (IUCN 2019). The first IUCN Red List for Threatened Species, published in 1964, was for mammals and birds, while the first comprehensive Red List for plants only followed thirty-four years later, in 1998 (Walter and Gillett 1998). This milestone was completed at the World Conservation Monitoring Centre (WCMC) at Cambridge, England. When the CMC had moved from the Royal Botanic Gardens, Kew to Cambridge to become the WCMC there was a zoological team of sixty-five—and one botanist (Walter pers. comm. 2019).

While plants provide the foundation for both landscape and fauna conservation, the titles of early conservation societies illustrate that a commitment to plant conservation lagged behind that for animals. For example, the Society for the Preservation of the Wild Fauna of the Empire was founded in 1903 (sometimes referred to at that time as the "Penitent Butchers") and only explicitly acknowledged flora in its name when it became the Fauna and Flora Preservation Society in 1981. The Society was renamed in 2001 as Fauna and Flora International (UIA 2021). The gulf in membership between plant- and animal-focused conservation organizations continued into the twenty-first century (Heywood 2017: 323). Plantlife International, established in Britain in 1989, claims eleven thousand members and supporters (Plantlife International 2019). By contrast, Bird Life International, founded in the United States as the International Committee for Bird Preservation (ICBP) in 1922, claims ten million members and supporters (Bird Life International 2019). The relative neglect of plants in conservation science echoes their neglect in the humanities (Marder 2013).

PART OF THE LIVING WORLD

While environmentalism has progressed in the modern era, this era is referred to, in part or wholly, as the Anthropocene (Laurance 2019). Human environmental impact is now apparent on a planetary scale with climate change the most significant immediate threat to human well-being. The range of challenges to our long-term well-being and survival are effectively articulated in the United Nations Sustainable Development Goals (SDGs) (UN 2015, 2019). The seventeen SDGs require people to collaborate effectively—two of the SDGs focus on governance—no. 16 Peace, Justice, and Strong Institutions and no. 17 Partnerships for the Goals. These two may be both the most critical and difficult of the SDG challenges. However, the SDGs also require people to collaborate effectively with plants. Plants underpin sustainability across the majority of the goals through their contribution to food production, ecosystem management, water conservation, and climate mitigation (Antonelli et al. 2019: 1101).

The reconciliation of people and plants requires a holistic view of the environment and of culture, society, and the economy. Holistic perspectives of the Earth's biotic and abiotic composition are apparent across written history and cultural traditions. *Oeconomia Naturae* of Carl Linnaeus (1707–78) and the Gaia theory of James Lovelock (b. 1919) are examples from different periods that embrace both a scientific and a spiritual perspective of our global environment (Bilberg and Linnaeus 1749; Lovelock 1979). Such doctrines acknowledge plants as the foundations of life on Earth. Plants have underpinned our past and present food, water, and climate security, while plant products have underpinned physical security and well-being. Plants will inevitably define our future.

The nature of our spiritual relationship with our environment is critical here. The seminal paper, Lynn White Jr.'s (1907–87) "The Historical Roots of Our Ecologic Crisis," remains relevant in identifying the nature of ideological and cultural factors underlying the environmental crises characterizing the modern era (White 1967). In editing a volume of essays on the Lynn White Thesis at fifty, Todd Le Vasseur and Ann Peterson summarized his key claims: "First, he argues that ideological and cultural factors, especially religion, are the root causes of the 'ecologic crisis' facing contemporary humans. Second, he identifies Western Christianity as particularly influential in creating environmentally destructive attitudes. Third, he suggests that, just as the fundamental causes of ecological destruction are religious, so too must their solution be religious" (2016: 2). Indeed, environmental ethicist Dale Jamieson pointed to White's clear assertion that "the environmental crisis is fundamentally a spiritual and religious crisis, and that its ultimate solution would have to be spiritual and religious" (Jamieson 2008: 20). Christianity may well be linked to the modern era's environmental crisis. However, the other major religious traditions appear to have fared no better in this regard. Pope Francis' 2013 encyclical *Laudato si'* endeavored to directly address this crisis (Francis 2015).

The acknowledgment that humans cannot be separated from nature is critical to environmental reconciliation. Rachel Carson's observation is eloquent in this regard: "So nature does indeed need protection from man, but man too needs protection from his own acts, for he is part of the living world. *His war against nature is inevitably a war against himself*" (Carson 1962).

The instrumental value of plants to provisioning humans is largely taken for granted. However, the value of growing plants as direct interventions for individual and global environmental reconciliation as well as for their instrumental value has been well articulated in the modern era. The Trillion Tree Campaign and Wendell Berry's social criticism, environmental activism, and writing illustrate two compatible but divergent perspectives on these issues.

As a global endeavor, the Trillion Tree Campaign (TTC) acknowledges the role of trees as a Nature-Based Solution (NBS) to lock up carbon in the world's forests, grasslands, and wetlands. TTC is a World Economic Forum initiative supporting the UN Decade on Environmental Restoration 2021–30 and is led by UNEP and FAO. Science writer Ben Guarino (2020) outlined some of the precursors for the TTC, influenced by the powerful example of Kenyan Nobel Laureate Wangari Maathai (1940–2011), who in turn inspired TTC champion Felix Finkbeiner (b. 1997). He included a quote, "There is a saying in forestry: It is not about trees, it is about people. If people will find trees important, then they will look after the trees." Effective advocacy for trees is critical for activating Nature Based Solutions (NBS) for climate change. The TTC observes that the "Latest science suggests that NBS can provide up to one-third of the emissions reductions required by 2030 to meet the Paris Agreement targets."

At a local level, the agrarianism of American farmer, writer, and social critic Wendell Berry (b. 1934) speaks to both the utility and beauty of plants, to their influence on us, and perhaps even to our connection with the Divine. His writing and actions through the last fifty years of the modern era have been significant to changing perspectives of our relationship with plants and with one other. In his influential 1969 essay "Think Little," reprinted in *The Last Whole Earth Catalog* in 1970, he argued for the value of growing food (a project that can be made inclusive through community gardens and containerized gardens). His advice is straightforward:

> Odd as I am sure it will appear to some, I can think of no better form of personal involvement in the cure of the environment than that of gardening. A person who is growing a garden, if he is growing it organically, is improving a piece of the world. He is producing something to eat, which makes him somewhat independent of the grocery business, but he is also enlarging, for himself, the meaning of food and the pleasure of eating.
>
> (2002: 88)

CHAPTER ONE

Plants as Staple Foods

Feast and Famine in Global Food Systems

DEREK BYERLEE

INTRODUCTION

Where once humankind depended on wild harvests of edible plants, fruits, seeds, animals, and fish, today cultivated crops provide nearly all the world's food supply. Even livestock are now mostly fed from cultivated grains and forages.[1] Although there are thousands of plants that provide edible foods, only 150–200 cultivated crops are used by humans.[2] A subset of these can be classified as global food staples defined as foods consumed on a regular basis to provide the bulk of the world's dietary energy, protein, and fats. Strikingly, a handful of only six crops provide the bulk of these three nutrients in global diets. Four of them are from the *Gramineae* or grass family—rice (*Oryza sativa*), wheat (*Triticum aestivum*), maize (*Zea mays*), and sugarcane (*Saccharum officinarum*).[3] Soybeans (*Glycine max)* is from the *Fabaceae* or legume family, and oil palm (*Elaeis guineensis*) is from the *Arecaceae* or palm family. Figure 1.1 shows the origins of these six crops along with other food staples that are often locally important, and Table 1.1 provides more detail on the production and nutrient composition of their main products.[4]

The overwhelming importance of these six crops in feeding humanity is evident in Table 1.2. They produce between half and two-thirds of the world's supply of energy, protein, and fats from cultivated plants. However, not all these nutrients are consumed directly as foods, since major shares are fed to animals and some are not eaten at all but used for seed or biofuels, or they are wasted before being eaten. Using estimates by the United Nations Food and Agricultural Organization (FAO) of food uses of these crops and making some simple assumptions about conversion into animal products of crops used for feed, Table 1.2 also provides an estimate of the role of these crops in human food supply. Again, the predominance of the "big six" is overwhelming with a sharp drop to the next group of important food staples such as potatoes, sorghum, and rapeseed. The three cereals and sugarcane provide the bulk of the calories, and palm oil and soybean oil are the most consumed vegetable oils in global diets. In addition, maize and soybeans provide about 70 percent of the composition of prepared animal feeds and, together with sugarcane and palm oil, over 90 percent of the feedstuffs for biofuels.

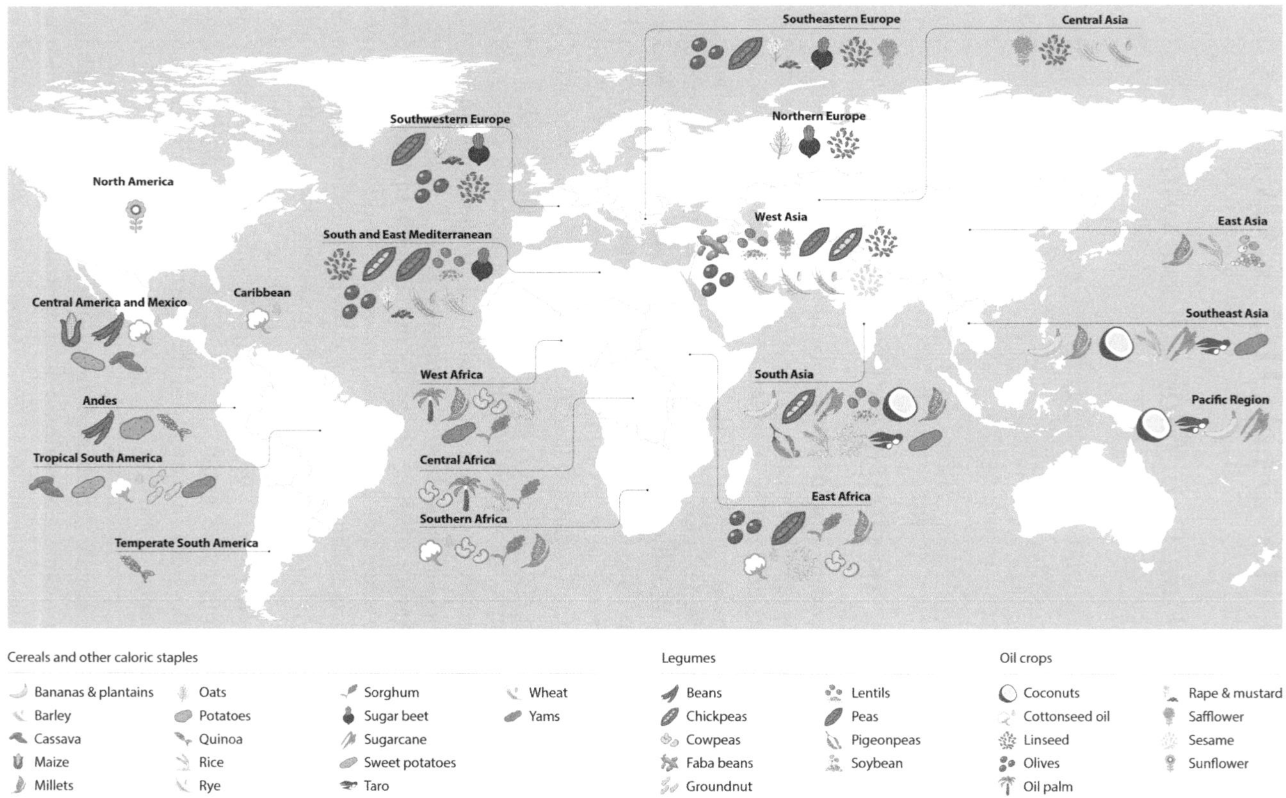

FIGURE 1.1 Origins and primary regions of diversity of staple food crops. Adapted from Khoury et al. 2016. Note: Cereals and other caloric staples are the major source of calories and are also important for protein supply. Legumes are important sources for protein supply and some (groundnuts and soybean) also provide significant amounts of edible oils.

TABLE 1.1 Scientific names, origin and adaptation, production, and nutritive values of the top six staples

	Rice	Wheat	Maize	Sugarcane	Soybeans	Oil palm
Family	*Gramineae*	*Gramineae*	*Gramineae*	*Gramineae*	*Fabaceae*	*Arecaceae*
Genus and main species	*Oryza sativa*	*Triticum aestivum*	*Zea mays*	*Saccharum officinarum*	*Glycine max*	*Elaeis guineensis*
Origin	Asia	West Asia	Mexico and Central America	New Guinea and Southeast Asia	North China	West and Central Africa
Climatic adaptation[a]	Mostly tropical and sub-tropical	Temperate to sub-tropical	Warm temperate to tropical	Tropical and sub-tropical	Warm temperate to tropical	Tropical
Calories / gram	3.7[b]	2.9	3.0	3.2[c]	3.4	8.7
Gram protein / gram	.07[b]	.09	.07	0[c]	.32	0
Gram fat / gram	.01[b]	.01	.03	0[c]	.22	1.00
Annual global production, 2010–14 (Mt)	487[b]	690	933	182[c]	271	52[d]
Average yield (t/ha)	3.0[b]	3.2	5.3	7.0[c]	2.5	3.0

a. Broadly defined by latitude north or south of the equator—temperate > 35°, subtropical 23° to 35° and tropical < 23°
b. Milled rice at 67 percent of rice paddy
c. Refers to sugar assuming 10 percent milling rate of sugarcane to sugar. Includes sugarcane produced for ethanol. Excludes sugar from sugar beet that supplies about 20 percent of the total market including for biofuels
d. Production of palm oil and palm kernel oil
Source: Computed from FAOSTAT at http://www.fao.org (accessed December 1, 2017).

This chapter reviews changes since 1920 in agriculture and food systems and cultures through the lens of these six staples. The changes in food systems during the century were profound and historically unprecedented, driven by three major forces on the demand side—rapid population growth, rising incomes and urbanization, and globalization. In 1920, the world fed about 1.9 billion people, mostly very inadequately, with regular occurrence of famines in Asia. After a series of major famines in India in

TABLE 1.2 World production of major nutrients from crops and amount used for food

	Calories	Protein (g)	Fat (g)
Total production from crops per day per person[a]	5,081	141	97
Share by crop			
• Rice	0.13	0.09	0.02
• Wheat	0.15	0.17	0.04
• Maize	0.22	0.19	0.10
• Soybeans	0.07	0.24	0.23
• Palm oil	0.04	0.00	0.23
• Sugarcane	0.05	-	-
Total share from six crops	**0.66**	**0.69**	**0.62**
Food supply only per day per person[b]			
All crops	2,328	49	41
All livestock	482	32	34
Total food supply from crops and livestock	2,810	81	75
Share food supply from six crops + livestock fed with maize and soybeans[c]	**0.62**	**0.50**	**0.46**

a. Total calories, protein, and fats produced by all crops (excluding fiber crops)
b. Food supply after subtracting feed, seed, waste, and industrial uses as estimated by FAOSTAT
c. Includes estimated share of production of beef, milk, pork, and poultry from maize and soybeans based on Mottet et al. 2017 and Byerlee et al. 2017.

the late nineteenth century in which tens of millions died, a British agricultural chemist, John Voelcker, traveled throughout India and prepared a voluminous report in 1893 that recommended stepping up investment in public institutions and infrastructure serving the food system to prevent future famines. Although many of these reforms were enacted in India and elsewhere and did reduce the incidence of famine, population growth in the twentieth century was relentless. In the first decade of the twentieth century, world population increased by about 10 million annually. With improvements in health and sanitation, this doubled to 21 million in the 1930s, more than doubled again to 50 million in the 1950s before peaking at 83 million in the 1980s (Figure 1.2). Just to keep up with population growth was a daunting challenge for the world food system, let alone making sufficient progress to reduce the incidence of hunger that affected as much as half of the world's population in the mid-twentieth century (Sukatme 1961).

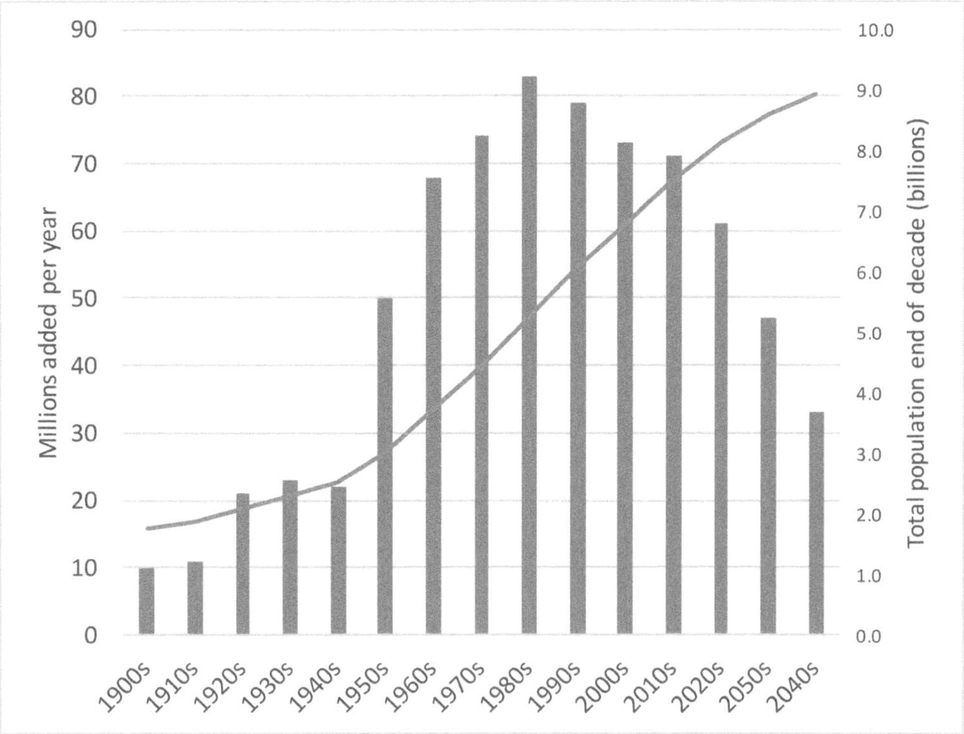

FIGURE 1.2 Annual population increase by decade represented by bars (left axis) and total world population at decade end represented by the solid line (right axis), actual 1900–2020, projected 2020–50. Calculated from FAOSTAT.

On top of the population explosion, incomes increased and the world steadily urbanized. This happened slowly in the first half of the century, a period of two world wars interspersed by the global economic depression, but then accelerated sharply in the second half of the century, led by North America, Europe, and Japan, followed by much of the rest of the world. In 1900, only 15 percent of the world's population lived in cities but by 2008, half of the world's population were living in cities. Rising incomes and urbanization have had profound impacts on farming and rural communities and on dietary patterns.

The third major force in the food systems was globalization. The first "golden era" of globalization based on free trade and improved steam transport and communications bloomed in the second half of the nineteenth century but was halted by the First World War followed by the economic depression of the 1930s that led many countries to raise import barriers. However, in the second half of the twentieth century, globalization in the form of better air and sea transport, gradual liberalization of trade, international migration, tourism and cultural exchanges, and the internet accelerated dramatically. Agriculture and food systems were deeply affected by these developments through increased trade, the rise of multinational food companies, homogenization of diets, and the rise of ethnic foods.

After briefly introducing each of the six dominant staples, this chapter reviews the enormous changes in farming from the dimensions of technological change, the organization of farming, and conservation of biodiversity on and off the farm. The chapter then turns to dimensions of changes in food consumption starting with trends in under-nutrition combined with the recent upsurge in obesity and its health-related issues, and the globalization of diets through trade, big business, and tastes. The final section concludes by noting the challenges and difficult trade-offs that today's food system actors must grapple with in the twenty-first century as the global community searches for sustainable and healthy solutions to feeding ten billion people by 2050.

SIX GLOBAL STAPLES IN A "NUTSHELL"

Wheat

Wheat is probably the oldest and most widely consumed staple in the world, particularly in temperate countries. The dominant species is *Triticum aestivum,* commonly known as bread wheat since it is mostly used for making leavened and unleavened bread. *Triticum durum,* commonly known as durum wheat and mostly used for making pastas and couscous, accounts for about 5 percent of today's wheat supply. Both species originated in the Fertile Crescent of the Middle East from ancestral wheat species (Harlan 1992) and then spread west to Europe and North Africa and east to South Asia and China, to become a major food staple in those regions. From the sixteenth century wheat moved with European settlers to temperate areas of the Americas and Australasia to become a staple of temperate parts of the New World, as well as a major export crop from the New World to the Old World.

Wheat is mostly grown under dryland conditions, but in some important producers, especially China, Egypt, India, and Pakistan, it is grown largely under irrigation. Although wheat cannot be successfully grown in the humid tropics, bread based on imported wheat is now widely consumed in tropical countries. For most of the twentieth century, wheat was the most important agricultural commodity in world trade, supplied by the US, Canada, Argentina, and Australia. However, just prior to the First World War, Russia and Ukraine (then part of Russia) were the most important exporters of wheat but in the second half of the twentieth century they became major importers, before again reclaiming first place in world exports in 2015.

Rice

Rice is the dominant food staple of most of Asia and an important staple in parts of Africa and Latin America. Globally it contributes equally to wheat in supplying dietary energy. Nearly all rice produced is *Oryza sativa,* which is further distinguished by *indica* types grown in the tropics and subtropics, and *japonica* types grown largely in temperate areas. In both types, there are many variations in preferences for grain type, fragrance, and stickiness. *O. sativa* is thought to have been first domesticated in southern China. In West Africa, another species, *O. glaberrima,* was domesticated and remains locally important.

About three-quarters of rice is produced under irrigated conditions (Fischer et al. 2014). The need for good water control and maintenance of irrigation systems has been associated with cultures that encourage more community cohesion and collective action compared to the more individualistic wheat cultures (Talhem et al. 2014). In much of Asia religion and traditional festivities until today revolve around rice farming.

Rice is consumed in many forms but mostly it is milled to remove the hull (about one-third of the weight) and then boiled to serve as white rice. Rice consumers typically have strong local preferences for grain qualities. Although widely traded, the volume of trade is small relative to total rice production and relative to wheat trade. In recent years, scented rices, such as basmati and jasmine rice, that may sell for twice the price of normal rice have been gaining market share.

Maize

Maize (*Zea mays*) is the only new world crop in our top six staples. It originated in Mexico and Central America and became the dominant food staple of the Americas before spreading to the rest of the world from the sixteenth century in what is known as the "Colombian Exchange" of crops. It became a major food staple in the eighteenth and nineteenth century in parts of southeastern Europe and the foothills of the Himalayas. In eastern and southern Africa, maize expanded rapidly after the First World War to replace traditional millets and sorghum and supply emerging towns and mining centers. Today, maize in much of eastern and southern Africa dominates dietary energy intake in the same way as in its home in Mexico and Central America. Many food cultures are built around maize as for example in Malawi where "Chimanga ndi moyo" or "maize is life" (Smale 1995). Among the indigenous groups of Mexico maize deities are part of mythology and religion and the Mayans believe that humankind was created out of maize.

Over the twentieth century, with the rise of intensive livestock operations, maize in the US (where it is known as "corn") became the dominant livestock feed, followed after the Second World War by Europe and Japan who imported much of their maize for livestock. With rising incomes, Latin American and Asian consumers joined this livestock revolution in the late part of the century. A further development was the rise of the ethanol industry in the USA in the early 2000s as part of efforts to replace fossil fuels. Today the US feeds its ethanol plants about three times more maize than the maize consumed as food in all of Africa.

With these diverse uses, maize is now the most important crop in the world in terms of tonnage and energy production—in 2014, world production exceeded one billion tons for the first time. It is grown widely from the equator to about 50° latitude, mostly under rainfed conditions in areas of good moisture supply. White maize is mostly used for food by making it into a flat bread (e.g. a tortilla) or a porridge, but there are many other food uses as well. One of the most controversial is the conversion of maize into high fructose corn syrup that is widely used in the processed food and beverage industry, accounting for up to half of consumption of sweeteners in the USA in 1990. Perversely, yellow maize that is high in pro-Vitamin A, a vitamin that is deficient in much of the world's diet, is mostly used for animal feed.

Sugarcane

Sugarcane (*Saccharum officinarum*) is generally not thought of as a staple crop. However, given its widespread regular consumption as sugar throughout the world and the fact that it accounts for nearly 10 percent of calories produced by plants, it fits many elements that define a staple. The main difference is that sugar is not consumed alone but as an additive to many foods and beverages. It is also ingested directly as sucrose rather than starch so that it is rapidly absorbed into the bloodstream and does not give a feeling of satiation provided by the cereal staples.

Sugarcane is a perennial crop that is believed to have been domesticated in New Guinea and then spread throughout Asia and parts of Africa, before its locus of production shifted decisively to the new world in the seventeenth century, especially the Caribbean. It is grown in the tropics and subtropics under both irrigated and high rainfall conditions. Another sugar crop, sugar beet (*Beta vulgaris*) has been grown in the temperate areas since the 1800s to produce the identical sugar product. Sugar beet made up nearly half of world sugar supply in 1900 but this has fallen to only about 20 percent today given the inherent efficiency advantages of sugarcane.

Sugarcane is closely associated with the shameful era of transportation of slaves from Africa to work on the new world sugar plantations throughout the tropical Americas (Mintz 1985). As sugar consumption rose in Europe, sugar became a major commodity in world trade in the eighteenth and nineteenth century. With the emancipation of slaves in the nineteenth century, the plantation systems of the Caribbean collapsed and Java (now in Indonesia) and Cuba, together with a handful of smaller producers (Mauritius, Fiji, Australia), became the major suppliers in world markets. In the late twentieth century, Brazil decisively emerged as the dominant player in the world sugar market, aided by its national program to promote sugarcane-based ethanol. India is the world's second largest sugarcane producer but its harvest is mostly used to supply its domestic market.

Soybean

Soybean (*Glycine max*) only emerged as a major world crop in the second half of the twentieth century. Soybean has its origins in northern China where it is consumed in a variety of forms, especially tofu, making it an important source of dietary protein. It evolved into a global crop in Japanese-occupied Manchuria (now in China), where large areas were cultivated to supply soybean oil to Japan and world markets, including the US. A by-product of oil extraction was soymeal that was exported as fertilizer for sugarcane in China and rice in Japan. Beginning before the Second World War and accelerating after the war, soybean emerged as a major crop in the US to feed livestock (along with maize). Soybean meal, which makes up 80 percent of the weight of soybeans, is together with maize now the major feedstuff for the poultry, pig, and dairy industries around the world. World soybean production has increased twenty-fold since 1950 with 90 percent destined to livestock feed and oil. Production has also shifted decisively to South America, especially Brazil, now the world's largest producer and exporter. Despite its origin in high latitudes, the major "soy baskets" today are in tropical central Brazil.

About two-thirds of world soybean production is destined for international markets and soybean and its products emerged around 2000 as the most valuable agricultural commodity in world trade, exceeding that of wheat. China, the origin of soybeans, is by far the largest importer as its struggles to satisfy its burgeoning appetite for meat.

Oil Palm

African oil palm (*Elaeis guineensis*) is a tree crop of West and Central African origin, producing two types of oil, palm oil from the fleshy fruit and palm kernel oil from the inner palm nut, as well as a small amount of high protein palm kernel meal. A related oil palm species, *E. oleifera,* originating in the Americas is locally harvested for oil but not commercially produced.

In Africa, oil palm was traditionally produced in a semi-wild state and processed in artisanal mills into red palm oil as an integral part of local cuisine. A thriving trade in palm oil and palm kernels also developed in Nigeria and other West African states to supply lubricants and fuel to Europe. The invention of margarine in the Netherlands and Germany as a cheap substitute for butter resulted in a surge in demand for palm oil and palm kernel oil in the early 1900s. Lever Brothers, a predecessor of today's Unilever, set up large mills for processing palm oils in the early part of the century in what was then the Belgian Congo (now Democratic Republic of Congo) to meet this new demand.

Adrien Hallet, a Belgian national who had worked in the Congo, transferred oil palm cultivation to Sumatra (then part of the Netherlands East Indies and today Indonesia). He was "extremely lucky" in being able to use African oil palms already being grown there for ornamental purposes for commercial cultivation since they turned out to have good oil yields (Martin 2005). During the 1920s, the introduction of large-scale milling technologies in Sumatra and the exportation of palm oil to Europe in bulk provided standardized palm oil suitable for many food uses.

From 1970, oil palm has emerged as the most dynamic industry in world agriculture, doubling in size in each decade since then (Byerlee et al. 2017). It is a huge industry in Indonesia and Malaysia that together supply about 90 percent of the world market, with India, China, and the European Union as the largest importers. As the world's cheapest edible oil, it has rapidly substituted for other oils such as coconut oil, groundnut oil, and even soybean oil to dominate edible oil use. It is now the third most valuable product in world trade and will likely overtake wheat for second place after soybeans within a few years.

Palm oil and palm kernel oil are used in margarine, cooking oil, and a wide variety of processed foods. Edible oils, led by palm oil and soybean oil, accounted for at least a quarter of the increase in food dietary energy from 1990, reflecting their growing availability and high energy density of nine calories per gram of oil compared to about three calories per gram of the cereals. However, palm oil is high in saturated fats that are regarded as being deleterious to health if consumed in excess. Palm oil and soybean oil are also used widely for non-food consumer products such as cosmetics and soap, in many industrial processes, and in the biofuel industry.

FARMING THE STAPLES

A Scientific Revolution

Over the past two centuries, the world has periodically rediscovered the famous prediction of Thomas Malthus in his 1798 treatise, *An Essay on the Principles of Population*, that population growth would outstrip food supply. Indeed, the twentieth century began with another such gloomy forecast by a distinguished British scientist, William Crooks, in his book, *The Wheat Problem*. As it turns out, the twentieth century was marked by an unprecedented pace of technological change that led to spectacular increases in crop yields around the world that outstripped population growth. For the first time in history, world food supply in the twentieth century was based largely on yield increases rather than expansion of the land area sown.

From the late 1800s crop breeding was carried out in France and a few other industrializing countries especially for wheat (Perkins 1997). One of the earliest successes was around 1900 when Australian breeder William Farrer (1845–1906) crossed Australian wheat varieties with Indian varieties to produce rust-resistant and early maturing wheat varieties for dryland areas that greatly expanded Australia's wheat frontier. Farrer's image was immortalized in the Australian $2 currency note in 1966, the only plant breeder to achieve such fame (Figure 1.3).

FIGURE 1.3 The Australian two dollar note issued in 1966 and replaced by a coin in 1988. The image is of William Farrer, Australia's most famous wheat breeder, who bred many widely accepted varieties around 1900. Photo by Ben Molyneux/Alamy Stock Photo.

Yield progress accelerated after the rediscovery of Mendel's laws of inheritance around 1900 that provided a scientific basis for crop breeding programs. In the early part of the century, cereal yields took off in several countries, as a result of these efforts. Wheat yields in England and rice yields in Japan are witness to the success of these efforts (Figure 1.4). In the US, the breakthrough occurred with the development of hybrid maize by inbreeding parents to exploit heterosis or hybrid vigor. Hybrid maize was rapidly adopted in the 1930s and 1940s in the US, aided by the rapid rise of private seed companies since farmers had to buy seed each year to maintain hybrid vigor. The first large-scale success was in Iowa where Henry Wallace, later vice president of the US, established the Pioneer Hi-Bred company in 1926, now part of a giant agro-chemical company, Corteva. The high intrinsic yield of the maize plant (due to its C4 photosynthetic pathway) and huge private investments in R&D have enabled maize to out-yield all other cereals by the end of the century. US yields marked a milestone in 2014 when average national yields exceeded 10 t/ha.

In the tropics and subtropics, much of which was under European colonial regimes until the middle of the twentieth century, the early emphasis in plant breeding was on export crops to supply raw materials such as rubber, and the emerging tastes for beverages and sugar in the imperial powers. Research, for example, was carried out on sugarcane in the early 1900s by the Dutch in Java, then a major sugar exporter, and the British in Barbados. After 1925, Javanese varieties were widely exploited in Cuba, the major sugar producer, indicating that scientific discoveries could travel long distances and across empires (Bosma and Curry-Macado 2013).

Synthetic fertilizer was the second major element in the scientific revolution. Following John Lawes pioneering development and experimentation with superphosphate fertilizers at what is now Rothamsted Research in the UK, superphosphate was rapidly adopted. Most notably, it took off in Australia around 1900 based on scientific experimentation,

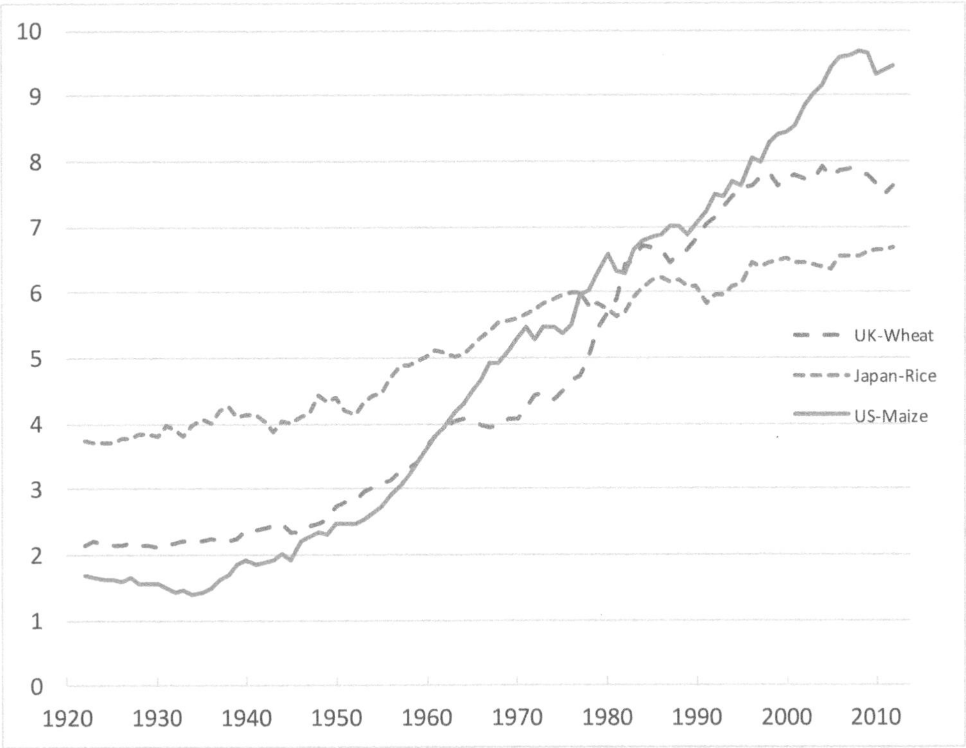

FIGURE 1.4 Yields (t/ha) of wheat, rice, and maize in selected industrialized countries (five-year moving averages) 1920–2014. Adapted from Brown 1966 and FAOSTAT.

farmer innovation, and strong support from the state (Byerlee 2021). Also, by 1920, the Haber-Bosch process was developed for the manufacture of nitrogenous fertilizer using fossil energy to convert nitrogen from the air. This process ushered in a period of declining fertilizer prices. Countries with scarce land resources, especially western Europe and Japan, gradually adopted this technology even before the Second World War. Adoption accelerated dramatically from 1940, especially in the US, contributing to record yield.

By contrast in the immediate post-Second World War period, most of Asia and Latin America were experiencing a rapid rise in population, and declining per capita food production. These trends were truly scary to the analysts and world leaders at the time, who were nurtured in the Malthusian views of the early twentieth century. It is not surprising that many popular books, including Paul Ehrlich's *Population Bomb*, Georg Borgstrom's *The Hungry Planet*, and the Paddock brothers' *Famine 1975!*, all published in the 1960s, echoed a deep Malthusian pessimism about looming world famines.

However, the simultaneous development of short-statured rice and wheat varieties carrying dwarfing genes in the 1960s provided much better response to fertilizers, especially in irrigated areas, that led to an abrupt yield take off in Asia and Latin America—faster even than those experienced in the temperate world. This research was led by the Rockefeller Foundation that had started supporting food crop research in Mexico in 1943 and then extended the effort to several other countries. Norman

Borlaug led the work on wheat in Mexico and in the 1950s began crossing lines he had developed in Mexico to a short-statured variety developed in Japan. When Borlaug traveled to the Middle East and South Asia for the first time in 1960 he recognized the potential of these new varieties to transform wheat production around the world. Following the success of wheat, short rice varieties were produced by the International Rice Research Institute (IRRI) in the Philippines established by the Rockefeller and Ford Foundations in 1960.

In what became known as the green revolution, hundreds of millions of farmers rapidly adopted the new wheat and rice varieties along with fertilizer, reversing the downward spiral in per capita food production in Asia and very likely averting massive famine. Borlaug was recognized with the Nobel Prize for Peace in 1970 in one of a handful of Nobel prizes awarded for discoveries in the plant sciences. His growing international program was institutionalized in the form of the International Maize and Wheat Improvement Center based in Mexico (CIMMYT) as a sister center to IRRI in 1966. These experiences also convinced many governments of the developing world to build their own research capacity in food crops aided by a new system of international research centers to share germplasm and knowledge for a range of food staples, known as CGIAR.

The intensification of agriculture over large areas was not without costs. In particular, fertilizer use eventually reached very high levels across much of the world, especially in China, now the largest user. High nitrogen use has increased nitrogen runoff and produced algae blooms downstream such as that in the Gulf of Mexico that are visible from space. Even more damaging was the wide and indiscriminate use of pesticides in the post-Second World War period that was exposed by Rachel Carson, an American ecologist, in her highly influential book, *Silent Spring,* published in 1962. Carson, an avid bird watcher as well as a scientist, associated the use of pesticide with declining bird populations.

Many systems have also increased specialization in one or two crops as seen in the vast areas of the rice-wheat system in the Indo-Gangetic Plains and the maize-soybean rotation in the US "Corn Belt" and the Brazilian "Soy Basket." Greater specialization resulted in increased use of agro-chemicals to control pests and maintain soil fertility as well as the supply of less diversified foods.[5] Specialization was also often encouraged by government subsidies and price supports biased toward major staples. By the 1980s, the world was beginning to espouse more diverse agro-ecological approaches to improve sustainability of food production systems. Practices such as zero tillage, integrated pest management, and a renewed emphasis on nitrogen-fixing legumes in crop rotations are becoming more widely accepted.

A further milestone in crop technology was the discovery of DNA in 1953 and the subsequent biotechnology revolution that enabled the insertion of genes from other species in crop varieties through genetic engineering. The first so-called genetically modified (GM) crops were adopted in the US in 1996 for maize and soybeans that quickly spread to the major producers in South America. GM crops were and continue to be controversial for several reasons including perceived health effects, possible negative impacts on the environment, ethical concerns about "playing with nature," and their association with growing concentration in the seed industry. However, after twenty years, the scientific evidence strongly indicates that while the yield effects of GMOs were modest, they are as safe as conventional technologies for consumption, and they have helped reduce costs and application of toxic pesticides. Even so, GM products (mainly maize and soybeans) are still largely destined for animal feed, biofuels, and industrial uses and have yet to be released for rice and wheat, the premier cereals used directly for food.[6]

The most worrying trend associated with GM crops is the growing concentration of the seed industry in the hands of a few large multinational companies. Interpretation of patent laws in the USA in 1980 to include genetic engineering processes have enabled the largest companies to control a high percentage of all such technologies through mergers and acquisitions (Heisey 2015). This combined with the 1992 Convention of Biological Diversity giving countries sovereignty over their genetic resources has severely curtailed the "open source" approach to plant breeding, and free exchange of germplasm across the world and among farmers that supported the rapid gains in crop yields for much of the twentieth century (Byerlee and Dubin 2009).

Finally, Africa was left behind by the scientific revolution in agriculture in the twentieth century for a number of reasons. First, a much wider array of staple crops is grown in Africa including sorghum, millet, cassava, and yams, so breakthroughs in the global staples were not enough to transform food economies of the region. Colonial priorities on export crops also resulted in underinvestment in research on these African staples. Second, nearly all of sub-Saharan Africa agriculture is rainfed, reducing yields and increasing climate risks of input use. Third, poor infrastructure has constrained the intensification of input use, especially fertilizers. Still, improved maize and rice varieties are now widely grown in the region, and in the early twenty-first century, African yields are finally growing.

The Family Farm and Corporate Agriculture

Agriculture the world over in the twentieth century was characterized by family farming where families own and operate farms and provide most of the labor. Family farming, linked into tight-knit rural and village communities, is integral to strong rural cultural traditions and community progress. It is often characterized as the basis for cultural values of hard work, independence, and egalitarian rural development, as, for example, in Thomas Jefferson's portrayal of the "yeoman farmer." Family farms are also the building blocks for the variegated rural landscapes of fields of staple crops, garden plots of fruits and vegetables, pastures for livestock, and natural areas that we see across the world from East African highlands to European valleys.

A remarkable fact is that while nearly all manufacturers and many service industries moved during the nineteenth and twentieth centuries from family businesses to large national and multinational corporate structures, farming of food staples remained overwhelmingly family owned and operated.[7] Even in the twenty-first century some five hundred million family farmers produce most of the world's food (Lowder et al. 2016). The resilience of the family farm in the modern world reflects the diversity and seasonality of agricultural operations that do not lend themselves to standardization and specialization of labor tasks as in manufacturing industries. Family farmers with their in-depth knowledge of local soil and weather conditions accumulated over many generations also have distinct advantages in tailoring management to specific climatic and soil conditions.

To be sure, in rich countries family farms have grown steadily in size through mechanization responding to rising wages and the need to provide farmers income levels comparable to those in the non-farm sectors. Many family farms in rich countries are now quite specialized in one or two crop-livestock enterprises and manage highly sophisticated businesses but they still depend largely on family management and labor. The average family farm for cereal-sheep production in South Australia today operates about 1,500 ha, more than ten times the average of about 130 ha in 1920.[8]

By contrast in the developing world over the twentieth century, family farms became smaller, especially in Asia, as population growth pushed up against land scarcity and families subdivided farms into ever smaller units for their children. In Bangladesh with sixty million people living on farms, farm size halved in two decades from 1.3 ha in 1977 to 0.6 ha in 1996 and it is questionable whether many tiny farms today can be efficient producers of staple foods (Rapsomanikis 2015). Fortunately, wages in much of Asia are now rising rapidly spurring migration off farms to manufacturing and service jobs and relieving pressure on the land (Wiggins and Keats 2014). Farm size has been increasing steadily for many years in Japan and this trend is now being seen in China and some other countries.

Family farms are not universal, however, and especially toward the end of the century, a new style of large corporate farm emerged for the production of some staple foods in units of tens and sometimes hundreds of thousands of hectares. In Russia and eastern Europe these farms with wheat as their major crop had their roots in the failed Soviet experiment in collective and state farms implemented from the 1930s. After the collapse of the Soviet Union in part because of poor agricultural performance, the state farms were privatized and often merged into large corporate “agro-holdings” sometimes with the injection of foreign capital and management to upgrade them to modern farming standards.

In Argentina and Brazil similar types of very large corporate farms emerged for producing soybeans often rotated with wheat or maize. These farms emerged from the unequal landownership in the region and poorly functioning financial markets and other services to support family farms. Today’s mega-farms in Latin America are able to manage the huge challenge of supply logistics and supervision of operations over large areas by employing state-of-the-art information technologies (Deininger and Byerlee 2012). The adoption of zero tillage using GM varieties resistant to herbicides has also reduced the number of farm operations and facilitated the effective management of very large areas.

Finally, two of our staples, sugarcane and oil palm, are often grown on plantations—a special form of farm organization akin to industrial companies. These crops have year-round production and labour needs that provide permanent jobs with a high degree of specialization and standardization, similar to many industrial firms. Another feature is that production is integrated with large mills that must process the raw materials quickly after harvest. The very large scale of modern plantations and the ability to make “every farm a factory” (Fitzgerald 2003) has encouraged large companies to control this sector. This as well as the mono-crop nature of plantations has also resulted in very different rural societies—immigrant labor often from long distances resides on the plantation in an enclave community. They also produce different landscapes characterized by large contiguous areas uniformly planted to a single crop.

In sugarcane, after the demise of slavery in the latter part of the nineteenth century, indentured labor was employed, often from far away. In particular, Indian labor was employed in the Caribbean, South Africa, Fiji, and Mauritius from the late nineteenth century; Japanese and other immigrants were employed in the Hawaiian industry; and Chinese in the Cuban industry. Similarly, today the great majority of laborers working on oil palm plantations in Malaysia are immigrants from surrounding countries, especially Indonesia. These waves of immigrant labor have left a multi-ethnic heritage in many countries of the tropical world today.

Up to about 1900, the area of sugarcane required to meet mill capacity was quite small, and most sugarcane farming operations were family owned. This changed drastically from

1900, with a revolution in processing technology that provided economies of scale in milling. In Cuba, the area per mill jumped from about 500 ha in 1904 to 6,000 ha in 1929 (Dye 1994). This trend has continued to the present day; an average sugarcane-ethanol plant in Brazil today, usually owned by a large corporation, requires a supply area of 20,000 ha, and the most recent mills may require up to 70,000 ha with a total investment of $1 billion or more (Deininger and Byerlee 2012). Today, sugarcane is also the only plantation crop that allows full mechanization of production operations (including harvesting), thereby largely solving labor supply problems. Similarly, over the century, oil palm mills have increased in size from 2000 ha in 1934 to around 10,000 ha today and huge companies, mostly Asian owned, have emerged that invest in oil palm plantations around the world (Byerlee et al. 2017). The implications of these transformations to very large plantation and processing operations on rural cultures and landscapes are still poorly understood.

Biodiversity on and off the Farm

The widespread adoption of improved varieties across the world in the twenteith century has led to growing concerns about the loss of biodiversity on the farm as well as reduced dietary diversity as a few modern varieties displaced a large number of traditional varieties. This loss is seen as a threat to food security in an age of climate change although the story is more complex—modern varieties themselves often have diverse genetic makeup as breeders incorporate new sources of genetic diversity into varietal pedigrees with each generation, and farmers periodically update varieties sown (Smale 1997). Major efforts have been mounted to preserve remaining diversity in "gene banks" around the world. For example, Paul Mangelsdorf at Harvard University and colleagues around Latin America put together a remarkable collection of highly diverse local maize varieties and wild relatives in the 1950s that are now conserved in several sites and have become a major resource for breeding of tropical maize varieties and hybrids (Mangelsdorf 1974). One unique site is the backup vault for food and agricultural germplasm opened at Svalbard in the Arctic by the government of Norway in 2008 that now stores nearly one million seed samples.

Who controls genetic resources has also been one of the big debates of the late twentieth century. On the one side, most scientists have favored the principle that germplasm for food crops is the heritage of humankind and should be freely exchanged and accessed. On the other side, there have been moves to "privatize the commons" through treaties on intellectual property protection for private breeding under the World Trade Organization, and to recognize countries' sovereignty over their genetic resources as enshrined in the UN Convention on Biological Diversity. A major accomplishment in 2001 was an agreement on the International Treaty on Plant Genetic Resources for Food and Agriculture to recognize core germplasm collections as the heritage of humankind. Still only three of our staples (rice, wheat, and maize) are included in this treaty. Sugarcane and oil palm were regarded as industrial crops and soybean was withdrawn from the agreement when China, its center of origin, wished to protest against a US spy plane incident at a critical juncture in negotiating the agreement (Hodgkin et al. 2013).

Off the farm, the expansion of the wheat frontier in the Americas and Australia to supply European markets in the last half of the nineteenth century continued into the early part of the twentieth century, resulting in the loss of most of the remaining natural grasslands in temperate areas suited to farming. However, in the 1930s, drought coupled with poor land management practices and movement of wheat into marginal growing environments led to an environmental catastrophe in the form of immense dust storms in the Great Plains of the US and in much of Australia's wheat areas. Despite the experience of the US and Australian Dust Bowls of the 1930s, Nikita Khrushchev, premier of the Soviet Union, launched his Virgin Lands program in the 1950s ploughing up over 40 Mha (million hectares) of the steppes in Central Asia (today mainly Kazakhstan) to produce wheat on huge state farms (McCauley 1976). Eventually, with the advent of dry years and low yields, much of this area was abandoned—with lasting damage to the environment. In the early 1960s, dust storms caused by wind erosion from this area affected cities as far away as eastern Europe.

Despite these ecological disasters and associated human costs the story in much of the rich world, and more recently China, has been reforestation or re-vegetation of marginal cropland made possible by intensification and rapidly rising crop yields in more favored areas. For example, from 1950 to 1980, the area of farm land declined by 50 Mha in the thirty-one eastern states of the US (Williams 2003). These trends were driven by economic forces, especially declining prices of food staples, but also the growing recognition in the second part of the twentieth century of the value of natural areas in conserving biodiversity. Greater areas of natural vegetation and more sustainable agricultural practices have resulted in a revival of much wildlife throughout the rich world.

In much of the tropical world the story has been the opposite. Deforestation has steadily increased over the twentieth century. A major part of this loss was due to rapid population growth as small farmers steadily encroached into forest land for both subsistence foods and cash crops. For example, Thailand cleared 28 percent of its forests from 1955 to 1975 partly due to rising population but also to supply export markets (Feeny 1984). In the 1950s, Thailand, already the top rice exporter, became one of the world's largest exporters of maize, an almost completely new crop to the country, as well as sugarcane.

By the late twentieth century, tropical forest loss was mostly driven by large-scale commercial agriculture for global markets, except in Africa (Rudel 2007). Two of our staples, soybeans and oil palm, featured in the transformation of millions of hectares of tropical forests and savannahs (oil palm for forests in Indonesia and eastern parts of Malaysia and soybeans for savannas in Latin America) (Byerlee et al. 2017). Rising demand for these crops occurred at a time when the world was increasingly focused on the global value of tropical forests for humankind first as a store of rich biodiversity and more recently as a store of carbon to reduce greenhouse emissions. These values were enshrined in international conventions such as the Convention on Biodiversity in 1992 and the Paris Climate agreement of 2016. Tropical deforestation and clearing of other natural areas for cropland have drawn intense scrutiny from global environmental organizations, civil society, and many governments.

In Brazil, one of the last remaining tracts of woodlands and savannahs has been steadily cleared in the Cerrado for the world's "soy basket"—an area of 200 Mha of infertile soils in central Brazil bordering on the Amazon. Some encroachment on the Amazon forests occurred but this was successfully addressed by a series of measures by the Brazilian government and a moratorium by international grain traders against buying soybeans originating in recently deforested areas in the Amazon biome (Byerlee et al. 2017). However, clearing of savannah woodlands continues until today and recent political changes in Brazil are again threatening efforts to conserve the Amazon forests.

Indonesia now has the fastest rate of tropical deforestation and the highest emissions of greenhouse gases from land-use changes, due to the explosion of oil palm as well as other commercial activities especially in peat lands dense in soil carbon. With the loss of its habitat the orangutan has emerged in much of the rich world as the iconic species for biodiversity loss leading to boycotts and campaigns against palm oil that has been labeled as the world's "most hated crop" (Yan 2017). In 2015, international traders committed to a moratorium against buying palm oil from recently deforested land but there is as yet little sign of a reduction in deforestation rates in Indonesia. Global concern about continuing deforestation was expressed in the ambitious international commitment in 2015 to a sustainable development goal to halt all deforestation by 2020.

EATING THE STAPLES

Overcoming Malnutrition—a Glass Half Full

There is little doubt that the twentieth century was a turning point in feeding the world due to the upsurge in cereal yields that started in the rich countries (mainly wheat and maize) before the Second World War and expanded after the war to Asia and Latin America (mainly wheat and rice). Per capita grain production that had been falling in much of the world through the first part of the century rose steadily so that by 2015 it was more than one-third above that in 1961–5. As one of its first priorities when it was established, the FAO conducted a number of world food surveys that estimated that one-third to one-half of the world's population went hungry in 1960 (Sukatme 1961). By 1990, some 20 percent of the world was undernourished, declining to about 11 percent today.[9] The spectacular successes in rice and wheat in Asia undoubtedly helped the world stave off a major food crisis and very likely famines in that region in the second half of the twentieth century. The strong performance of maize and soybeans from the 1980s in Latin America enabled better-off consumers to diversify diets to animal products.

Indeed, not only did world grain supply outpace population growth by a considerable margin, world prices in real terms trended downward over the twentieth century, albeit with periodic spikes due to climatic or other shocks or demand surges. By 2000, world wheat prices were only about one-third of prices in 1900, despite the sharp fall in per capita land area devoted to cereals.

Still, the glass is only half full. An estimated 800 million people today are hungry and 155 million children (22 percent of the total) are stunted, impairing cognitive potential. These are shameful statistics in a world of ever-growing prosperity. The United Nations has periodically put a spotlight on the hunger challenge through a series of world food summits, starting in 1943, and most recently in formulating consensus around the Sustainable Development Goal of ending hunger by 2030. The biggest challenge is in sub-Saharan Africa that benefited little from the Green Revolution and where over 20 percent of the population remain undernourished, most of them small farmers. Even in 2017, some twenty million people faced the threat of famine in conflict areas of Africa, and another fifty million required direct food assistance, according to FAO.

Meanwhile, the number of people who are "over-nourished" rose sharply in the second half of the century. An estimated 2.1 billion people in the world were overweight in 2013, up from 857 million in 1980 (Global Panel 2016). This process is most advanced in the US and some other rich countries where close to one-third of the population is now classified as obese. Many middle-income countries, notably Egypt, Mexico, and Hungary, have closely followed the same path (Figure 1.5). Even poorer countries such as India are experiencing multiple burdens of malnutrition, with sizable numbers of overweight people along with the continuing burden of large numbers of people with inadequate

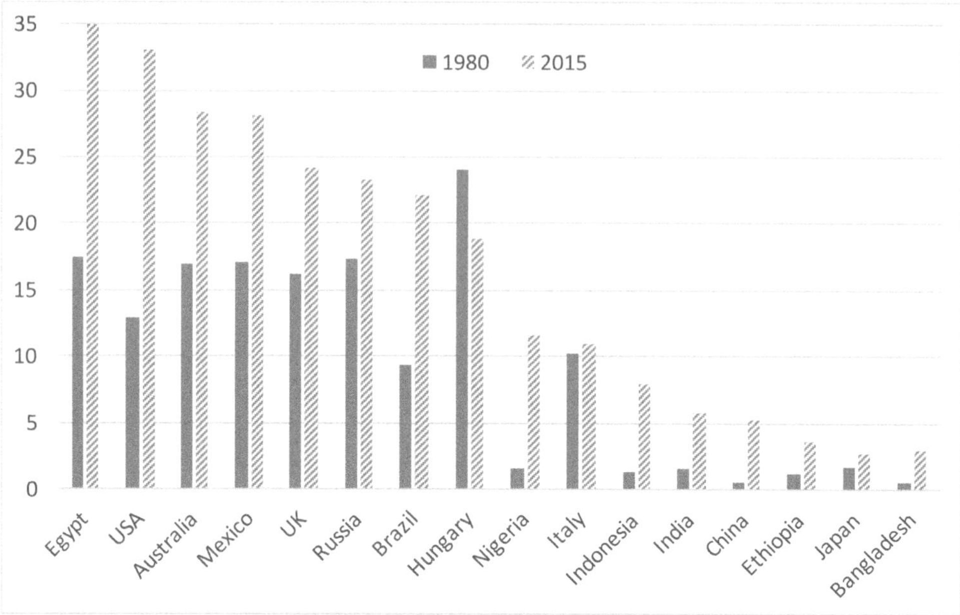

FIGURE 1.5 Prevalence of obesity as percent of adults, selected countries, 1980 and 2015. Calculated from Global Burden of Diseases 2015 Obesity Collaborators 2017.

energy intake or micronutrient deficiencies. Obesity in turn is closely associated with an epidemic of non-communicable diseases, notably diabetes and cardiovascular disease that incur high health costs and reduce life expectancy. In the US, health costs of diabetes now exceed those of tobacco.

The problem of obesity relates to many causes including more sedentary lifestyles in urban areas, but diet is the most important causal factor (Global Burden of Diseases 2015 Obesity Collaborators. 2017). Dietary changes favoring obesity include excess consumption of carbohydrates, sugars, oils, and animal fats—all closely related to the direct and indirect consumption of our six staples—the theme of the next section.

Finally, an estimated two billion people lack essential micronutrients that are delivered in the diet through legumes, nuts, and fruits and vegetables that have often been neglected at the expense of the major staples. Diversified diets are the best solution to this so-called "hidden hunger." There have also some recent success in bio-fortification of staple crops, such as development of wheat varieties with enhanced zinc content that have started to be grown in South Asia.

Changing Diets

Over the past century, there has been a steady convergence in diets across the world with our six staples taking center stage in what is known as the "nutrition transition" (Khoury et al. 2014; Popkin 2001). The main drivers of this convergence have been rising incomes and urbanization especially after the Second World War. These drivers together with increased participation by women in the labor force have put a premium on convenience foods to reduce cooking time and fuel.

There are several common drivers toward the homogenization of diets globally. First, rice and wheat, and to a smaller extent maize, have substituted for other cereals and for starchy roots. In the case of rice, this increase is most apparent in regions that largely consumed roots such as cassava, yams, and potatoes, as in tropical Latin America and more recently in West Africa. Almost everywhere, wheat consumption has increased with urbanization since bread is prized as a convenience food, especially white bread that has been deprived of important micronutrients. Many processed foods, such as breakfast cereals and snacks, are also wheat-based convenience foods. Wheat is now the most important staple in tropical Latin America despite the fact that little is grown in the region. The role of maize has also increased dramatically in Africa over the century where it has displaced the traditional cereals, sorghum and millet, again largely based on ease of processing for a growing urban and mining population.

Consumers everywhere have increased their consumption of animal products, meat, milk, eggs, and fish, as incomes rose in what has been termed the "livestock revolution" (Delgado 2003). The composition of animal products in the diet varies but the common denominator has been the rise of poultry meat. Poultry was a luxury food even in the Western world up to the Second World War but its price has plummeted relative to other meats, through technological progress in poultry breeding and feed efficiency (Silbergeld 2016). Almost all poultry today is raised in intensive systems based on feeds prepared largely from maize and soybeans. In the mid-1990s poultry meat exceeded beef consumption and will soon overtake pork as the world's most important livestock product.

Rising incomes, declining prices, and the demand for processed foods have also driven demand for vegetable oils, led by palm oil and soy oil that have substituted for traditional sources of fats. Much cheaper margarine made from vegetable oils displaced butter up until quite recently. This was despite protests from dairy farmers everywhere who supported policies to discourage margarine, including outright bans on production and imports in Canada and New Zealand and a host of measures to regulate its appearance relative to butter. For example, federal regulations in the US up to the 1950s forbade margarine to be sold with a yellow color and five US states actually required that margarine be colored pink (Snodgrass 1930)! However, after the health risks of transfats produced by the hydrogenation process for manufacturing margarine became apparent, margarine consumption began to fall.

More recently, palm oil has displaced traditional oils in cooking in many countries. In Indonesia, coconut oil provided over 90 percent of edible oil supply in 1965—by 2010, palm oil provided over 90 percent of the supply (Byerlee et al. 2017). Likewise, India is now the world's largest importer and consumer of palm oil as traditional oils such as groundnut oil have failed to keep pace with demand.

Finally, as incomes increase, the consumption of processed food and beverages has risen sharply with the rise of the food manufacturing industry and supermarkets (Figure 1.6). Processed foods are made up of many ingredients but the harvests of our six crops are by far the most important. Beverages sweetened with sugar are now pervasive throughout the world, and are a major source of "empty calories" contributing to the obesity epidemic. Consumption of sugar is now thought to be especially important for obesity due to its rapid absorption into the blood stream (Taubes 2016). Mexico is one of the world's highest consumers of sugar and also has one of the highest incidence of diabetes. In the US, one third of sweeteners are consumed as high fructose corn syrup in non-alcoholic beverages, produced by processing maize.[10]

The figures on dietary provision of calories in France in the early twentieth century and China since 1980 illustrate these trends (Figures 1.7 and 1.8). China has yet to

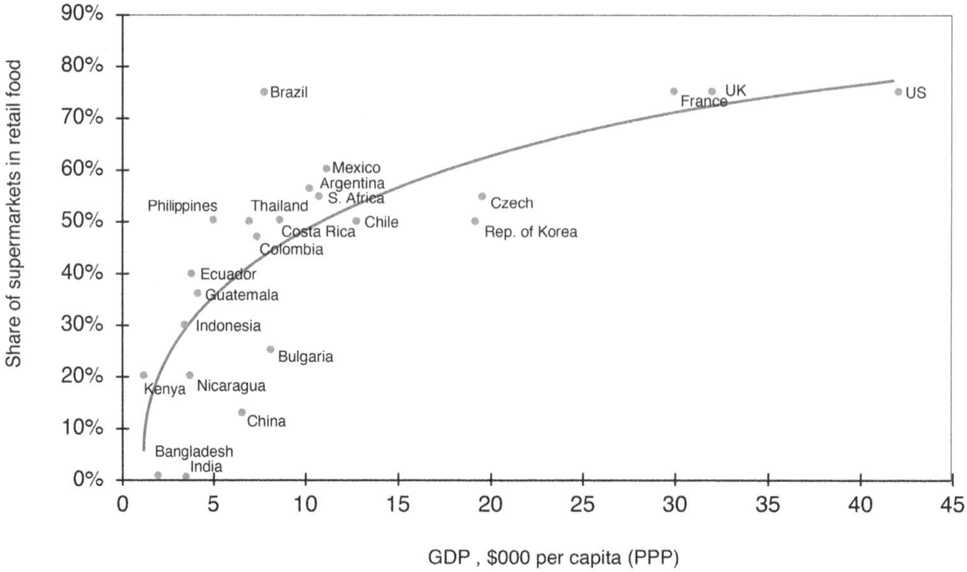

FIGURE 1.6 Share of supermarkets in food retailing. Reardon and Berdegue 2006.

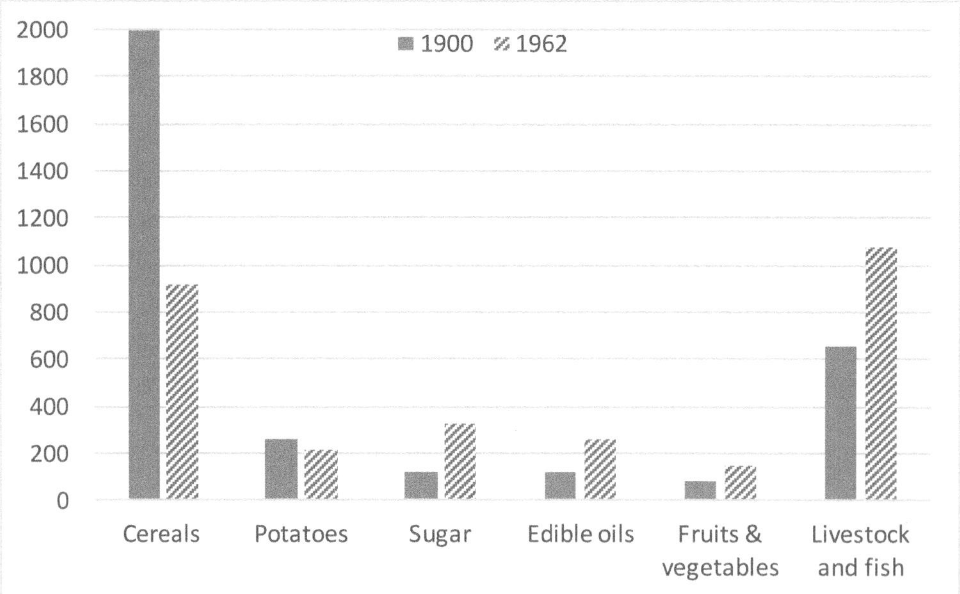

FIGURE 1.7 Changes in food consumption patterns in France in the first part of the twentieth century (calories per person per day). Adapted from Toutain 1971. Note: Cereals is mostly wheat but in 1900 included considerable amounts of other cereals, such as rye and buckwheat (*Fagopyrum esculentum*—not a true cereal). Minor food groups such as pulses and nuts are not included. Livestock includes dairy products and eggs.

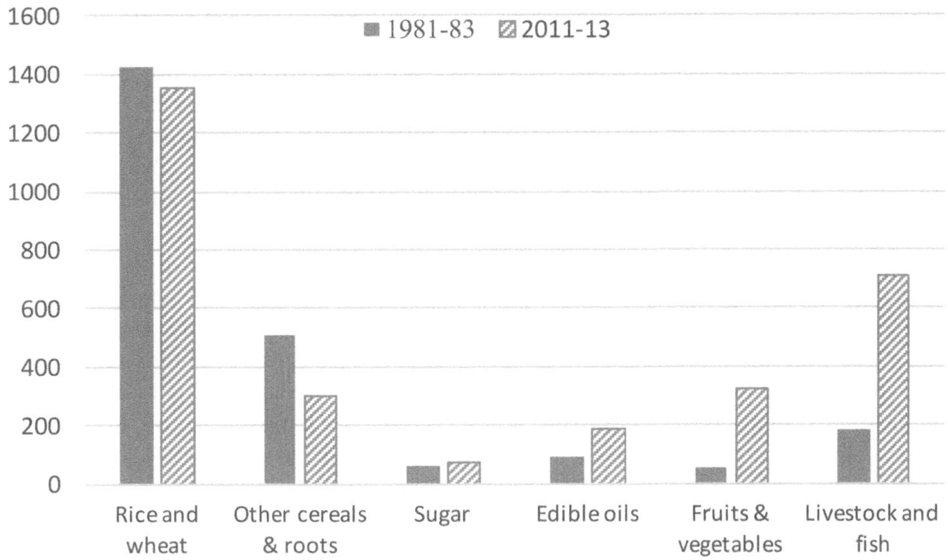

FIGURE 1.8 Recent changes in food consumption patterns in China (calories per person per day). FAOSTAT. Note: Minor food groups such as pulses and nuts are not included. Livestock includes dairy products and eggs.

see a sharp decline in per capita cereal consumption but it has already started for rice in the past decade. The biggest changes have been increases in edible oils, fruits and vegetables, and livestock products in diets following the path of France nearly a century earlier. China still consumes relatively low amounts of sugar per capita and one of the big questions for Chinese health is whether it will follow the Western model or Japan or Korea, which consume less than half the sugar per person of the French.

GLOBALIZATION

Staple food crops have long moved around the world especially since the Colombian Exchange from the sixteenth century when "old world" crops such as wheat and sugarcane moved to the Americas and "new world" crops such as maize moved to the rest of the world. On average, about two-thirds of crop production in a country derives from crop species of foreign origin (Khoury et al. 2016) and this share has increased over time. During the twentieth century, two of our six crops have decisively shifted their geographical center of production out of their region of origin. Starting around 1920 soybeans moved from China to the Americas and oil palm from Africa to Southeast Asia.

Food consumption patterns have also been affected by growing international trade. After the first period of globalization ending with the First World War, the economic depression of the 1930s led to protectionism and isolation that sharply reduced trade. After the Second World War, agricultural trade has been slowly but surely liberalized although agriculture remained until very recently a highly protected sector. Trade has enabled many countries to consume crops that they do not grow. In particular, wheat is a temperate crop but is now widely consumed in tropical countries based on imports. During the latter part of the twentieth century, many tropical countries encouraged bread consumption through subsidies and cheap foreign exchange, supported by policies of wheat exporting countries such as food aid in the form of wheat (Byerlee 1987). In contrast, palm oil is produced only in the tropics but is now consumed mostly in temperate and subtropical regions based on imports.

Production of both oil palm and soybean is heavily concentrated in two to three major countries who produce to supply consumers across the world. Wheat, maize, and sugar exports are also quite concentrated although a lower share of those crops ranging from 10–30 percent is traded. In general, trade in grains and oilseeds tends to flow from relatively land abundant regions such as the Americas to countries experiencing acute land or water scarcity, such as East Asia and the Middle East. Africa, a net importer of the products of all six crops, is an exception to this rule in large part due to low productivity of its domestic agriculture.

The growing dependence on a few countries and crops for global food supplies has ignited a controversy between the "globalizers" who favor policies of providing cheap food and the "locavores" who favor supporting local food cultures by directly linking farmers and consumers such as through farmers' markets. Locavores are also concerned about "food miles" and the emissions of greenhouse gases from long-distance transport. These concerns apply more for richer consumers willing to pay premium prices, since the local hinterland is often not very efficient in providing basic food staples although local farmers may be efficient for high-value horticultural and livestock products. There is also evidence that local production may sometimes lead to higher emissions of greenhouse gases than food transported over long distances (Desrochers and Shimizu 2012). Climate change will only increase the need for global trade as an adaptation mechanism for food staples.

Beyond trade, there are other factors playing out in globalization of food systems. One has been growing international migration and the rise of ethnic foods especially given the trend towards eating away from home. Diets in Europe and the US, for example, have been greatly enriched by rice-based Asian dishes and maize-based Latin American cuisines. The increased flow of information through the media, including social media, has further increased awareness and appreciation of "foreign foods" around the world.

Finally, food systems beyond the farm are increasingly dominated by multinational agribusinesses, food manufacturers, and fast-food chains that tend to homogenize diets across the world (Table 1.3). These companies influence diets not only through standardization of products but also through aggressive advertising. On the retail end (not shown in Table 1.3), supermarkets including giants such as Walmart have become common in urban areas and everywhere supermarkets account for a growing share of food retailing. Fast food chains, such as McDonalds, are also ubiquitous in urban areas, based largely on our six crops—wheat for the bun, maize and soybeans for producing the meat, and edible oil for frying, accompanied by a sugared beverage. Although many of these multinational companies originated in the north, a growing number have their homes in the emerging economies of the south.[11]

TABLE 1.3 Sales of the world's ten largest food and beverage companies (excluding retail), 2015

Rank	Company	Main product	US$ billion 2015
1	Nestle	Diversified	72
2	PepsiCo	Beverages and snacks	67
3	JBS	Meat	53
4	Coca-Cola	Beverages	46
5	Archer Daniels Midlands	Grain-based processed foods	43
6	Tyson Foods	Meat	38
7	Kraft Foods[a]	Diversified	38
8	Cargill	Grain-based processed foods	34
9	Mars	Confectionery	33
10	Unilever	Diversified	29

[a] Includes merger with Mondelez Int in 2015
Source: Global Panel on Agriculture and Food Systems for Nutrition, 2016.

MOVING BEYOND THE AGRI-CULTURE WARS[12] OF THE EARLY TWENTY-FIRST CENTURY

The major trends in the world's food system described in this chapter have led to a lively debate in the early twenty-first century about the way forward that has been acted out in a number of high-profile global reports. On the one side are those who espouse the high productivity of the current food systems in major food-producing countries and the central role of science in feeding the world and achieving better environmental and health outcomes (e.g. Foresight 2011; World Bank 2007). On the other side are those who believe the food system is "broken" and a fresh start is needed that espouses principles of organic farming that minimizes use of agro-chemicals and emphasizes small farmers linked to local markets (e.g. International Assessment of Agricultural Knowledge, Science and Technology for Development 2009). Behind these debates are a multitude of themes around the role of GM crops, agro-chemicals, animal welfare, agribusiness, small farmers, and self-sufficient versus globalized food systems. As a lead author of the *World Development Report*, I clearly belong in the first camp mainly because organic agriculture cannot sustainably feed the world due to the large amounts of land that would be required to provide sufficient nitrogen for crop production (Connor 2013). However, I recognize that the performance of existing food systems needs to be improved in several dimensions—diversity, sustainability, food security, equity, and health and nutrition.

Just as science was the driving force for food systems in the twentieth century, there is little doubt that science will play a key role in not only meeting the challenge of feeding ten billion in 2050 but also in ensuring that this is achieved through more sustainable and healthy food systems. The private sector is playing an ever-greater role in delivering the science for food and agriculture. However, the growing concentration of ownership of genetic technologies in a handful of companies in the food system and especially in the seed system is a major concern for ensuring competitive and diverse technologies and more diverse diets. Increased public investment in science will be essential to supplying healthy foods and meeting social and environmental goals.

Returning to our six staples, there is nothing intrinsically unhealthy for people or the environment in any of them, except possibly for sugar. The problems arise on the production side with weak governance and poor policies that encourage overuse of agro-chemicals as in much of Asia for rice or clearing of forests and savannahs as for oil palm and soybeans. On the consumption side, the problems arise when products are refined into less healthy products such as white rice and bleached palm oil, or converted into less nutritious processed foods, such as white bread, or are consumed in excess, especially sweeteners and fats. The heavy dependence on six crops to feed the world is also a wake-up call to seek more diverse food systems that could also be more resilient against climate change. One priority would be to encourage greater production and consumption of legumes due to their value in fixing nitrogen and in enhancing protein supplies in diets.

Both the larger and more specialized commercial farmers and small farmers of the world are needed to meet future food supplies (Herrero et al. 2017). In a globalized world, larger farmers often have an advantage in supplying food staples to distant markets while small farms that produce for their own consumption and for local markets have an advantage in higher value products and diversifying diets. Over time, small farmers will inevitably increase in size as they become more commercialized to meet growing urban markets and seek higher incomes. An unprecedented migration of hundreds of millions of small farmers to cities is already underway in much of the developing world creating space for farm consolidation.

The most urgent task for humankind is to end hunger, as recognized in the Sustainable Development Goal set in 2015. Nowhere is this challenge greater than in sub-Saharan Africa which is now home to the largest number of undernourished people, most of them farmers. Looking to 2050, Africa will account for over half of the world's population growth, reaching some 2.5 billion people, nearly as large as China and India combined today. Rapid intensification of food systems in Africa is imperative. In a region where farmers use very low levels of inputs, increased use of external inputs especially fertilizers will be needed. Maize and to a lesser extent rice and palm oil are food staples in Africa but progress will be needed on many foods given the diversity of staples in the continent. Elsewhere, with climate change there is little room for complacency in ensuring global food supplies but fortunately there are many opportunities for further intensification in a more sustainable way as outlined in Fischer et al. (2014).

Meanwhile obesity is, or is becoming, endemic with huge costs to society in the form of rising incidence of non-communicable diseases such as heart disease and diabetes. The war on obesity will require progress on many fronts. Some countries, such as Mexico, have already imposed a "fat tax" to curtail the overconsumption of foods and beverages high in sugars, with some success.[13] The big food manufacturing and retailing companies must also do much more to deliver and promote healthier and more environmentally friendly foods and governments need to better regulate the industry. However, ultimately it is consumers that must decide on the type and quality of food they eat aided by much better public education on nutrition. The greatest challenge is to ensure that the hundreds of millions of poor of the world will have access to cheap food at the same time that food systems deliver diverse and healthy foods to all to help address obesity and micronutrient deficiencies.

As we conclude this review, consensus is building around two closely related challenges. The first is the need to shift to healthier diets by moving from animal-based protein to plant-based protein, and to diversified diets high in micronutrients, such as nuts, fruits, and vegetables (Willett et al. 2019). The second challenge is to sustainably produce these foods and reduce food losses so that food systems remain within planetary boundaries of water, land, biodiversity, and greenhouse gas emissions (Searchinger et al. 2019). Both challenges require a shift away from our six staples and from meat based on intensive livestock production, especially in high- and middle-income countries. Such a shift requires major changes in consumer behavior that will necessarily be evolutionary rather than revolutionary.

ACKNOWLEDGMENTS

I thank Colin Khoury and Carlos Navarro for adapting Figure 1.1 and Ken Cassman, Greg Edmeades, Jessica Fanzo, Tony Fischer, Steve Forbes, and Colin Khoury for valuable comments on earlier drafts.

CHAPTER TWO

Plants as Luxury Foods

Affordability in an Environmentally Uncertain Future

FREDERICA BOWCUTT

Luxury is a contested and divisive concept. What constitutes a luxury plant food varies historically, culturally, and individually. Frequently, expensive, rare, and/or finely crafted items meet the criteria. Aesthetics factor into the construction of what is luxurious and can include ephemeral experiences, not just material objects used to establish one's class status. In Japan, the term *wabi sabi* speaks to this concept of savoring unique, transitory experiences which often fetch a high price or, depending on your aesthetic, cost nothing. Although luxury plant foods typically go beyond basics and necessities, there is a tension in capitalism caused by the growth imperative that drives efforts to increase sales by "democratizing" access, hence the term "affordable luxuries" common in commodity and food history literature. Thus, a high-status item in one century like white sugar can later become a staple through heightened production that makes it no longer rare and thereby reduces consumer costs. In turn, this drives a perpetual search for new foods that appeal to the aesthetic that access to true luxuries is restricted to the most affluent in society, hence one of its commonly divisive qualities which can contribute to conflict and even social movements.[1] Put bluntly, "entrance barriers" to purchasing luxuries such as high prices and "selective and exclusive distribution" exclude access to most consumers, while luxury brands depend on being "desired by all but consumed only by the happy few" (Kapferer 1997: 255).

Over the history of capitalism, access to food in the dominant commercial centers of the world has been improved by making food cheaper through technological innovation, agricultural advances, the exploitation of new arable lands and the farmers who work them. Not only have staples become more abundant with greater affluence, the lower classes could increasingly afford a variety of plant foods previously reserved for society's elites. For example, during the seventeenth century luxuries like sugar (*Saccharum officinarum*), tea (*Camellia sinensis*), and black pepper (*Piper nigrum*), among other rare spices, such as cloves (*Syzygium aromaticum*) and nutmeg (*Myristica* spp.), became more widely available in the Dutch Republic as wealth increased in part due to domestic land reclamation projects and trade. Dutch commercial shipping linked the small republic to Chinese ports and tropical colonies (Schama 1997). The Dutch also established colonial plantations of luxuries like cacao (*Theobroma cacao*) and coffee (*Coffea* spp.) (Brockway 2002: 51, 53). By the eighteenth century and even earlier, "colonies served as a fertile ground for the procurement and production of tropical plants that would not grow in harsh European climates" (Schiebinger 2004: 6). Botanists as bioprospectors aided in the search for "green gold," by seeking desirable edible and medicinal plants that could provide "renewable profits long after gold and silver ran out" (Schiebinger 2004: 7).

During the eighteenth and nineteenth centuries, the price of tea and sugar steadily dropped in England as it urbanized and industrialized, making at least lower grades of both accessible to wage laborers while more nutritious foods became less accessible (Mintz 1985: 114–22). Increased production globally of luxury foods, like sugar, tea, coffee, spices, and tropical fruits, at more affordable prices, depended on slavery and other forms of servitude in the early history of capitalism (Busch et al. 1995: 19–20; Mintz 1985; Soluri 2005). Indeed, politicized British housewives led boycotts of slave-produced sugar beginning in 1791 advocating for abstention or purchase of more expensive sugar produced with free labor thereby helping to cultivate domestic support for abolition (Midgley 1996). By the twentieth century, the British and white settlers in Britain's colonies and former colonies increasingly gained greater access to tropical foods including various perishable fruits through territorial conquest, exploited labor, deforestation, and improvements in transportation networks (Collingham 2017).

During the 1900s, an increasing number of plant foods was added to the diet of middle- and lower-income consumers in the United States (US) and other industrialized regions of the world, foods that had historically been reserved for elites. Former high-status foods, like avocado (*Persea americana*), chocolate, citrus (*Citrus* spp.), coffee, pineapple (*Ananas comosus*), sweet banana cultivars (*Musa acuminata*),[2] and white bread, became common consumables over the century within these core commercial areas. Some luxury plant foods that have remained luxuries might be reasonably viewed as perverse. For example, single specimens of rare Japanese melon cultivars cost more than a year's pay for the poorest people on the planet, and the most expensive coffee beans are retrieved from the droppings of caged and force-fed civet cats (*Paradoxurus hermaphroditus*) by poor people in the Global South (primarily in Indonesia) for affluent consumers in the Global North, primarily in Japan and the US (Bale 2016; Berthon et al. 2009: 48; Money 2007: 50).

Despite the historic pattern in wealthy nations of expanding access for the middle and lower classes to cheap staples and former luxuries, continued access to both may be threatened in the twenty-first century for multiple reasons: a foreseeable end to new agricultural lands, increasing drought and water costs caused in part by climate change, a rise in unresolvable insect and pest problems exacerbated by globalization and climate change, evolution of herbicide resistant "superweeds," and a growing number of social movements opposed to agribusiness as usual (Friedmann 2004; Moore 2010). Also linked to climate change, rising sea levels and more numerous extreme weather events pose serious threats to producing enough food globally (Chakraborty and Newton 2011; Lobell et al. 2006). According to the Intergovernmental Panel on Climate Change (IPCC), increases in food prices globally for wheat (*Triticum aestivum*), corn (*Zea mays*), and soy (*Glycine max*) during the 2010s were linked to extreme weather events and associated drops in production (Vermeulen 2014: 3). Late spring frosts, flooding, high winds, and unusual gaps in rainfall can devastate crops (Creasy and Creasy 2018: 91).

Many agricultural scientists argue that increasing crop variability through genetic engineering will reduce the negative consequences of climate change (Dicenta et al. 2017: 184) and other threats to food production such as diseases and pests (Hefferon and Herring 2017). "Some advocates claim that GMOs [genetically modified organisms] could result in more sustainable food crops, such as drought-tolerant or saline-tolerant wheat" and that they "could assist with widespread nutritional deficiencies" by engineering foods with higher nutritional value such as beta-carotene enriched "golden rice" (Ankeny 2012: 468–9).

With the turn of the twenty-first century, great hopes, along with large sums of capital, have been invested in biotechnology including genetic engineering herbicide-resistant crops. However, thus far, the returns on those investments have overall fallen short of expectations (Moore 2010). Many productionist scientists believe more capital needs to be invested in agricultural technologies to be able to produce 50 percent more food globally in order to feed the projected world's population by 2050 (Chakraborty and Newton 2011). This perspective stems from the belief that a lack of food causes hunger; however, historically famines can often be linked in large part to political, economic, and social factors.[3] Despite potential positive outcomes of crop genetic engineering including greater food security, controversy over its safe and ethical use has stymied its adoption (Ankeny 2012: 468). Hefferon and Herring (2017) argue that investment in modern agricultural biotechnology might be improved by "removing obstacles of regulation and the stigma of the GMO from genome-edited crops."

While various environmental challenges threaten to revert multiple crops that were formerly luxuries back into high-status plant foods,[4] a variety of contemporary food movements are working towards a different kind of future food system that includes a reframing of luxury foods. In contrast to the common definition of rare and hence expensive, delectable items consumed by an affluent minority as an expression of their class-privilege, many contemporary food activists argue having "educated taste" includes avoiding "cruel foods" produced through exploiting people and other animals, plants, and ecosystems (Strong 2011).[5] In this alternative reality, those with good taste support the production and consumption of crops that foster cultural and biological diversity and wellness, e.g. through agroecological methods and fair trade (Nabhan et al. 2016; Salmón 2012).

With a focus on the US, this chapter explores how access to former luxury plant foods increased from 1920 to the present. Simultaneously, Americans significantly increased their consumption of processed foods made from cheap staples like wheat, rice (*Oryza sativa*), corn, sugar, soy, and hydrogenated palm oil (primarily from *Elaeis* spp.). Despite wider access to many nutritious foods, these cheap staples contributed to a multitude of "affluence diseases" which commonly appear in populations that shift from largely plant-based diets to ones with more dairy, grains, meat, and processed foods (McMichael 2013). The contemporary obesity epidemic in the US and the many chronic diseases associated with the modern industrial diet are now sparking a lucrative expansion of "wellness foods" marketed as health promoting (MacKendrick 2018; McMichael 2013; Winson 2013). Current market trends suggest that a healthy diet may be a luxury, accessible primarily to the affluent (Birch and Bonwick 2019: 1478). Many "superfoods" have become novel luxuries favored by discriminating consumers and celebrity chefs although efforts are under way to make them cheaper, following a centuries-long trend.[6] This chapter considers how contemporary food movements, focused on fundamentally changing the global food system to make it more socially just and sustainable, embrace novelties like regional indigenous staples and heirloom fruit and vegetable cultivars. Ironically, by helping to popularize new plant foods that then become globally traded and marketed as exotic superfoods, these movements can unintentionally contribute to culinary colonialism, the privatizing of botanical food resources held in common. Alternatively, they can achieve some of their equity and sustainability goals, for example, through organic and fair-trade certification programs. Using a transdisciplinary approach, evidence is drawn from agroecology, economic botany, plant pathology, political economy, sociology, and historical accounts including commodity histories.

THE INTERWAR YEARS

Economic growth in the food sector radically expanded consumption options in the interwar years. The end of the First World War in 1918 marked the beginning of a period of major cultural, political, economic, and social change in the US and Europe. For example, suffrage victories swept the US and Europe during the early twentieth century. Simultaneously, economic opportunities increased for many women. As more women moved into paid work in the early twentieth century, a new market emerged for processed foods that reduced the labor to put food on the table. Processed foods often mimicked luxury foods that had been formerly out of reach to the middle and working classes, such as white bread. The motto "the best thing since sliced bread" captured the sentiment of a generation of women whose mothers had been expected to bake bread from scratch to meet the daily needs of their households. "WW I helped move the baking of bread from the home to commercial bakeries as more women took work outside the home" (Newman 2013: 220). Marketing of foods increased rapidly during the interwar years and typically targeted women as the primary decision-makers regarding food choice. With "the discovery of vitamins in the twenties," manufacturers of processed foods began to appeal to "nutrition anxiety" with claims of scientifically improved foods that met daily requirements of "adequate nourishment" (McFeely 2000: 65).

The 1920 US census captured another important turning point "when urban residents first outnumbered rural" (Mendelson 2013: 221). Populations that had been dominated by farmers relatively quickly became city dwellers often without easy access to fresh homegrown fruits and vegetables. California, Oregon, and eastern Washington State became important industrial agricultural producers of fruits and vegetables for the entire US. Late-nineteenth- and early-twentieth-century horticulturalists searched for new fruit and vegetable species and cultivars that could be industrially produced as staples or luxuries. For example, as an agricultural explorer with the US Department of Agriculture, David Grandison Fairchild (1869–1954) introduced many new crops into the US, including avocados, dates (*Phoenix dactylifera*), mangos (*Mangifera indica*), nectarines (*Prunus persica*), pistachios (*Pistacia vera*), and soybeans (Petruzzello 2020). He had great hopes for mangosteen (*Garcinia mangostana*, from the Malay Archipelago), which he called "the queen of tropical fruits" and believed it would "bring fancy prices" (Stone 2018: 190).

Increased productivity and efficiency resulting from industrialization, scientific agriculture, improved transportation infrastructure, and new business models transformed many formerly exclusive high-priced foods into affordable luxuries. Shortened transit times combined with new refrigeration technologies reduced losses. Developing the capacity to transport perishable produce over long distances was a key achievement of twentieth-century capitalists (Julier 2005). By 1920, United Fruit emerged as one of the largest US corporations with "a vertically integrated network of plantations, refrigerated steamships, and railroad cars to produce, transport, distribute, and market bananas" (Jenkins 2013: 113). United Fruit established "banana republics" in several Latin American countries and controlled vast tracts of land used for banana plantations. The resulting abundance enabled the invention of numerous new dishes like the iconic banana split ice cream treat, which became a mainstay of American soda fountains by the 1920s (Jenkins 2013: 114). By the twentieth century, sweet bananas had "lost their status as a luxury food and were transformed through low price, year-round availability, and abundance into comfort food" (Jenkins 2013: 114).

During much of the nineteenth century, citrus fruits were consumed as a rare treat for most people living in frost-prone regions of the US. However, many wealthy Americans with a greenhouse grew their own oranges (*Citrus X aurantium*) and lemons (*Citrus X limon*) (Karp 2013: 418). As a crop that favors a subtropical to Mediterranean climate, "large-scale commercial cultivation began in Florida and California in the 1870s and 1880s" (Karp 2013: 418; see also Figure 2.1). Shipping costs to transport fresh fruit to

FIGURE 2.1 During the late nineteenth and early twentieth century, southern California agriculturalists established monocultures of luxury foods like citrus which were promoted as more efficient modes of food production. Sunkist marketed grapefruit picturing Indigenous peoples as tacitly condoning this transformation of their dispossessed lands as "beautiful," while simultaneously misrepresenting North American tribal diversity by picturing the man and his "belle" with beaded moccasins and other Plains Indian apparel. Indian Belle Grapefruit, lithographic label produced in Porterville, California, *c*. 1920. Photo by Transcendental Graphics, courtesy of Getty Images.

the east coast spurred California growers, in particular, to develop year-round products to recover their capital investments in infrastructure. 'Valencia' oranges proved important as a popular source of fresh orange juice (Figure 2.2). By the late nineteenth century, California 'Valencia' orange groves were bearing enough to saturate the domestic market (Farmer 2013: 252). Cooperative marketing associations like Sunkist attempted to manage supply and demand to avoid periods of over-production which proved costly for growers. They increased demand in part through branding and marketing. By the 1930s, Sunkist was advertising orange juice as a healthy alternative to artificial soft drinks (Sackman 2007: 101).

During the early twentieth century, smaller-scale horticultural producers of citrus gave way to agribusiness plantation growers, further facilitating production management. "As

FIGURE 2.2 An upper-class, white, male consumer pictured as a connoisseur of the best orange juice. Branded for the epicure, 'Valencia' oranges were marketed by the Santiago Orange Growers Association in Orange County California, *c*. 1930–40. Conditional copyright permission granted by Villa Park Orchards Association, image file courtesy of the Orange Public Library & History Center.

of WWI, 30 acres was still considered a large grove, and the majority of growers owned just 5 to 10 acres ... [But] by 1940, roughly 4 percent of California's orange growers owned 40 percent of the total acreage" (Farmer 2013: 267). Surplus production allowed Americans to ship frozen orange juice concentrate to armed forces and Second World War allies (Sackman 2007: 291). There emerged a powerful horticapitalist minority that contributed to the rise of pervasive advertising culture. During the Second World War, "Sunkist [even] lobbied to place oranges on the list of essential foods," foods that would help win the war (Sackman 2007: 291).[7] After the Second World War, the advent of commercial frozen orange juice concentrate sparked "explosive growth of the citrus processing industry" (Karp 2013: 418).

Despite efficiencies gained through managerial control and other shifts in business practices, twentieth-century agricultural workers, such as those producing and harvesting oranges in California, continued to suffer from insufficient compensation and grueling working conditions, as well as dangerous pesticide exposures and violent union busting (Sackman 2007). "During the early 1970s, thousands of banana workers were rendered sterile by" the widely used fungicide dibromochloropropane (DBCP); however, "historically the banana companies have not accepted responsibility for the health and safety of their workers, the community, or the environment" (Vandermeer and Perfecto 2005: 5–6). Exploited labor, combined with greater efficiencies resulting from monocultures and industrialized production, meant some enjoyed luxuries previously unattainable, but at the expense of others (Figures 2.3, 2.4, and 2.5).

FIGURE 2.3 Sorting and drying coffee berries on a Kenya coffee plantation in 1925. Kenya was a British colony from 1920–63. Courtesy of Alamy.

FIGURE 2.4 A 1938 coffee house in Prague, Czech Republic. Sueddeutsche Zeitung. Courtesy of Alamy.

THE SECOND WORLD WAR AND AMERICAN INDUSTRIAL SCIENTIFIC AGRICULTURE

By the late nineteenth century, regions of the US with fertile soils and a semi-tropical or Mediterranean climate had already begun to be transformed into important food production zones. By the Second World War, investment in industrial scientific agriculture yielded significant productivity gains in both staple and specialty crops contributing to an abundance of food for the US and its European allies. Access to high-status foods like canned pineapple, white sugar, and coffee remained during the Second World War, and even increased for some Americans, due to federal food policies and higher employment rates. After the earlier deprivations during the Great Depression (1929–33), Americans willingly "spent more on food" and many more were able to "afford food that otherwise would demand exorbitant prices," in part due to government price controls and rationing (Bentley 2013: 227).

During wartime rationing, the US Department of Agriculture advocated one domestic meatless day a week to enable the sending of more meat and dairy to the front line (McFeely 2000: 74–5). Nuts offered an alternative plant-based source of protein. However, most were consumed primarily in fall and winter as seasonal specialty crops. Access was also significantly constrained regionally. Prior to 1930, consumption of

FIGURE 2.5 A Cadbury's milk chocolate advertisement promoting cacao grown within the British Empire and claiming their candy is food, due to its high cocoa butter content. As such, the advertisement claims, it is a good source of fat, given wartime dairy butter rationing. From a 1940 UK magazine. Contributed by John Frost Newspapers, courtesy of Alamy.

various pecan (*Carya illinoinensis*) dishes, including pecan pie, was primarily restricted to the southeastern US where pecans are native. Through the application of agricultural science, particularly high-yielding grafted cultivars, pecan production began to exceed demand in the 1930s, provoking the US Department of Agriculture to promote it as a nutritious alternative protein source, especially during the Second World War. This federal agency even provided American housewives with recipes including for savory nut loaves to take the place of meatloaf (McWilliams 2013).

During the Second World War, various federal agencies increasingly shaped public opinion and buying habits through education and food policies often influenced by corporate lobbyists. For example, the government promoted oranges and grapefruit[8] as a means to obtain optimum health through the daily consumption of these and other Vitamin C rich fruits. The Food and Nutrition Board sought to educate housewives through the establishment of recommended daily allowances (RDA) of various nutrients (Bentley 2013: 230). Although more efficient and productive food systems nearly eliminated starvation in industrialized countries, it was discovered in the 1930s that a majority of "the well-fed middle class" suffered from "hidden malnutrition" even in superpowers like the UK (Collingham 2017: 267). In 1941, President Roosevelt called for a National Nutrition Conference for Defense which "established standards for a healthy diet" based on a greater diversity of foods including fruits and vegetables and less reliance on meat and potatoes (McFeely 2000: 84). The credibility of nutrition science increased in the 1940s, and it began to be used to affect public policy.

MID-CENTURY CONSUMER CULTURE

The post-war period, with its greater access to former luxury foods, particularly in Europe and the US, is described by many historians as a period of food democratization. For example, industrial California orange producers claimed their efforts "democratized the fruits of Eden to create a landscape of abundance that could be enjoyed by all" (Sackman 2007: 7–8).[9] Mid-century US foreign policy included distributing agricultural surpluses to regions in the world with strategic importance so as to increase American global influence (McMichael 2013: 5). Once signed into law by President Dwight D. Eisenhower, the Agricultural Trade Development and Assistance Act of 1954 became known as Public Law 83–480 (Riley 2017: 183). The law, abbreviated to PL-480, resulted in excess US-grown grains and other staples being routed to select, anti-communist Third World states as food aid (McMichael 2013: 5).

Although American foreign food assistance functioned in large part to dispose of domestic agricultural surplus, in the 1960s the humanitarian purpose of Food for Peace policies was increasingly emphasized by President Robert F. Kennedy and others (McDonald 2017: 143–8). Food aid enabled efforts to spark industrialization and an expansion of the global economic system including "the penetration of international capital into previously self-sufficient agrarian societies" (Friedmann 1982: S251). Recipient countries embraced "the U.S. model of national agro-industrialization" including the use of Green Revolution technologies (McMichael 2013: 5), and shifted their own agricultural production to fewer staples and more cash crops for export like bananas, cacao, and coffee. Caught in a cycle of "ever-rising productivity, falling prices, shifts among commodities, and the search for profit through the launch of new products," agribusiness repeatedly "[turned] luxury consumption very gradually into mass consumption" (Freedman 2007: 340).

After the Second World War, the rapid expansion of US suburbs, the post-war baby boom, and the rise of mass media all contributed to a dazzling expansion of American consumer culture. By mid-century, pre-made and processed foods came to symbolize the scientific, sanitary, and modern choice (Figure 2.6). And these foods were increasingly available in the rapidly expanding grocery store chains. Supermarkets "became a symbol of American political, economic, and cultural dominance in the Cold War era" and came to be called "shining food palaces" (Bentley 2016: 12). Although economic growth occurred as the American military-industrial complex ramped up during the war, post-war economic growth in the US was unprecedented. The Servicemen's Readjustment Act of 1944 (aka the G.I. Bill) facilitated reintegration of American veterans back into civilian life. However, Euro-Americans benefited disproportionally from low-cost mortgages and

FIGURE 2.6 Woman shopping at the Los Angeles Farmers Market bakery with a wide selection of pre-made pies, Los Angeles, California *c.* 1950. Photo by Transcendental Graphics, courtesy of Getty Images.

other federal financial benefits. Throughout the US, racially exclusive suburbs expanded rapidly, amplifying hetero-normative and racist values, as well as a consumer culture.

During the post-war baby boom, many women left paid work to focus on child-rearing. From 1946–64, American women gave birth to over seventy-five million babies (US Census Bureau 2015); however, family size at roughly three children was still smaller than the period before the First World War with an average of five children (US Bureau of Labor Statistics 2006: 15). Convenience foods became one of "mommy's little helpers." The expanding generation of suburban, white, middle-class Americans could also now afford modern kitchen appliances. After the deprivations of the Depression and the Second World War, a heightened demand for durable goods was unleashed. New refrigerators with built-in freezers led to a sharp rise in frozen food options (Figure 2.7). While "fresh-squeezed orange juice remained a luxury through the early 20th century, ... by the early 1950s [frozen] orange juice accounted for 20 percent of the frozen food market" (Smith 2013b: 748).

Readily available, affordable baked goods and other processed foods relieved mid-century housewives of having to cook a full meal. TV dinners "first marketed in 1954" offered a welcome relief from the demands of motherhood (McFeely 2000: 32). However, when women did cook from scratch, the stakes seemed higher. "The sustained crises of Prohibition, the Depression, and WWII ... left Americans eager to demonstrate recovered morale and prosperity through ample indulgence of appetite" (Mendelson 2013: 226). Good living journalism helped to unleash "the fantasy of giddy indulgence" (Mendelson

FIGURE 2.7 Advertisement of a huge open refrigerator stuffed with food, with a little girl reaching for a pink cake, 1955. Screen print from a photograph. Photo by GraphicaArtis, courtesy of Getty Images.

2013: 226). Launched in 1941, *Gourmet* magazine set new standards for the culinary skills of housewives, promoted the use of unfamiliar ingredients, and inspired food-based travel. By the 1950s, many middle-class Americans were joining gourmet clubs and embracing wine tasting as a pastime (Lovegren 2013a: 237). French cooking became popularized in the US in large part by Julia Child (1912–2004), who published her first book *Mastering the Art of French Cooking* in 1961. The interest in *haute* French cuisine went hand-in-glove with a new interest in cultivating wine connoisseurship. Child became one of the first celebrity chefs, as did her contemporary James Beard (1903–85), who opened a cooking school in 1955 that promoted American food culture.

Cookbooks during this period reflect shifting trends. Rising interest in gourmet cooking extended to foods from other regions of the world. Avocados were in large part popularized by California growers through decades of promotion beginning in 1900. In 1966, the first cookbook devoted to their preparation appeared that wasn't industry produced. Guacamole played a key role in rising American consumption of the Central American fruit (Smith 2013a: 100). As "large-scale, capital-intensive farms dedicated to single cash crops" expanded in the early twentieth century, California farmers sought out "horticultural crops" in the form of fruits and vegetables that satiated specialized market niches (Mendelson 2013: 225). As a result, by mid-century, a wider diversity of vegetables, such as artichokes (*Cynara cardunculus*), broccoli (*Brassica oleracea*), fennel bulbs (*Foeniculum vulgare*), and zucchini (*Cucurbita pepo*) fed an expanding enthusiasm for Italian cuisine and other ethnic culinary traditions.

INCREASING DEMAND FOR HIGHER QUALITY FOOD AND DRINK IN THE 1960S AND 1970S

Trend-setters play an important role in sparking demand for luxury goods. Beginning in the 1960s, California became a hub of culinary innovation. Peet's Coffee opened in Berkeley in 1966, at first selling only freshly roasted beans. Founder Alfred Peet (1920–2007) was a Dutch immigrant who was appalled by the quality of the coffee Americans were drinking and set out to remedy the problem (Marshall 2007).[10] The neighborhood of his original shop became a food culture mecca with the celebrated restaurant Chez Panisse established in 1971. Founding chef Alice Waters (b. 1944) was an early promoter of local sourcing of produce, as was her nearby contemporary Deborah Madison who co-founded Greens, a vegetarian restaurant that opened in 1979 as an affiliate of the San Francisco Zen Center and its Green Gulch Farm. Less than a block away from the original Peet's Coffee, a gourmet chocolate shop called Cocolat opened in 1976. The owner Alice Medrich (b. *c.* 1950), who first tasted chocolate truffles in Paris, began selling a handmade version from her Berkeley home in 1973. According to a *New York Times* article, on their first day of business the Cocolat shop sold every truffle and cake available in three hours (Barron 1988: 31). Medrich is credited with launching a truffle craze among Americans, most of whom at the time had not tasted French-style chocolate confections before.

Increased demand for higher-quality coffee and chocolate in the US contributed to an expansion during the 1960s and 1970s in select post-colonial regions of specialty crops grown for export. That shift in agricultural priorities was made possible in part by US food aid beginning in 1954. Access to cheaper staples in recipient Third World states came at the expense of self-sufficiency as traditional agricultural practices waned, including an emphasis on greater crop diversity (Friedmann 1982: S267–S272). Also, during the 1970s, economic stagnation in the US contributed to significant shifts. To afford the consumptive, suburban "American dream," an increasing number of middle-class housewives were re-entering the paid labor force to offset inflation and falling wages of the primary earner. From 1960 to 2000, the number of American women working outside the home rose sharply from 40 percent to nearly 70 percent (Lovegren 2013b: 241). As a result, the new normal became store-bought, prepared food consumed at home, as opposed to homemade meals.

LATE-TWENTIETH- AND EARLY-TWENTY-FIRST-CENTURY FOODIES

In the 1980s, "'Foodie' became a new word, used to describe someone who is passionate or knowledgeable about cooking or eating" (Lovegren 2013b: 244). Rarity often contributes to desirability. An extreme example of a rare specialty export crop is civet coffee (aka Kopi Luwak coffee), which is among the most expensive in the world (Figure 2.8). The beans are retrieved from civet cat feces after they eat the fleshy coffee fruits (Bale 2016; Money 2007: 50). To connoisseurs, they yield a distinctive and desirable brew worthy of the high price. Despite contemporary concerns about omnivorous caged civet cats being force-fed coffee beans, demand remains high (Carder et al. 2016). The Densuke black-skinned watermelon (*Citrullus lanatus*) and Yubari King melon (*Cucumis melo*) are expensive, premium fruits from the Japanese island of Hokkaido and sought after for gift giving, a cultural practice called *omiyage* (Bosker 2017; Forbes 2016b). The Japanese Agricultural

FIGURE 2.8 Civet cat feces pictured with visible coffee beans in Bali, Indonesia, the beans being retrieved and processed to make some of the most expensive coffee in the world. Photo by Pakawat Thongcharoen, courtesy of Getty Images.

Standards Act of 1947 requires that the place of origin of perishable foods be provided to consumers. With a 2004 amendment to the Japanese Trademark Act, regional producers now brand and market their agricultural products based on geographical origin (Kojo 2006). Ethiopia trademarked several popular coffee varieties despite opposition initially from Starbucks, which later raised their prices on products containing the regional specialty crops claiming sales aided poor farmers (Vaidhyanathan 2017: 5). Connoisseurs pride themselves in recognizing the unique taste of the land also known as the *terroir*.

Although not technically plant products, wild foraged fungal delicacies also fetch high prices as exotic vegetables. Ironically, many can be collected by knowledgeable foragers for nothing, being in abundance under the right conditions and in the appropriate habitats, but they are purchased only at a premium. For example, while related species in the genus *Sparassis* are “a staple item in the monsoon diet of many Himalayan villagers,” the cauliflower mushroom (*S. radicata*) found in old growth coniferous forests in western North America is considered an “elegant” and “exceptional” fungus for “a special occasion” (Arora 1986: 657).[11] Some prized and economically significant fungi, like chanterelles (species of *Cantharellus* and allied genera) and matsutake (*Tricholoma matsutake*), can only be collected in the wild because successful cultivation of them eludes us. By the late twentieth century, depopulation of the Japanese countryside reduced traditional management of pine forests and thus the optimum growing conditions for matsutake. As production dropped, Japan began importing the wild foraged fungus from western North America (Tsing 2015). Matsutake and many other fungal species are claimed to be medicinal, which increases their desirability (Voeks 2018: 160).

Industrial agriculture homogenized food systems globally concentrating on high-yielding cultivars and greater uniformity (Busch et al. 1995: 21). The "focus on particular improved varieties has narrowed the genetic base of our food supply" (Pilcher 2016: 44). For example, 'Lisbon' lemons became the only lemon in most US grocery stores during the twentieth century. Thin-skinned lemon cultivars like 'Meyer' do not transport well over long distances because they are prone to rot more quickly than the thicker-skinned 'Lisbon' is. They are now marketed as a specialty crop along with many other unusual citrus. Many perishable delicacies have become novelties for urban epicures who seek them out in boutique grocery stores and farmers' markets specializing in unusual fruits and vegetables. While foodies often embrace expensive and novel ingredients, an increasing number of shoppers began seeking out organic, fair-trade, and locally sourced products because of their health and social justice concerns.

TURN OF THE CENTURY FOOD MOVEMENTS

Grassroots responses to the negative impacts of the industrialized global food system began proliferating in the late twentieth and early twenty-first centuries. Due in part to less expensive food, the percentage of income spent on food by Americans dropped from over 40 percent in 1920 to less than 15 percent by 2002 (US Bureau of Labor Statistics 2006: 69). However, "greater concentration, standardization, and globalization" of food systems occurred contributing to "world production [that] ... increased at a staggering rate, yet these gains have been won at a heavy cost to the environment, and to human and animal welfare" (Pilcher 2016: 44). American consumers increasingly used educated choices of foodstuffs in an attempt to protect their health. "The 1990s marked a major turning point in the American regulatory landscape, establishing consumer self-protection as the standard response to government inaction ... precautionary consumption has become a necessary default response to a regulatory system that consistently puts industry profit ahead of public health" (MacKendrick 2018: 38). By emphasizing local, organic production, small-scale artisanal and craft foods and beverages producers sought to provide alternatives to agribusiness that reduces costs in part by increasing productivity with pesticides and chemical fertilizers while disregarding environmental and human health risks. European Union (EU) policies established in 2007 reduced the personal burden on European consumers to be informed and selective, by maintaining stricter safety standards than those established in the US, in order to reduce health hazards from the industrialized global food system such as pesticide residues (MacKendrick 2018: 31, 38).[12]

Nutrient-poor foods produced by industrialized food systems, as well as chemical hazards, have contributed to a global health crisis (Winson 2013: 36). With chronic degenerative diseases now "the leading cause of death worldwide, including cancer, diabetes, ... and neurological disorders, ... prevention offers the advantage of being far less costly compared with treatment" (Santana-Gálvez et al. 2019: 179). Consumption of more fruits and vegetables reduces the incidence of chronic disease, although the reasons for this are poorly understood (Nestle 2018: 76). Plant-based diets can be effective in treating common health conditions. For example, multiple clinical studies indicate a vegan diet can prevent or reverse diabetes (Davis and Melina 2014: 62–4). Gourmet plant-based meals are increasingly featured in fine dining venues as luxuriously decadent for those who can afford it. For example, Millennium, an elite vegan restaurant in the San Francisco Bay Area, strives "to elevate vegetable-based cuisine to the sublime" (Millennium Restaurant 2019). Advocates of organic, plant-based diets often claim they are paying their health costs forward by reducing their risks for problems later in life.

As diet-related diseases increasingly became global epidemics, sales of food marketed as medicinal expanded. Between 2002 and 2007, the global nutraceutical market increased from \$46.7 to \$74.7 billion, with the greatest growth in the US, UK, and Japan (Das et al. 2012: 174). The term nutraceutical was coined in the late twentieth century as "any substance that is a food or part of a food and provides medical or health benefits, including the prevention and treatment of disease" (DeFelice 1995: 59). Many plant foods are now marketed as nutraceuticals or similarly health-promoting, nutrient-dense functional foods or superfoods. For example, maca is promoted as an aid to stamina including sexual, and lucuma (*Pouteria lucuma*) as a healthy sweetener for diabetics. In some cases, their reputed healing properties are confirmed through phytochemical and clinical research (Voeks 2018: 160).

Many of the globally traded superfoods were until recently little-known native species used by Indigenous people primarily from the Americas (see Table 2.1; see also Figure 2.9). Market forces drive a perpetual hunt for rare and precious commodities that have the potential to become profitable. Advocates for the industry realize that,

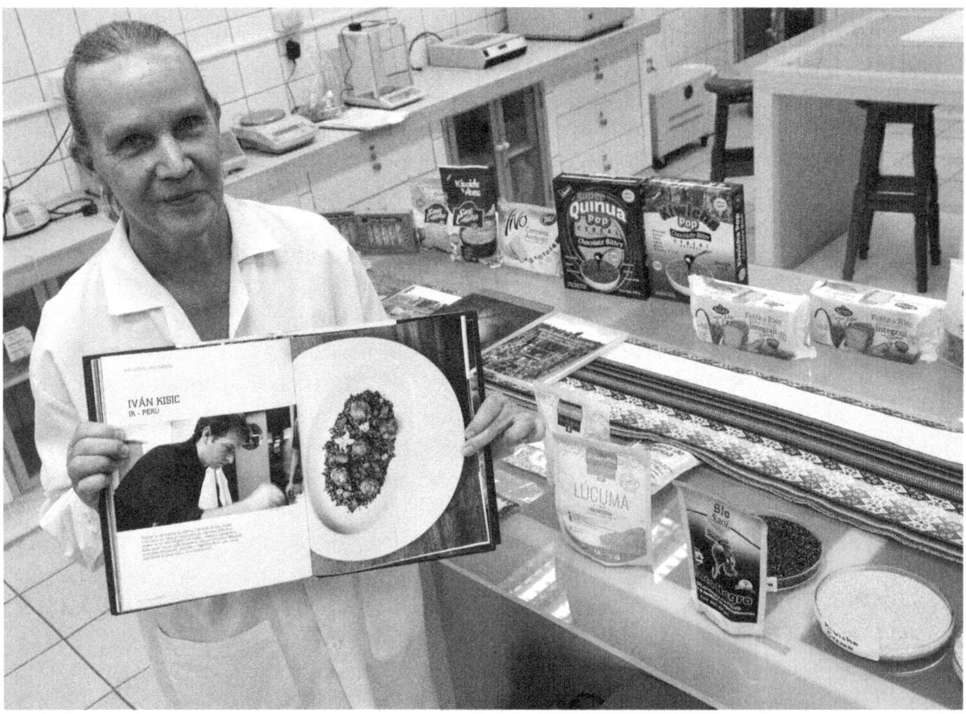

FIGURE 2.9 Finnish food chemist Ritva Repo-Carrasco-Valencia, pictured in the food lab of the National Agrarian University La Molina, in Lima, Peru in 2018. She is showing a photo of the Peruvian celebrity chef Ivàn Kisic (d. 2012) using superfoods in a trendy culinary creation. Peru is a major producer and exporter of superfoods, which have been part of the diet of Indigenous people in the Andean highlands for hundreds if not thousands of years. Photo by Cris Bouroncle/AFP, courtesy of Getty Images.

to maintain their credibility, proving health claims is vital (Birch and Bonwick 2019; Papetti 2019). However, according to nutrition scientist Marion Nestle (2018: 76), the moniker of superfood "is an advertising concept," not a scientifically defendable term. Nevertheless, in many developed countries aging people who seek to remain active with a healthy lifestyle provide an economic incentive to continue to market medicinal foods.[13]

For a variety of reasons, testing the various health claims made about health foods is challenging. Clinically based, double-blind tests on the impact of foods are difficult to perform due to confounding variables and funding for researching alternative medical interventions through nutrition is highly limited. Nonetheless, even optimistic, inconclusive results can boost use and sales, particularly when difficult-to-treat conditions are involved. For example, some individuals living with HIV or AIDS consume avocado and a range of other tropical fruits "to maintain and strengthen their immune system" (Voeks 2018: 160). Foods high in antioxidants have been promoted for years as protective against Alzheimer's disease, an assertion based on research results that suggest "total antioxidant intake *may* influence cognitive decline with age through the neuro-protective action of antioxidants" (Tapsell et al. 2006: S13). Blueberries (*Vaccinium* spp.) in particular have been successfully branded as a superfood for this reason. Claims that wild foraged blueberries contain even higher amounts of antioxidants than the more common commercial highbush blueberry (*Vaccinium corymbosum*) does contributed to increased wild collection in the eastern US and neighboring areas of Canada. Heightened demand sparked productivity enhancement efforts to the point that the market was saturated and prices dropped (Nestle 2018: 77–8). Antioxidant-rich foods continue to be marketed as an effective means to address oxidative stress, which may contribute to a variety of diseases often associated with aging (Das et al. 2012: 177, 183; also see Table 2.1).

Because of the cost, for many consumers it is a luxury to consume healthy, organic, non-GMO, wild-foraged, and fairly traded foods, a reality that many food activists seek to push back against (Figure 2.10). Critics claim a focus on rare, nutritious foods can manifest as an elitist lifestyle rooted in gender-, class-, and race-based privilege. Some critics condemn the common assumption that women are going to prepare all this time-consuming healthy food as a part of their unpaid "care work" expected of them in patriarchal societies (Cairns et al. 2015: 177–201). Other critics condemn what they call "culinary colonialism" (Johnston and Baumann 2015: 89–92). According to Enrique Salmón, author of *Eating the Landscape*, "preserving an heirloom tomato [*Solanum lycopersicum*] that can be purchased for an outlandish price at swanky farmers markets" misses the point of current efforts to revive endangered vegetables, fruits, and other plant food species and varieties (2012: 150). It is about revitalizing small-scale agriculture as a means to improve the quality of our food and food security by supporting family farmers as "dedicated stewards of foods" who are capable of being more adaptive and resilient than industrial farmers in the face of rapid environmental changes, including climatic (Salmón 2012: 11). Relying on a wider range of plant foods is a key element of food security.

Agriculture responsive to local community needs may be better positioned to increase access to healthy food for people. Some small producers can even profitably harvest and sell nutritious weeds from their farms, such as dandelion (*Taraxacum officinale*) and nettle greens (*Urtica dioica*) as seasonal gourmet specialties. Increasingly in Indigenous communities, traditional diets are promoted as a legitimate response to the diabetes epidemic and a means to renew an "identity connected to responsibility toward one's place in a community within a landscape" (Salmón 2012: 150). Rates of diabetes rose rapidly in

TABLE 2.1 Select superfoods marketed as health promoting

Common Name	Scientific Name	Place of Origin	Reputed Therapeutic Uses
acai	*Euterpe oleracea*	Trinidad to tropical So. Am.	antioxidants, healthy fats, weight loss
aronia	*Aronia melanocarpa*	e. No. Am.	antioxidants, nutrient dense
bitter-melon (balsampear)	*Momordica charantia*	tropical Asia and Africa	blood sugar regulation (anti-diabetic)
blueberries	*Vaccinium corymbosum*	e. No. Am.	antioxidants (anti-aging)
buffaloberry	*Shepherdia canadensis*	No. Am.	antioxidants
cacao	*Theobroma cacao*	Mexico to n. So. Am.	antioxidants
calafate	*Berberis microphylla*	Chili	antioxidants
camu berry	*Myrciaria dubia*	tropical So. Am.	antioxidants, immunity booster (high Vitamin C)
chia	*Salvia columbariae* *S. hispanica* and other *S.* spp.	s. Mexico to Ecuador	antioxidants, anti-anxiety, mild laxative, healthy fat, hepatoprotective
goji berry (wolfberry)	*Lycium barbarum*	n. and central China	antioxidants, immune booster and nutrient dense
goumi	*Elaeagnus multiflora*	China, Korea, Japan	antioxidants
hemp hearts	*Cannabis sativa*	se. Europe, Russia to nw. China	nutrient dense, protein source
kiwicha	*Amaranthus caudatus*	Ecuador to nw. Argentina	gluten-free pseudograin
lucuma	*Pouteria lucuma*	Colombia to n. Chile	blood sugar control, detoxification
maca	*Lepidium meyenii*	s. Peru to nw. Argentina	aphrodisiac, energy booster, hormonal balance
Matcha (green tea powder)	*Camellia sinensis*	probably China	antioxidants, energy and mood enhancer, cognitive performance
quinoa	*Chenopodium quinoa*	Ecuador to nw. Argentina	protein-rich, gluten-free pseudograin
sacha-inchi	*Plukenetia volubilis*	Windward Islands to tropical So. Am.	nutrient dense, protein source
saffron	*Crocus sativus*	India, the Balkans, e. Medit. region	anticancer, antioxidants, nutrient dense, may improve neurological conditions
yacon	*Smallanthus sonchifolius*	w. South Am.	blood sugar regulation, detoxification, prebiotic (FOS)

Sources included "Plants of the World Online" hosted by KewScience (http://www.plantsoftheworldonline.org) and Thomson Health Care, *PDR for Herbal Medicines* (Montvale, NJ: Thomson, 2007).

FIGURE 2.10 A protest against GMOs at New York City's Whole Foods Market on May 25, 2013. The upscale American multinational supermarket chain brands itself as a source of healthy foods produced through sustainable agriculture. According to their website on July 31, 2020, they "require that all non-GMO label claims be third-party verified or certified." Courtesy of Alamy.

the twentieth century and continue to rise among Americans who eat an industrial diet, with particularly high rates among Indigenous peoples (Winson 2013: 233). However, "the prevalence of type 2 diabetes tends to be much lower" among Native people who eat their traditional diet (Winson 2013: 233).

Various organized food activists seek alternatives to the negative impacts of transnational corporate agriculture on food systems (Robinson 2014: 223). For example, the Slow Food Movement's Ark of Taste project promotes "agricultural biodiversity and small-scale, family-based food production systems [that] are in danger throughout the world due to industrialization, genetic erosion, changing consumption patterns, climate change, the abandonment of rural areas, migration, and conflict" (Slow Food Foundation for Biodiversity 2019). However, marketing "edible treasures" and other gourmet specialty crops to affluent consumers risks negative consequences for the communities that have ensured their survival.

Fair-trade branding became for some producers a marketing strategy which co-opted the principles and instead promoted elitist consumerism (Ankeny 2012: 472, 473–5). By partnering with non-profit organizations, some companies use "cause-related marketing" to enhance their public image while making insignificant contributions to change that addresses social inequities and unsustainable production (Ankeny 2012: 470–1). For example, Starbucks misled consumers into believing they "pioneered the fair-trade model" while sourcing "only three in a hundred Starbucks beans ... from fair-trade sources" (Jaffee 2007: 203). According to director emeritus of the Institute for Food and Development Policy, Eric Holt-Giménez (2017: 172), believing in market-based efforts focused on transformation "lulls us into the magical belief that somehow we can change the food system without changing the capitalist system in which it is historically embedded. This is the political fetishization of food."

Novel indigenous foods have become trendy among foodies, which introduces the risk of food gentrification. Chefs of some fashionable restaurants purchase wild foods from professional foragers (Goldstein 2013: 310), such as California bay nuts (*Umbellularia californica*), Devil's club shoot tips (*Oplopanax horridus*), and Siberian miner's lettuce (*Claytonia sibirica*). Menus featuring wild plants and mushrooms can set a restaurant apart from its competitors (Strand and DiStefano 2010). Although many foragers are ethical, some overharvest on public and private lands or gather from roadsides potentially impacted by elevated lead levels in the soil, pesticide residues, and other pollutants. The potential environmental and social impact of wild foraging and commercial production of novel indigenous foods also raises concerns about continued cultural appropriation, biopiracy, and reduced access for Native people. In her essay entitled "Indigenous Food Sovereignty in Canada," Priscilla Settee (2018: 183) explains that "rapacious resource extraction based on colonial models have forced many land-based peoples into cities. In urban areas deprived of traditional foods we experience food deserts, extreme hunger, and the onset and entrenchment of preventable diseases leading to early death." Traditional native foods contribute to the affirmation of cultural identity key to survival.

Given the risks of commercializing indigenous foods in an attempt to capitalize on "gastronomic multiculturalism," some scholars argue that food sovereignty should include "the right to hold gastronomic capital back from the market" (Grey and Newman 2018: 717–18). Increased demand can raise the economic value of traditional indigenous foods to the point that industrial producers appropriate them, negatively impacting Indigenous peoples in the process. For example, industrially produced "wild" rice (*Zizania* spp.) cultivated in California fetches lower prices than the wild harvested grains of *Zizania palustris* (LaDuke 2005: 171–4). The Anishnaabeg prevailed legally in setting some limits on the marketing of industrial *Zizania* to protect the livelihood of community members who traditionally harvest a surplus of *manoomin*[14] as a source of income (Grey and Patel 2015: 439–40). Similarly, from 2011 to 2014, Indigenous Peruvian farmers benefited financially as the Andean protein-rich pseudograin quinoa "took consumer markets by storm," but the short-lived boom ended with a devastating collapse of prices when foreign producers entered the market (McDonell 2018). Frequently, a focus on growing cash crops for export can also cause reduced production of other food and medicinal plants and thus promote less self-sufficiency and more market dependency. Many Indigenous activists emphasize that cheap industrially produced food sabotages decolonization struggles (Grey and Patel 2015: 433–4).

CONTEMPORARY ENVIRONMENTAL THREATS: CLIMATE CHANGE, PLANT DISEASES, AND PESTS

Climate change is already impacting the production of chocolate, coffee, grapes (*Vitis vinifera*), and many other crops. According to the authors of *Climate Leviathan*, "the luxurious life of the capitalist global North is desiccating West Africa and scorching South Asia" (Wainwright and Mann 2018: 76). For chocolate consumers, this should be unsettling news, since two-thirds of cacao production occurs in West Africa, primarily Cameroun, Ghana, Côte d'Ivoire, and Nigeria (Money 2007: 69). "Market analysts pay constant attention to West Africa's seasonal weather and longer-term climate change, as well as its politics" (Leissle 2018: 6). In addition to its direct impacts, climate change can also increase cacao disease and pest outbreaks (Leissle 2018: 12).[15] Diseases cause a loss of up to 38 percent of global cacao production annually (Marelli et al. 2019: 1331).[16] Scientists predict the losses will increase with climate change. The severity of the current threat posed by cacao diseases has largely been masked temporarily, "production has only kept pace through unsustainable rainforest clearing and new plantings" (Marelli et al. 2019: 1340).

By comparison with annual crops, perennials like trees, shrubs, and woody vines are less adaptable to climate change. Based on computer models, "climate change in California is very likely to put downward pressure on yields of almonds [*Prunus amygdalus*], walnuts [*Juglans* spp.], avocados, and table grapes by 2050" (Lobell et al. 2006). Significant geographical shifts in production zones have already occurred. For example, vineyards in the western US have expanded northward including within the Willamette Valley of Oregon and eastern Washington. California's Central Valley, with its Mediterranean climate, produces over 80 percent of almonds traded globally and, based on computer models, will continue to be suitable for almond production through the twenty-first century (Parker and Abatzoglou 2018: 211–12). However, almonds require large amounts of water to thrive, making them a controversial crop in a warming world. Almond milk and other nut milks are capturing an increasing share of the milk market with more consumers adopting a plant-based diet. But rising water costs for growers have contributed to significant price increases for almond milk, almond butter, and other almond-based products, especially for their organic certified versions. Based in part on the rising cost of irrigation and in some cases public policies which restrict water use, some fear that "avocados are toast" as a global commodity because they may "become too expensive to grow profitably" (Sternberg 2015).

Allowing market demands to drive agricultural practices can result in unintended negative outcomes that reduce the resilience of the food system. For example, Chinese consumer demand for American pecans illustrates the influence that a rising middle class can have. China imported less than 1 percent of the US crop of pecans in 2005. By 2011, that jumped to 27 percent. Between 2005 and 2010, the Chinese increased their Mexican pecan imports from 0.6 to 15 million pounds (McWilliams 2013: 143). In China the brain-looking nuts are believed to reduce the risk of Alzheimer's disease, contributing to their popularity (McWilliams 2013: 143–4). However, the Chinese prefer a limited number of cultivated varieties, which raises concern about the loss of genetic diversity in the US (McWilliams 2013: 146–51). Given that the Chinese are applying scientific agricultural methods to enhance pecan production in southeastern China (Zhang et al. 2015), many people fear that, after American growers have shifted to disease risky monocultures of fewer cultivars, the Chinese will reduce US pecan imports. In the meantime, "the dominance of the Chinese market drove up the price of pecans" in the US and, as a result, Americans are eating fewer of the North American nuts. Domestic industries "were being priced out of the pecan market" and increasingly forced to drop the native nut from their ingredients list (McWilliams 2013: 144–5). Thus, pecans have made a full circle in the US from a regional specialty to being widely available and affordable, then to an expensive luxury.

In the late twentieth and early twenty-first centuries, agro-industrial production expanded and free-trade policies resulted in increased long-distance movement of food products. As a result, fossil fuel consumption increased exacerbating climate change in order to produce and transport food stuffs (Belasco 2008: 108–9). Globalization also resulted in more frequent movement of plant diseases and pests through trade in food as well as nursery stock plant material and wood products (Bowcutt 2015: 105–12). A focus on monocultural production, as opposed to biodiverse agricultural models, heightens biosecurity risks of disease spread and pest outbreaks once pathogens and pests arrive in a new location. As a result, many former luxury tropical plant foods, now widely commodified, are threatened. The agricultural risks taken to improve productivity are coming home to roost. Many agricultural scientists predict virulent plant diseases will in the near future cause significant consumer price increases for avocado, banana, chocolate, coffee, oranges, and others.

Commercial citrus production in Florida may be over because of the currently incurable citrus greening disease (aka huanglongbing caused by *Liberibacter asiaticus*). Between 2005 and 2019, the disease caused a 75 percent reduction in orange production and an 85 percent reduction in grapefruit output (Nelson 2019). Citrus greening has also killed groves in Asia, Brazil, and the Dominican Republic and has started to spread in southern California. According to Neil McRoberts, a plant disease epidemiologist and advisor to the Citrus Research Board of California, "citrus in the Central Valley [of California] might be spared because of the climate—that the winters are too cold and summers too hot for the insects that transmit the disease. … But … a few degrees warmer in the winter and cooler in the summer could change that. I think we're at a tipping point" (Nelson 2019). The popular dessert banana clone that is currently widely sold, 'Cavendish,' may soon be a thing of the past as a cheap globally traded staple because of Panama disease (*Fusarium oxysporum*) (Ordonez et al. 2015).

CONCLUSIONS

Throughout this chapter the term luxury plant foods has been interrogated by placing such foods in both historical and cultural contexts. Put simply, "the problem with the word 'luxury' is that it is at once a concept (a category), a subjective impression and a polemical term, often subjected to moral criticism" (Kapferer 1997: 251–2). In an economic system built to undercut prices through efficiency and disregard of the social and environmental costs of production, what meets the criteria of a luxury is not only debatable but perpetually shifting. Given the long history of converting high-status foods into affordable commodities, it remains unclear whether current food movements have the capacity to shift luxury food culture fundamentally. The rising popularity of artisanal foods and craft beverages in the Global North could reflect mounting critiques of capitalism's homogenizing effects and tendency to produce more food at the expense of quality and diversity, regional differences, labor rights, local control, and environmental sustainability. In the late twentieth and early twenty-first centuries, these commodities are often marketed as contributing to community-based initiatives focused on "taking back the food system." As market shares in carbonated soft drinks decline due to their correlation with diabetes, corporations producing these drinks buy small-scale kombucha producers and vertically integrate them. Now even discount American purveyors of foods sell cheap kombucha along with hemp hearts, maca, and quinoa. In capitalism, what was once on the fringe can become mainstream, and in the process the original associated critique and emancipatory visions can be eroded.

Currently, the use of genetic engineering and other technological interventions are increasingly embraced in an attempt to control threats to ongoing production of luxury and former luxury plant foods (Dale et al. 2017; Marelli et al. 2019).[17] However, it remains to be seen whether these interventions can effectively address global threats to a food system dependent on industrial production through monocultures of a limited number of plant species and cultivars. These homogenized agricultural systems are increasingly showing signs of strain. While "some believe that genetic modification and other biological innovations ... will feed the planet into the foreseeable future, others argue that sustainable biodynamic farming ... could replace factory farms" (Pilcher 2016: 44–5). Either way, without significant political and economic changes, a diet rich in organic, fresh, local produce with a diversity of nutrient-dense plant foods will continue to be a luxury out of reach for most people worldwide. Reaching environmental limits caused in part by climate change, diseases, and pests may force change that addresses the numerous critiques of industrial agriculture.

ACKNOWLEDGMENTS

I am grateful for comments on earlier drafts from Savvina Chowdhury, Stephen J. Forbes, Susan Frankel, Rachel Friedman, David Mabberley, and Melissa K. Nelson. Their thoughtful feedback helped me to polish my argument. I am especially indebted to political economist Jeanne Hahn who read two drafts and gave extensive feedback. The chapter benefited from her wise counsel and her high standards for academic writing. Another colleague from The Evergreen State College, sociologist Prita Lal, like Hahn, met with me to discuss the piece at length and suggested excellent books and articles to read. Thanks to Miko Francis, Aidan Linder, Paul McMillin, and Liza Rognas for their skilled assistance at The Evergreen State College Library. Keisha Loidolt at the Orange Public Library & History Center facilitated use of the Epicure Brand orange crate label. I am also grateful for the adept aid of two reference librarians at University of California, Berkeley's Bioscience Library: Becky Miller and Elliott Smith.

CHAPTER THREE

Trade and Exploration

The Impact on Plant Diversity

SARA OLDFIELD

Plants are essential for life on earth and for the well-being of humankind, providing a wide variety of goods and services. Plants provide both subsistence needs for billions of rural people and the basis for financial livelihoods at a local, national, and global scale. The importance of plants as major commodities in world trade has increased dramatically since the beginning of colonial exploration by the Spanish and Portuguese in the fifteenth century. Global plant trade in both cultivated and wild collected plant material has and continues to shape whole economies and societies. The world has changed dramatically during the twentieth century with technical, cultural, and economic shifts on an unprecedented scale and natural resources, including plants located, traded, and consumed worldwide to an extent not previously possible. In this chapter we explore the importance of plants for production and trade concentrating on a few key sectors—timber, food, and medicine. The roles of scientific knowledge and understanding of traditional ecological knowledge relating to plants are considered. At the same time as the expansion of global trade in plant products there has been a growing realization that plant diversity is not an inexhaustible resource. The decline of plant diversity, caused both by direct consumption and indirect effects, is considered and the actions required to secure plant diversity for a sustainable future are discussed. While plant exploration continues along with the commodification of novel crops in the modern era, the major story for this period is the conflict between the accelerating exploitation of forests and the development of agricultural lands for cropping accompanied by an accelerating destruction of biodiversity. The response of the international community to this crisis has seen the development of institutional frameworks in an endeavor to stem biodiversity loss.

Extending back into prehistory, humans have exerted an influence on biodiversity initially causing very slow and gradual changes at the local level. Globally, the pace of environmental change only began to accelerate rapidly in the early nineteenth century as the human population passed one billion, coinciding with the beginning of the industrial revolution (Blackmore 2009; Steffen et al. 2007). With European colonization and expanding international trade, rapid environmental change spread around the world, with particularly significant loss of biodiversity in the species-rich tropics. By the end of the twentieth century the impact of humanity on the biosphere was considered so great that we have entered a new geological era in the history of the planet known as the Anthropocene (Crutzen 2002; Steffen et al. 2011). Although not fully recognized in the political arena, humankind has become a global geological force in its own right through factors such as the amount of concrete used in buildings and the amount of plastic deposited at sea. The consequences for the Earth's future are unprecedented.

In the twenty-first century Earth's biological resources are being utilized at an ever-increasing rate even though knowledge of biodiversity remains incomplete. As discussed further below, the total number of flowering plants remains unknown with certainty and the properties and values of relatively few plants have been studied and cataloged. Plant species are, however, clearly of immense value as a source of food, medicine, timber, fuel, fibers, gums, resins, and ornamentals. At least thirty thousand plant species have a documented use as recorded in major databases (RBG Kew 2016). The true figure is likely to be considerably higher taking into account geographical areas that are poorly documented, informal use of plants, and the acknowledged paucity of data on harvest and trade in so-called Non-Wood Forest Products (FAO 2010). The knowledge of the distribution and abundance of individual plant species and their conservation status also remain poorly understood and documented, limiting the potential for rational and sustainable utilization. The situation is perhaps most acute in tropical regions given the concentration of plant diversity and rapid ongoing transformation of ecosystems.

As noted in Global Biodiversity 4, a synthesis report showing the status and trends of global biodiversity, overall use of natural resources is projected to continue to increase in absolute terms until 2020. Humans are appropriating between 30 and 40 percent of the entire planet's plant production, more than double the amount appropriated a century ago (Secretariat of the Convention on Biological Diversity 2014). As we continue to consume and destroy plant resources, it is widely recognized that the causes of most environmental problems are closely related to the generation and distribution of wealth, knowledge, and power, and to patterns of energy consumption, industrialization, population growth, affluence, and poverty (Greene 2005).

TIMBER PRODUCTION AND TRADE

Timber was an early commodity traded internationally and remains the most valuable commodity traded from the wild. Currently about 15 percent of the timber produced globally is sourced from tropical forests and the rest from boreal and temperate regions (Arets et al. 2010). The World Bank estimates that the trade in timber products contributes some $468 billion annually to global GDP. Total global wood production was estimated at 3.53 billion m^3 in 2012 of which 46 percent was for timber and paper and 54 percent for wood fuel. Industrial roundwood production is slowly recovering following the economic recession starting in 2008. Roundwood production is dominated by the US, Canada, Brazil, China, and the Russian Federation which account for 55 percent of the total (FAO 2016). A recently compiled list of internationally traded commercial timbers documented 1,575 timber taxa (Mark et al. 2014). In total there are around sixty thousand tree species (Beech et al. 2017) and many more tree species are used locally as a source of timber or as a source of fuel. Globally, wood provides over 9 percent of primary energy supply, and in 2005, the total reported value of harvested fuel wood amounted to $7 billion (FAO 2010). Plantations account for about 20 percent of global wood supply.

The forests of the Caribbean region were exploited for timber for use in Europe from the time of Columbus, with mahogany *Swietenia mahagoni* a prized resource. Exploitation of West African timber for the European market began later and can be traced back at least to 1672, when the Royal African Company received a charter from

King Charles II of England to trade in African mahogany (*Khaya* and *Entandrophragma* spp.). By the early eighteenth century, timber from coastal forests of West Africa had replaced that from the Caribbean region (Sayer et al. 1992). Trade in African timber increased significantly with European colonialism at the end of the nineteenth century and evolved rapidly after 1945, still controlled by European interests, to supply utility grade wood. In 1962, approximately 30 percent of the tropical hardwood imported into Europe was sourced in West Africa (Sayer et al. 1992).

In Southeast Asia, forest resources in Peninsular Malaysia have been managed for production and trade since the early twentieth century with exploration, selection, survey and demarcation of forests for both production and protection in the early 1900s (Burgess 1989). The timber export industry initially developed to supply UK needs. International demand for Malaysian timber became significant in the 1950s with a rapid increase in demand after the end of the Korean war in 1953 and with agricultural expansion after Malaysia's independence in 1957 making large quantities of logs available (Hing and Salleh 1989). In Indonesia, teak plantations developed in Java under Dutch colonial rule formed the basis for commercial forestry until after the Second World War. The vast forests of Indonesia and East Malaysia were opened up in 1965 for the international market, replacing, for example, about half the imports of tropical hardwoods into Europe (Sayer et al. 1992). In the Philippines intensive logging of the dipterocarp forests took place from the end of the Second World War until the 1970s. In 1968 timber accounted for 33 percent of all foreign exchange earnings, falling to 5 percent by 1986 because of lower global prices, forest depletion, and conservation policies.

By the end of the twentieth century the first global assessment of the extent of logging in the tropics indicated that 20 percent of the tropical forest biome was either

FIGURE 3.1 Illegal and legal timber extraction remains a threat to forests globally—timber here being removed from Omo Forest Reserve, Ogun, Nigeria. Photo by Peter Martell/Getty Images.

actively logged or allocated to logging concessions between 2000 and 2005 (Asner et al. 2009). About half of this area had already been heavily impacted by land-use change having previously lost over 50 percent of its potential forest cover. Commercial timber extraction has now reached into the center of Amazonia, is expanding rapidly into new areas including Central Africa and Papua New Guinea, and could soon reach the last remote tropical forests (Wright 2010). Timber production and trade remain major global economic activities. Sustainable management of the natural forests from which most of the wood is sourced remains a major challenge. Progress has been made in many countries with certification helping, to a varying degree, to ensure that forests are managed for the benefit of people, biodiversity, and recognizing the importance of forests for carbon storage (Rametsteiner and Simula 2003). Nevertheless, weak governance and growing demand for speciality timbers continue to threaten both tree species and the forests of which they form part—as discussed further below.

THE EXAMPLE OF MAHOGANY

Mahogany has been traded internationally for over four hundred years and remains an important commodity harvested from the wild. Despite its importance, management of natural stocks remains weak on a global level. There are three species of true mahogany, in the genus *Swietenia*, all native to the Americas, growing in forests from Florida to Brazil. Mahogany was first traded internationally in the sixteenth century when stocks of *Swietenia mahagoni* were exploited in the Caribbean region. In the eighteenth century, mahogany from the Caribbean islands and Central America was greatly valued for furniture production in the UK. Now big leaf mahogany *Swietenia macrophylla* is the main species in international trade with the US the major importer. In addition to *Swietenia*, various other tropical hardwood species are also traded as mahogany, with timber traders making use of the traditional cachet of the name (Melville 1936).

Swietenia macrophylla has a wide geographical range, distributed from Mexico southward into western South America and extending over a crescent-shaped area across southern Amazonia. It grows naturally in tropical dry and tropical wet forests on a wide variety of soil types. General deforestation and logging pressures have led to concern about the conservation status of *Swietenia macrophylla* over the past thirty years and to calls for action at an international level. Following a study of this species and other so-called mahogany species in trade undertaken in 1983, recommendations for listing on CITES (the Convention on International Trade in Endangered Species of Wild Fauna and Flora) were made (Knees and Gardener 1983) and the first formal listing proposal for *S. macrophylla* was put forward to CITES in 1992). *Swietenia macrophylla* was listed as globally Vulnerable by IUCN in 1998. However, despite the fact that *Swietenia humilis* and a few other trees were already covered by CITES, opposition to the idea of listing major commercial timbers was strong (Oldfield 2013). It was not until ten years later in 2003 that a proposal to list *Swietenia macrophylla* on CITES Appendix II was finally accepted.

Brazil was the major exporter of *Swietenia macrophylla* in the twentieth century. Production of mahogany between 1971 and 2001 is estimated to have been valued at US$4 billion, with 75 percent exported to the US and European high-end furniture and construction markets (Grogan, Barreto, and Veríssimo 2002). In 2001, prior to the CITES listing, the Brazilian government introduced a ban on extraction and trade in the species. This unfortunately resulted in an increase of illegal trade with mahogany stolen from conservation areas, private land and Indigenous land and exported under other names. It has also, tragically, been linked with an increase in violent deaths (Chimeli and Soares 2017).

Following the ban on Brazilian mahogany, the major exporting countries as reported by CITES (2002–11) were Bolivia, Guatemala, Mexico, and Peru (Ferriss 2014). Ensuring sustainable and legal trade in *Swietenia macrophylla* as required by CITES is challenging. CITES requirements need to be considered alongside national forestry policies and legislation. The Convention requires decisions based on scientific findings that trade will not be detrimental to the species in the wild and if levels of overall trade are considered to be potentially damaging international action can be taken. The working partnership between ITTO (International Tropical Timber Organisation) and CITES on mahogany and other timber issues is helping to support practical solutions. Sharing of experiences between countries is facilitated through workshops bringing together the different range states.

In general, managing timber production is complex in tropical regions. Independent forest certification and chain of custody certification for timber such as those under the umbrella of the Forest Stewardship Council (FSC) is another approach to verifying the sustainability of timber production. CITES, although legally binding, arguably sets less exacting standards than required by FSC and the two approaches are generally considered to be broadly complementary.

FOOD PRODUCTION AND TRADE

Plants are essential for food production. Unlike the timber industry which still relies to a very significant extent on wild-harvested materials, the majority of the world's food needs are supplied by cultivated and improved crops many of which have been grown for millennia. A small number of agricultural crops have major global importance. Seven plant species that provide wheat, sugar, and rice are among the most significant contributors to per-capita calorie intake in 90 percent of countries (Khoury et al. 2014). Coffee, wheat, maize, oats, rice, soybeans, cocoa, orange juice, and sugar are the main plant-based agricultural commodities traded internationally. Palm oil is another global commodity traded on the futures market and is discussed further below. In addition, thousands of plant species are grown locally for subsistence or traded in local markets, some of them scarcely or only partially domesticated, and many more species are gathered directly from the wild (Heywood 2011). Food plants which are still gathered from the wild for international trade include various species of nuts, bamboo shoots, palm hearts, and edible oils.

Traditionally rural communities made use of a wide variety of wild and cultivated plants in their diets, especially in tropical regions. It has been noted that, in tropical Asia, colonial administrations introduced highly profitable cash crops which led to an impoverishment in both the number of cultivated species and varieties grown locally and subsequently through intensive breeding and vegetative propagation, an impoverishment also of genetic variability (Ashton 1981).

Globalization of food production in the twentieth century developed especially rapidly after 1945 with the establishment of a global food regime based on the incorporation of local systems into a global system of food production. Local subsistence food producers have become increasingly involved in cash crop production with transnational corporations controlling production and trade in food commodities. At the same time local subsistence farmers dislocated from land developed for industrial scale farming have migrated to cities. The development of the global food regime was stimulated by the US which was producing large food surpluses at the end of the Second World War. Cheap imported food provided an incentive for industrialization in developing countries and for large-scale mechanized agriculture. The "Green Revolution" in the 1960s and 1970s resulted in massively increased crop yields and industrialized agriculture (Thomas 2006).

Changes in food production over the past century have had a profound impact on biodiversity through land clearance as discussed further below. Reliance on growing fewer and fewer high-yielding crop varieties brings risks to livelihoods and food security. The globalization of food production has, for example, contributed to the global spread of weeds, pests, and diseases. Some of these could potentially have devastating impacts on genetically uniform crops, recalling the potato blight of the nineteenth century. Climate change is also having an increasing effect on agricultural food production and food security. The loss of genetic diversity within crops and species diversity across crops weakens resilience to the impact of climate change on food security.

Crop wild relatives (CWR) are wild plant taxa that have relatively close genetic relationships to crop plants. This enables their use in plant breeding to improve agricultural varieties whether for yield, disease, or drought resilience. The global value of the introduction of new genes from CWR to crops is estimated to be $115 billion annually (Pimentel et al. 1997). Currently the contribution of CWR to the production chains of rice, wheat, potatoes, and cassava is worth US$25 billion, and could potentially be worth US$73 billion in the future. For the twenty-nine crops identified as being of major importance to global food security and listed in Annex 1 of the International Treaty on Plant Genetic Resources for Food and Agriculture these figures rise to current and potential values of US$42 billion and US$120 billion respectively. A recent global inventory of CWR includes 3,546 species (Vincent et al. 2013).

THE EXAMPLE OF OIL PALM

Global demand for oil palm together with modern production methods to supply that demand provide examples of the impact of international trade in an agricultural commodity on biodiversity. Oil palm, *Elaeis guineensis,* is native to the tropical forests of West and Central Africa where it was traditionally grown as part of mixed farming practices. Now oil from the fleshy fruits and kernels has a wide variety of uses in global food production and as a major ingredient in household products and cosmetics. Palm oil is used, for example, in the manufacture of margarine, cooking oil, ice cream, and coffee whitener. Palm kernel oil is extracted for use in the manufacture of glycerine soaps and as a substitute for cocoa butter in chocolate. Oil palm is also used in plastics, industrial chemicals, and biodiesel production. The oil palm provides one of the leading vegetable oils produced globally, accounting for one-quarter of global consumption and approximately 60 percent of international trade in vegetable oils (World Bank 2010). The plant is cultivated on approximately 15 million ha worldwide. Palm oil has risen in global importance in the past several decades, with world production rising from 14.7 million tonnes in 1990 to over 57 million tonnes in 2014 (FAOSTAT 2016). Global demand for palm oil is expected to double by 2020 (UNEP 2011). Indonesia is now the world's leading producer of palm oil, supplying approximately half of the commodity globally, and is itself driving increased palm oil consumption through a domestic biofuel policy. Together with Malaysia, the two countries account for more than 80 percent of global production (Pittman et al. 2013).

FIGURE 3.2 Oil palm plantations in northeastern Borneo, state of Sabah, Malaysia. Recently planted oil palms can be seen in the cleared grassy areas and a tiny bit of natural rainforest still struggles for survival farther away. Courtesy of Vaara, Getty Images.

Today, most oil palm production is being developed as an industrial-scale monoculture, imposing significant environmental risks through the replacement of rich tropical forest ecosystems and intensive use of fertilizers and pesticides. Production also impacts on local societies, particularly people with limited economic means (Colchester 2010). Modern oil palm cultivation is generally characterized by large plantations of uniform age structure with a low canopy and sparse undergrowth (Fitzherbert et al. 2008). The oil palm tree generates fruits from the third year, with yield per tree increasing gradually until it peaks at approximately twenty years (FAO 2002). Oil palm plantations are typically destroyed and replanted at twenty-five- to thirty-year intervals (Wahid et al. 2005).

MEDICINAL PLANTS AND TRADE

Plants have been used as sources of medicine for millennia and often the consumption of plants as food, flavoring, and medicines overlapped. Traditional plant-based systems of medicine continue to be extremely important today. An estimated fifty thousand plant species are used medicinally around the world (Schippmann et al. 2002), and in 2006 the estimated demand for medicinal plants was approximately \$14 billion per annum according to the World Health Organization (Booker et al. 2012). In China, 90 percent of hospitals have a traditional medicine department alongside "modern" medical departments. Around eight thousand plant species within the country are considered to have medicinal value. They are used in traditional Chinese medicine, practiced especially by the Han Chinese, and in ethnic minority medicine used by the fifty-five recognized ethnic minority groups. Most medicinal plants used in China are collected from the wild (Pei and Huai 2013). The Chinese *materia medica*, 80 percent of which is based on plants, is of huge economic importance with an estimated output of US$83.1 billion in 2012 (Sharrock et al. 2014).

Around the world, it is often quoted that 80 percent of people in rural communities rely on medicinal plants for their healthcare (Lambert et al. 1997). In addition, traditional and complementary medicine based on plants is still favored in developed countries. Currently, a hundred million Europeans are thought to use such remedies and there is a big market for natural health products in the US with an estimated value of over US$14 billion (Sharrock et al. 2014).

Exploration of the medicinal properties of plants continues with plants playing a major role in the development of new drugs. Between 1981 and 2002, 61 percent of new pharmaceutical drugs, including 67 percent of drugs used to treat cancer, had novel components that were derived from or inspired by natural chemicals (although not all derived from plants), with a quarter of prescription medicines derived directly from flowering plants or modeled on molecules they contain (Cragg and Newman 2005).

THE EXAMPLE OF *TAXUS*

Taxus is a small genus of coniferous trees and shrubs. The taxonomy is complicated but twelve species are generally recognized. In 1967 an anti-cancer drug was isolated from a North American species, the Pacific yew *Taxus brevifolia*. This was part of research by the US National Cancer Institute which screened a wide range of wild species for medicinal products. Subsequently, wild populations of Pacific yew experienced significant declines due to overexploitation for the production of cancer treatments. This exploitation has now generally stopped as alternative methods for Taxol production have been developed and exploitation shifted to Asian yew species.

Together with the North American *Taxus brevifolia* and the European yew *T. baccata*, various Asian species including *T. wallichiana*, *T. chinensis*, and *T. cuspidata*, are the source of taxanes from which the drug paclitaxel can be derived chemically. The young shoots, leaves, and bark of *T. wallichiana* have been used for their medicinal properties for centuries. Across most of its range through the Himalayas and western China, *Taxus wallichiana* has been heavily exploited for its leaves and bark which are used to produce the anti-cancer drug paclitaxel or similar chemicals. Declines of up to 90 percent have been reported in India and Nepal while in western China declines of more than 50 percent have been reported. The extent of exploitation in other countries is uncertain. This species is now listed as Endangered on the IUCN Red List (Thomas and Farjon 2011).

Concern about the international trade in *T. wallichiana* led to its listing on CITES Appendix II in 1995. Subsequently, four additional Asian yew species were included in CITES Appendix II: *T. chinensis*, *T. cuspidata*, *T. fuana*, and *T. sumatrana*. Problems with CITES implementation for these species include the generally low level of implementation effort for medicinal species, and problems with identification of the materials in trade (Mulliken and Crofton 2008).

Over the past fifty years, production of cancer treatments based on *Taxus* has seen a transition from direct extraction from the Pacific yew tree to industrial production based on plant cell culture. Bridging these two production strategies was a semisynthetic route that initially allowed larger-scale access to the Taxol compound. The application of metabolic and process engineering is now expected to improve production methods. Engineering of microbial production hosts is also being developed (Li et al. 2014). The shift from harvesting to industrial production has been accompanied by attempts to conserve *Taxus* species in the wild but six species remain of international conservation concern at least in part because of exploitation. According to Mulliken and Crofton (2008) a strong economic incentive for wild harvesting and the purchase by manufacturers of wild-harvested products is likely to persist even within countries such as China which ban such harvests.

BOTANICAL EXPLORATION

The eighteenth and nineteenth centuries are often thought of as the age of botanical exploration with exciting new discoveries showcased by botanic gardens and driving international trade. Botanical expeditions and the production of floras continued throughout the twentieth century. Data from Index Kewensis, which catalogs all newly published plant names, shows that an average of over two thousand new species were described each year during the last decade of the twentieth century mainly based on recent field collections. This indicates the ongoing discovery of new species with many more species being described from the tropics, especially the Americas, but also with new species still being discovered in China and North America (Prance 2001).

As noted above, about thirty thousand plant species have a documented use and many more are used locally. This is less than 10 percent of known plant species. A basic scientific inventory of the world's plant species has not yet been fully compiled. Understanding the historical, current, and potential uses of plants remains a huge area of research. There is a wealth of background material and knowledge that can be tapped in biocultural collections such as herbarium specimens with information on plant use, and museum collections of natural products and cultural artefacts. The importance of collating biocultural information and managing biocultural collections, which has perhaps been overlooked in the rush for scientific advancement, is increasingly acknowledged (Salick et al. 2014). At the same time botanical exploration still continues, driven both by the need for botanical inventory for biodiversity conservation planning and to an extent for the discovery of new products. Certain areas of the world remain poorly explored botanically with Papua New Guinea being probably the least known (Prance et al. 2000).

Exploration of the natural world for the development of new sources of chemical compounds, genes, micro-organisms, macro-organisms, and other valuable natural products is generally known as bioprospecting. This entails the search for economically valuable genetic and biochemical resources, usually for use in medical treatments, and has come to mean in effect exploring ways to commercialize biodiversity. More recently exploration and research on Indigenous knowledge related to the use and management of plant and animal resources has also been included in the concept of bioprospecting. Bioprospecting has fundamental relevance to the conservation and sustainable use of biological resources and to the rights of local and Indigenous communities.

Early approaches to botanical discovery and also to protection of natural resources generally took little account of local knowledge and practices of Indigenous peoples who were often seen as noble but primitive communities that needed to be rescued from their ignorance and their destruction of nature (Payyappallimana, Fadeeva, and O'Donoghue 2013). The views of Indigenous peoples were overlooked with negative consequences as noted by Beatriz Schulthess (1994), President of the Indigenous Peoples Ancestral Spiritual Council, who wrote that "it feels to me as if an enormous bulldozer, from essentially northern states and northern companies, is about to destroy both the last available resources on Earth and indigenous peoples simultaneously." It was only in the second half of the twentieth century that a shift in policy occurred internationally with the recognition of the links between ecology and society, the need for local communities to engaged in policy development, and new multilateral perspectives to be formed on cultural diversity and traditional knowledge.

THE MAIN THREATS TO PLANT DIVERSITY

At a global scale, the major threats to biodiversity in general are considered to be habitat loss, overexploitation, and the impacts of invasive species, pollution, and climate change (Millennium Ecosystem Assessment 2005). These major threats vary in their intensity and impact between different regions, ecosystems, and species, usually acting in combination. These global threatening factors all result from, or are influenced by, human behavior and to a large extent are intermingled with drivers of international trade. Human activities far outweigh the natural threats to plant species and their habitats (Brummitt and Bachman 2010). It is now generally recognized that human pressures on biodiversity will increase as the global population continues to grow, tastes in food change with the increase in meat consumption and consumerism generally increases. At the same time, production land, which is able to provide food and other commodities, is expected to decrease in extent and quality. Relating specifically to plant species, it has not yet been possible to carry out comprehensive analysis of which of the major threats to biodiversity have the greatest impact. However, information on threats recorded for over twenty thousand plants included in the IUCN Red List indicates that agriculture is the most commonly reported threat (Rivers 2017).

Agriculture

The clearance of natural habitats for farming has taken place for thousands of years, as a fundamental process in the development of many human cultures. Globally, by the end of the twentieth century 38 percent of the land area was agricultural land (http://data.worldbank.org/indicator/AG.LND.AGRI.ZS). Adverse impacts of agriculture on biodiversity can be summarized as conversion of natural ecosystems, management of agricultural landscapes in ways that limit the existence of wild plants and animals, and application of pesticides and other agrochemicals that damage natural biodiversity (Hazell and Wood 2008). In Southeast Asia deforestation is now mainly attributed to large-scale industrialized agriculture, especially oil palm plantations, while in other areas increased demand for land for local food production continues to be a major driver (Secretariat of the Convention on Biological Diversity 2014).

In Latin America, conservation efforts during the second half of the twentieth century focused on lowland deforestation for cattle grazing and slash-and-burn agriculture, but the relative importance of these drivers of deforestation is now declining. Today, soybean production, mainly to feed animals in China, is the major cause of loss of millions of hectares of seasonally dry forests in Brazil, Bolivia, Paraguay, and Argentina (Carvalho et al. 2019; Holmes 2019; Tyrrell 2018). At the same time, migration to urban areas is leaving marginal grazing and agricultural lands abandoned. In rural areas, an important

FIGURE 3.3 A truck passes a soy field in a deforested section of the Amazon rainforest near Ariquemes, Rondonia state, Brazil in 2017. Photo by Mario Tama/Getty Images.

conservation strategy has been to invest in community-based sustainable development projects. These projects have had limited success in improving socioeconomic outcomes (Fearnside 2017; Martinelli et al. 2017).

Overexploitation

The second most common reported threat to plant species as recorded on IUCN Red List is overexploitation (Rivers 2017). As noted above, timbers are a major category of species that are harvested from the wild, in some cases over centuries, leading to resource depletion and potential extinction. Examples of timber species that have become extinct through overexploitation have been hard to find but there are island endemic species that have been reduced to a handful of individuals because of centuries of commercial use (Oldfield 1988). The St. Helena redwood *Trochetiopsis erythroxylon* and St. Helena ebony *Trochetiopsis melanoxylon* are recorded as Extinct in the Wild and Extinct respectively as a result of deforestation and harvesting for timber. Another species is the Critically Endangered *Diospyros hemiteles,* an ebony found only on the island of Mauritius, which is now mainly threatened by invasive species and sugarcane plantations.

More recently there has been a dramatic increase in demand for hardwoods of the genera *Diospyros* spp. (commonly known as ebony) and *Dalbergia* spp. (commonly known as blackwood or sometimes ebony) for the Chinese market and there is a very real risk of extinction to heavily exploited species. In Madagascar, for example, the trade in timber of these groups, which has largely been unregulated, is threatening the survival of individual species which remain poorly known botanically with very limited data on distribution and abundance. Illegal logging of Madagascar's precious timbers has been a major problem for about twenty years and increased as a result of the political turbulence since 2009. This has resulted in unprecedented levels of illegal timber removals from protected areas particularly in the northeast of the island (Ratsimbazafy et al. 2016).

Medicinal plants harvested from the wild are also subject to overexploitation. African Cherry *Prunus africana* is one such species. This tree species has a wide range of local uses in Africa and yields a valuable medicinal product used to manufacture treatments for benign prostatic hyperplasia. The manufactured drugs are sold globally. The bark of *Prunus africana* is harvested in the largest quantity of any tree species and this has led to international concerns about sustainability. The international market is robust and projected to increase. Retail value of *Prunus africana* products is estimated at over $200 million annually and may be considerably more.

A third broad group of plant species that have been harvested from the wild at damaging levels are ornamentals. Exotic orchids have been cultivated in Europe since the seventeenth century and ownership of rare species has become extremely fashionable in the latter part of the nineteenth century. Enormous quantities of wild plants were imported especially to the UK with intense rivalry between collectors and between commercial companies (Jenkins and Oldfield 1992). Nowadays the vast majority of orchids traded worldwide are artificially propagated hybrids but demand for wild plants particularly of rare species continues. One orchid genus that has been severely impacted by overcollection is *Paphiopedilum*. Currently eighty-four species of this genus of Southeast Asian slipper orchids are included as threatened in the IUCN Red List. This represents nearly the entire genus. At the end of nineteenth century these were the most popular of all orchids with European collectors. Falling out of fashion, the discovery of *P. sukhakulii* in Thailand in the 1960s followed by other new species, revived interest in the genus and led to large quantities of wild plants appearing in trade. Likewise new discoveries in southwest China in the 1980s triggered intense demand. A recent study has shown the importance of social media as a means of trading in rare orchid species including those of *Paphiopedilum* (Hinsley et al. 2016).

Cacti and other succulent plants are also popular with collectors who often favor wild-collected plants for their authenticity and seek specimens of particularly rare species. A recent global assessment found that 31 percent of cactus species are threatened with extinction based on the IUCN Red List categories and criteria. Illegal trade in live plants and seeds for the horticultural industry and private collections are major threats to cacti, affecting nearly half of the threatened species (Goettsch et al. 2015).

Cycads are considered one of the most threatened groups of plant. There are 348 species currently known in two families, Cycadaceae and Zamiaceae (Calonje et al. 2017), over 60 percent of which are listed as threatened by IUCN (Zheng et al. 2017). Collection from the wild for ornamental horticulture is one of the major threats to these plants which generally have low seed germination and seedling survival rates in their natural habitats. In China there are twenty-three species, twenty-two of which are listed as Threatened or Near Threatened by IUCN. *Cycas revoluta* (the only Chinese species considered globally Least Concern by IUCN) and the Critically Endangered *C. szechuanensis* have both become virtually extinct in the wild in China due to excessive commercial harvesting (Zheng et al. 2017).

Invasive Species

Humans have been involved in the accidental and deliberate dispersal of plant, animal, and microbial species for millennia. The dramatic increase in the spread of non-native species globally roughly tracks the rise in human transport and commerce (di Castri

1989). In California, USA, which has a rich native flora, the first plant introductions began with the Spanish missionaries in about 1769, there were at least 134 established alien plant species by 1860, and by 1993 this number had risen to over a thousand, or 15 percent of the entire flora (Schwarz et al. 1996).

In the US, the spread of invasive and noxious weeds has been a major threat to native plants. Some species were deliberately introduced to improve pasture. European settlers in the Pacific Northwest introduced European grassland species on a huge scale. Grasses such as *Holcus lanatus*, *Anthoxanthum odoratum*, *Bromus hordaceus*, and *Lolium perenne* became established in response to the new practices of fire-suppression and year-round heavy grazing by cattle and sheep. *Bromus tectorum* (cheatgrass) is a species native to the Mediterranean region. Its original habitat was the decaying straw of thatched roofs. Accidentally introduced into the US in packing materials, ship ballast, and likely as a contaminant of crop seed, cheatgrass was first found in the US near Denver, Colorado, in the late 1800s (Whitson 1991). Cheatgrass spread explosively in the ready-made seedbeds prepared by the trampling livestock hooves of overstocked range lands. Disturbance associated with homesteading and cultivation of winter wheat also accelerated its spread and establishment. By the 1930s, cheatgrass was becoming the dominant grass over vast areas of the Pacific Northwest and the Intermountain West regions. Cheatgrass has developed into a severe weed in several agricultural systems throughout North America, particularly western pastureland, rangeland, and winter wheat fields. It is now estimated to infest more than 41 million hectares (101 million acres) in western states (Mack 2000).

In the US it is now estimated that invasive species are a threatening factor in 42 percent of threatened and endangered plants and the main threat for 18 percent. The annual cost of invasive species to the US economy has been estimated at $120 billion with one species, the yellow star-thistle, alone currently costing ranchers and farmers in California $17 million in forage and control activities and $75 million in water losses annually (Groffman et al. 2014).

Introduced species are a threat to 61 percent of recorded threatened species of plants and animals in Australia (Evans et al. 2011). Within the country, introduced weed species are a major problem costing millions of dollars annually. Of Australia's twenty worst weeds, sixteen were deliberately introduced (Low 2005).

Many of the world's worst invasive plants belong to relatively few families, predominantly the daisy family (Asteraceae) and grass family (Poaceae), together with the genera *Acacia*, *Mimosa*, and *Cyperus*. Despite the noted links between threatened plants and invasive plant species, as yet, there is a lack of evidence of actual plant extinctions resulting directly from invasive species, in part because of the lack of documentation (Downey and Richardson 2016). The introduction of mammals particularly to oceanic islands has had a more dramatic effect. Goats introduced to St. Helena Island in 1513, for example, almost certainly extinguished more than fifty endemic plant species, although only seven were scientifically described before their extinction (Groombridge 1992). Failure to tackle the issue of biotic invasions could result in major loss of agricultural and forestry resources in some regions, disruption of the processes that supply ecosystem services and the creation of homogeneous, impoverished ecosystems composed of cosmopolitan species (Mack et al. 2000).

Climate Change

The impact of climate change due to the increase in CO_2 emissions during the twentieth century is an increasing threat to plant diversity. In addition to emissions from burning fossil fuels, global deforestation has been one of the main drivers of climate change contributing to the release of the carbon stored in trees as CO_2 emissions. In turn, the impact of climate change on individual plant species is complex. Temperature and rainfall are fundamental to plant growth and climatic factors are a major determinant of plant distribution providing the so-called "climate envelope" for each species. Faced with climate change, species can adapt, migrate, or become extinct. With climate change, plant ranges are shifting both altitudinally and latitudinally. Widespread species with greater genetic variation are able to fare better than naturally rare species that may also have a restricted genepool. Restricted range plants, such as alpines which are often isolated at higher elevations, or coastal species, may simply have no suitable habitat to spread to.

A study twenty years ago indicated that with a mean global warming of 3 degrees Celsius, about 7–11 percent of North American plant species would be entirely out of their climate envelopes and thus vulnerable to extinction. Species already rare and/or under threat would be disproportionally affected with 10–18 percent out of their climate envelopes (Kutner and Morse 1996). A study of European plants indicated that 1,350 plant species face extinction by 2080 as a result of climate change (Thuiller et al. 2005).

HOW MANY PLANT SPECIES ARE THREATENED?

In response to the growing awareness of species extinctions, the documentation of the conservation status of plant species at a global level began in the 1970s. Robert Melville of the Royal Botanic Gardens, Kew produced a loose-leaf account of threatened Angiosperms, forming Volume 5 of the IUCN Red List series. This account recognized that for plants (and also invertebrates) so many species are rare or under threat and information about so many of them is so limited that the full treatment given to birds and mammals may never be possible (Scott et al. 1987). In 1978, the IUCN Plant Red Data Book was published providing 250 examples of the 20,000–25,000 vascular plant species considered likely to be threatened based on the IUCN Red Data categories in use at the time (Lucas and Synge 1978). Twenty years later a compilation of over thirty-three thousand globally threatened plant species was published by IUCN (using the same threat categories of Endangered, Vulnerable, Rare, or Indeterminate), with the authors stating that the number of threatened plant species recorded represented the tip of the iceberg (Walter and Gillett 1998).

The current IUCN Red List Categories and Criteria, introduced in 1994 with subsequent modifications, use five different criteria to measure symptoms of extinction risk. The criteria relate to biological processes underlying population decline and extinction. In a species assessment the species is evaluated against the five criteria: a) population reduction; b) geographic range; c) small population size and decline; d) very small or restricted population; and e) quantitative analysis. The five criteria each have quantitative thresholds and are further qualified by several sub-criteria. Based on the assessment process a species may fall into an IUCN threatened category of Critically Endangered (CR), Endangered (EN), or Vulnerable (VU). Some species are considered Data Deficient (DD) because of lack of appropriate information and/or taxonomic uncertainty. Species that do not meet the criteria for the threatened categories may be considered Near Threatened (NT), if they nearly do so, or Least Concern (LC).

At present approximately half of all species assessed as threatened on the IUCN Red List are plants (11,233 of 23,250) (IUCN 2016). This figure of 11,233 is significantly less than the thirty-three thousand included in the 1998 Red List because many species have not been reassessed using the more robust system introduced in 1994 which requires significant supporting documentation. Furthermore, the current IUCN Red List system measures risk of extinction rather than natural rarity. The majority of plant species are restricted in range (Joppa et al. 2013). The 1998 IUCN Red List included 14,504 species listed as Rare (Walter and Gillett 1998), a category no longer utilized by IUCN.

In total, only 6 percent of all plant species have been assessed globally and added to the IUCN Red List (Sharrock et al. 2014). Given the small number of global red list assessments for plants, it is difficult to estimate what proportion of the total number of known plant species are at risk of extinction. Of the plant assessments on the IUCN Red List over half are recorded as threatened but the plants currently included are not a representative sample of plants overall as many of the groups assessed to date are woody plants (building on a global survey carried out in 1995–8 and reported by Oldfield et al. 1998) and species previously assumed to be at risk. More recently there has been a move to also assess species that are not known to be at risk—and therefore the proportion of plants recorded as threatened on the IUCN Red List is actually declining over time (Rivers 2017).

In addition to the global assessment of plants using the IUCN Red List system, many countries have national approaches to the assessment of plants. Plants are currently by far the most common taxa represented in the National Red List database maintained by IUCN and the Zoological Society of London with ninety-eight countries having a national plant red list (National Red List 2016). Various different methodologies are used to categorize degree of threat, some of which are enshrined in national law. Whereas some 20,755 plant species have global IUCN Red List assessments (threatened, NT, DD, or Least Concern), a further 150,000 plant names have national, regional, or non-IUCN conservation assessments as recorded in a new database, ThreatSearch, maintained by Botanic Gardens Conservation International (BGCI). Not all these assessments relate to validly published plant names (with many being synonyms). According to a recent analysis of IUCN, national red lists and other assessments, 27,148–32,542 plant species (vascular plants and bryophytes) have been assessed at least once with a threatened category (Bachman et al. 2018).

Looking at knowledge of the threat status of useful plant species, it is clear that many timbers, medicinal plants, ornamentals, and CWR are threatened with extinction. Over a thousand tree species used for timber have been recorded as threatened (Oldfield et al. 1998). Detailed assessment of the conservation status of the fifty thousand medicinal plant species has not been undertaken but an estimated fifteen thousand may be threatened with extinction (IUCN/SSC MPSG 2007). While the majority of commercial material comes from cultivated sources, no more than a few hundred of the estimated 2,500 internationally traded medicinal plant species are thought to be commercially cultivated, so both international trade and unsustainable local use put wild medicinal plants under significant pressure. In India around 90 percent of the plants used by the country's medicinal plant industry are collected from the wild, and 315 of the 6,560 known medicinal species are threatened with extinction (Sharrock et al. 2014).

In a comprehensive assessment of the threatened status of European CWR, 572 European species relating to twenty-five economically important crops were assessed (Bilz et al. 2011; Kell et al. 2012). The study showed that at least 11.5 percent of the species are threatened, with 3.3 percent being Critically Endangered, 4.4 percent Endangered, and 3.8 percent evaluated as Vulnerable. A further 29 percent were recorded as Data Deficient, which is somewhat disappointing given the level of botanical knowledge in

Europe compared with much of the rest of the world. Over half the species assessed were considered to be Least Concern but of these around a third are threatened at national level (Kell et al. 2012).

GROWING AWARENESS AND CONSERVATION POLICY

Over the twentieth century there was a growing awareness of human impact on the environment and the need for policies relating to environmental protection. The first international treaty relating to flora, signed in Berne in 1889, was primarily concerned with preventing the spread of *Phylloxera*—a kind of aphid—which threatened to destroy European grape production. This was followed by further agreements in the 1920s and 1950s concerned with protecting the economic resources of cultivated plants by maintaining healthy stocks and preventing disease. The UN Food and Agriculture Organisation (FAO) was established in 1945 with conservation of natural resources included in its mandate (Greene 2005).

Also in the late nineteenth and early twentieth century, calls developed for the protection of wildlife for its own sake as well as to protect economic interests. These calls were partly driven by individual expressions of the need to celebrate and protect the natural environment for its scenic beauty and provide for the needs of iconic wildlife species. In the US, for example, the early conservation movement was partly aesthetic in nature influenced by writers and artists such as Henry David Thoreau (1817–62), Ralph Waldo Emerson (1803–82), and the painter George Catlin (1796–1872). These influential individuals articulated a desire to preserve areas of unique scenic beauty leading to the formation of Yellowstone, Yosemite, and Crater Lake National Parks. The national park movement was also partly an expression of the nationalistic spirit of the US with Catlin, for example, expressing a widely held sense that the destiny of the country was linked with its magnificent landscapes (Adams 2004). At the same time, rapid urbanization and industrial development following the Civil War led to calls for the preservation of wilderness values.

Contributing to the growing appreciation and acceptance of the need for conservation was the increasing scientific awareness of natural history through geographical and botanical exploration and the landscape planning movement. The waste of natural resources was recognized by George Perkins Marsh (1801–82), John Muir (1838–1914), and Frederick Law Olmsted (1822–1903). "By 1890 the ideas that there was an intricate and complex relationship between soils, water and forests was a matter of common knowledge among most of the American people" (Rakestraw 1955). There was general concern about the need to conserve timber resources in the US by the end of the nineteenth century, and legislation passed in 1891 allowed President Harrison to reserve public land with timber and undergrowth from settlement.

Elsewhere in the world, national parks were established in the 1880s and 1890s in the British Dominions of Canada, Australia, and New Zealand. Regulations were passed for the preservation of game and protection of animal species in various African colonies by the end of the nineteenth century (Adams 2004). In the UK, conservation concerns developed during the nineteenth century in response to the natural history movement, concern for animal welfare, and reaction to the rapid progress of industrialization and urbanization (Adams 2004).

The first international conservation organization was the UK-based Society for the Preservation of the Wild Fauna of the Empire established in 1903 by big-game hunters who realized that "they were in danger of running out of things to shoot" (Knight and Rose 2017). Initially the Society's main concern was to stop the decline of game animals in Africa by the establishment of sanctuaries. The eminent and influential founders were given the name of "The Penitent Butchers" by the press. Over the past century this organization has evolved to become a major force for the conservation of all threatened species and has been very influential in the international conservation movement. Now known as Fauna & Flora International, "Flora" was specifically added to the organizational name in the early 1980s (Knight and Rose 2017).

Globalization of the environmental movement began effectively after the Second World War. As mentioned above, FAO was established in 1945. There were also moves to form an international environmental organization focused on the conservation of wildlife rather than on the sustainable use of natural resources (Adams 2004). The International Union for the Protection of Nature (IUPN) was established in 1947 and changed its name to the International Union for Conservation of Nature and Natural Resources (IUCN) in 1956, reflecting an evolving and broader remit. The IUPN established a "Survival Service" with responsibility for cataloging threatened species of plants and animal. In 1956 this became a permanent Commission of IUCN and is now known as the Species Survival Commission. The Red Data book concept was devised by the leading British conservationist Sir Peter Scott in 1963 (Scott, Burton, and Fitter 1987).

From the 1960s onwards, environmentalism become increasingly popular in civil society. The World Wildlife Fund (WWF) was launched in 1961, initially to fund the work of IUCN. Friends of the Earth was founded in 1971, and Greenpeace was established in 1977. Growing environmental awareness around the world and popular campaigning led to concerted international efforts at a political level. The first UN summit on the environment, the United Nations Conference on the Human Environment, was held in Stockholm in June 1972. This placed environmental issues on the global political agenda with recognition of the need to increase awareness of the economic, social, and political effects of environmental problems. One outcome of the Stockholm Conference was the establishment of the United Nations Environment Programme (UNEP). The broad objectives of UNEP initially included to improve knowledge of ecological systems, to improve planning for development that took into account environmental considerations, and to build capacity for preservation and enhancement of the environment (Adams 2004).

During the 1970s efforts continued at national, regional, and international levels to catalog the conservation status of plants and animals, and to use this information to give legal protection to threatened species. In the US, for example, the Endangered Species Act came into force in 1973 with lists of Endangered and Threatened species subject to legal protection. Assessment of the conservation of plant species for the Act was undertaken by the Smithsonian Institution in response to a request from Congress. Internationally the Convention on International Trade in Endangered Species of Fauna and Flora (CITES) came into force in 1976. This Convention was designed to ensure that international trade in specimens of wild animals and plants does not threaten their survival. Appendices list species at risk through over-exploitation for international trade. When the Appendices were first developed with lists of species proposed by participating countries, there was an emphasis on protecting ornamental plants such as orchids and cacti together with iconic species of national importance. Over time the emphasis has shifted to include resource plants such as timbers and medicinal plants to address the threats described above.

The use of CITES to ensure sustainability in the timber trade was initially considered controversial (Oldfield 2013). The International Tropical Timber Organisation (ITTO), created in 1983 through the International Tropical Timber Agreement, considered the issue of controls on tropical timber species of conservation concern and were resistant to the idea of international legally binding action. Over time the position shifted with a joint CITES and ITTO program of work initiated in 2006 to strengthen sustainable management of timber resources and track the legality of CITES timber exports.

By the time of the second UN environmental summit held in Rio de Janeiro in 1992, there was global recognition of the inextricable links between environment and development expressed for example by the Brundtland Report (World Commission on Environment and Development 1987). A major outcome of the Rio meeting was the Convention on Biological Diversity (CBD) (de Klem 1993). The CBD is a framework agreement that comprehensively addresses biological diversity. The Convention's three objectives are: conservation of biological diversity, sustainable use of its components, and fair and equitable sharing of benefits arising from the utilization of genetic resources.

Another equally important outcome of the second UN environmental summit in Rio de Janeiro was the United Nations Framework Convention on Climate Change (UNFCCC). The ultimate goal of this Convention is to stabilize greenhouse gas concentrations "at a level that would prevent dangerous anthropogenic (human induced) interference with the climate system." It states that "such a level should be achieved within a time-frame sufficient to allow ecosystems to adapt naturally to climate change, to ensure that food production is not threatened, and to enable economic development to proceed in a sustainable manner." The historic UNFCCC Paris Agreement in 2015 stated "recognising that climate change represents an urgent and potentially irreversible threat to human societies and the planet and thus requires the widest possible cooperation by all countries." The central aim of the Paris Agreement is to strengthen the global response to the threat of climate change by keeping a global temperature rise in the twenty-first century well below 2 degrees Celsius above pre-industrial levels and to make efforts to limit the temperature increase even further to 1.5 degrees Celsius. Additionally, the agreement aims to strengthen the ability of countries to deal with the impacts of climate change.

There is now a suite of Multilateral Environmental Agreements (MEAs) including the CBD and CITES (both administered by UNEP) that help ensure the conservation of plant species. The MEAs provide a global framework for national legislation, policies, and action, and provide mechanisms for international cooperation. They have led to many small conservation successes and more significant advances, but they have not yet prevented the loss of plant diversity. Unfortunately, the conservation of biodiversity has tended to be perceived at governmental level as a matter of significantly less importance than the status of economic concerns. This is particularly the case with the conservation of the world's flora despite its fundamental importance to all life and to economic well-being. Successes in biodiversity conservation that have been achieved around the world frequently reflect the efforts of voluntary and charitable organizations. It has been suggested that this provides an indication of the importance that wider society places on the quality of life in a sustainable future, even when political priorities lie elsewhere (Blackmore and Oldfield 2017).

Major issues of climate change, food security, desertification, biodiversity loss, poverty, and human migration all have their roots in the degradation of the natural environment. The legacy of environmental concerns in the twentieth century can be seen in the outcomes of the United Nations Rio+20 summit held in Brazil in 2012. Governments have created a set of seventeen sustainable development goals (SDGs) adopted in 2015 (UN 2015, 2019). SDG 15 may be the most relevant goal in relation to the conservation of plant diversity. It aims to protect, restore, and promote sustainable use of terrestrial ecosystems, sustainably manage forests, combat desertification, and halt and reverse land degradation and halt biodiversity loss. SDG 2, with its focus on food security, is also directly linked to plant diversity as is SDG 11 which relates to making cities and human settlements inclusive, safe, resilient, and sustainable and SDG 12 on sustainable consumption and production.

THE BUSINESS SECTOR

The role of the business sector in promoting biodiversity conservation grew significantly in importance in the second half of the twentieth century. Various sectors including finance, travel, and retail together with primary production industries of energy, mining, agriculture, and forestry have all responded to some degree to international environmental concerns. Productive partnerships developed between businesses and the voluntary sector. In 1990, the Business Council for Sustainable Development (BCSD) was created with forty-eight business leaders to represent business at the 1992 Earth Summit in Rio. Subsequently this became known as the World Business Council for Sustainable Development (WBCSD) following a merger with the World Industry Council for the Environment (WICE).

In the forestry sector, calls for an international legal regime for control of forests and trade in timber were largely countered by the growth of voluntary schemes for certification of sustainable forest management supported by the private sector. The Forest Stewardship Council was created by business interests, environmentalists, and community leaders in 1993 after the Rio summit failed to agree a convention to stop deforestation. The mission of the Forest Stewardship Council (FSC) is to promote environmentally sound, socially beneficial, and economically prosperous management of the world's forests. Subsequently, national governments set up national forest certification schemes and these were brought together under the Programme for the Endorsement of Forest Certification (PEFC).

Currently over 440 million hectares of forest around the world have been certified by the FSC and PEFC, representing about 10 percent of the total global forest area (https://globalforestatlas.yale.edu/conservation/forest-certification). Unfortunately, certified forest is still very strongly concentrated in temperate and boreal regions which are less biodiverse than tropical regions.

Forest certification schemes have proved successful in linking production to the consumer of timber products and the trade in certified products has been particularly successful in areas with strong consumer interest in the environment. The same is true for agricultural certification schemes. These are growing in importance but still cover only a small proportion of farmed land. A specific certification scheme relating to agricultural production is the Roundtable of Sustainable Palm Oil (RSPO). The Roundtable grew from an informal co-operation among commercial companies such as Unilever and Migros (Switzerland's largest retail company), the Malaysian Palm Oil Association and WWF in 2002. It is currently the most broadly recognized framework reference for sustainability in oil palm production and defines standards for plantations, including environmental and socio-economic aspects (UNEP 2011).

Certification has also been developed for the sourcing of wild plants for medicinal, food, and other purposes. The FSC developed a working group to consider the certification of so-called Non-Wood Forest Products (NWFPs) in 1996 with chicle, the natural chewing gum derived from *Manilkara chicle*, the first product to be certified (Oldfield and Jenkins 2012). The Fairwild Foundation came into being in 2008 after lengthy discussions between business and environmental NGOs. The mission of the Foundation is "to enable transformation of resource management and business practices to be ecologically, socially and economically sustainable throughout the supply chain of wild-collected products" (Fairwild 2021).

BOTANIC GARDENS

Botanic gardens were initially established as centres for education and study of medicinal plants. Throughout their history, botanic gardens have been central to the exploration and cataloging of plants. During the era of colonial exploration and expansion they became extremely important in the development of tropical agriculture and the establishment of cash crops to new countries. The current, predominant role of botanic gardens in plant conservation has developed since the 1970s. Their institutional capacity to contribute to the conservation of plant diversity is supported by the documented collections of living plants, seeds, and herbarium specimens that are maintained together with staff expertise in plant taxonomy, identification, conservation exploration and assessment, and horticulture (reviews of the work of botanic gardens include Aronson 2014; Blackmore and Oldfield 2017; Chen et al. 2009; Donaldson 2009; Havens et al. 2006; Maunder 2008; Oldfield 2010).

As the biodiversity conservation agenda developed during the twentieth century, the focus was generally on threatened animal species and later ecosystems such as the tropical rainforests. Conservation of plant species generally received less attention, popular support, and financial resources than animal conservation. Botanic gardens have filled the niche, taking a lead in coordinating plant conservation work and providing outreach to the public on this issue. IUCN's Survival Service Commission set up the Threatened Plants Committee (TPC) in 1974 with its Secretariat based at the Royal Botanic Gardens, Kew. Developing the work of Melville, referred to above, TPC documented the global threat status of plants working closely with botanists at the Smithsonian Institution. Two international plant conservation conferences bringing together botanic gardens from around the world were held at the Royal Botanic Gardens, Kew, in 1975 and 1978 (Synge and Townsend 1979).

The Botanic Gardens Conservation Co-ordinating Body (BGCCB) was formed in 1979 in response to calls for closer collaboration between botanic gardens arising from the two conferences. BGCCB was set up as an initiative of IUCN with the aims to find out which plants identified as rare by IUCN are in cultivation and to keep members in touch and promote cooperation through a newsletter. In 1984, IUCN and WWF jointly launched a Plant Conservation Campaign and Programme—"To save the plants that save us." One of the objectives of the Programme was to work with botanic gardens, helping them to develop their conservation role. Arising from this, the Botanic Gardens Conservation Strategy was ultimately published in 1989, setting out how botanic gardens could effectively participate in plant conservation. The BGCCB evolved to became independent from IUCN and was established as a charitable organization, Botanic Gardens Conservation International (BGCI), in 1987.

The International Agenda for Botanic Gardens in Conservation was published by BGCI in 2000 (Wyse, Jackson, and Sutherland 2000), updating and broadening the Botanic Gardens Conservation Strategy. The International Agenda was important in contributing to the development of the Global Strategy for Plant Conservation (GSPC) a policy initiative within the framework of the CBD. The ultimate goal of the GSPC is to halt the continuing loss of plant diversity.

The GSPC resulted from a resolution at the XVI International Botanical Congress in 1999 which urged the world community to recognize plant conservation as an outstanding global priority. Responding to the resolution and with coordination from BGCI, leading botanic gardens, IUCN, and other national and international organizations worked together at the turn of the century to develop the new major plant initiative for plants within the CBD framework. The GSPC was adopted unanimously by all government parties to the CBD at the Sixth Conference of the Parties held in The Hague, Netherlands, in 2002 (Wyse, Jackson, and Kennedy 2009).

The GSPC includes sixteen output-oriented targets covering all aspects of the conservation and sustainable use of plants. It is remarkable that virtually all governments of the world agreed to a major initiative to protect and restore plant diversity, which if fully implemented would have major implications for land management, production of plant commodities, and international trade. In 2010, an updated version of the GSPC was agreed with the sixteen targets modified to reflect the growing awareness of the impacts of climate change on plant diversity. The rationale of the GSPC notes the urgent concern that "many plant species, communities, and their ecological interactions, including the many relationships between plant species and human communities and cultures, are in danger of extinction, threatened by such human-induced factors as, inter alia, climate change, habitat loss and transformation, over-exploitation, alien invasive species, pollution, clearing for agriculture and other development."

Implementation of the GSPC links to the overall UN Strategic Plan for Biodiversity 2011–20 with its CBD Aichi Targets agreed in 2010. These are in turn linked to the MDGs which are discussed above. The GSPC is primarily implemented at a national level and has additionally allowed monitoring of progress of plant conservation at a global level.

HOW EFFECTIVE IS PLANT CONSERVATION?

At a global level the two major international Conventions, CBD and CITES, as mentioned above, specifically address the conservation of plant species and provide frameworks for international collaboration and national implementation. CITES is arguably the most powerful of the international biodiversity conservation agreements because its provisions are translated into national laws in all the countries which sign up to it. Provisions of the Convention relate to over thirty thousand plant species listed in its three Appendices. CITES currently has 183 Parties—States that have agreed to be bound by the Convention. The Convention on Biological Diversity (CBD) is a framework agreement with broader objectives than CITES addressing conservation, sustainable use, and equitable sharing of the benefits of biodiversity.

The CBD is legally binding under international law and, to date, there are 194 parties and 168 signatories to the Convention. The CBD's objectives are met through the development and implementation of National Biodiversity Strategies and Action Plans (NBSAPS). Progress is assessed and monitored through the provision of National Reports.

The GSPC has been considered one of the successes of the CBD. Having clear targets allows monitoring of progress to be undertaken. Global progress in implementation of the GSPC was reviewed in 2009 (Secretariat of the CBD 2009) and in 2014 (Sharrock et al. 2014). This review highlighted the constraints of limited institutional integration at the planning level, for example, between agriculture, forestry, and plant conservation, and the limited financial and human resources allocated to the conservation of plant diversity. They also noted that the GSPC had galvanized coordinated action in plant conservation and that significant progress was being made towards certain targets. The lack of baseline botanical information, despite the progress during the twentieth century in botanical exploration and documentation hindered overall progress in the GSPC implementation. GSPC Targets 1, 2, and 5 as shown in Box 1 are effectively research and documentation targets without which progress in conservation and sustainable use of plants at the species and landscape level as required by other GSPC Targets cannot be fully measured internationally.

A consortium of botanic gardens and other scientific institutions is leading work on GSPC Target 1 which initially led to the first global checklist of plant species, The Plant List, and is now developing the World Flora Online (WFO 2020). This basic compilation of agreed plant names underpins assessment of the conservation status of plants together with overall planning for and monitoring of plant conservation action.

Progress towards GSPC Target 2 which calls for "An assessment of the conservation status of all known plants as far as possible, to guide conservation action" has been limited as described above.

Target 5 of the GSPC calls for the identification and protection of the most important areas for plant diversity. This process generally takes into account presence of threatened plant species, overall plant species richness, and unique species assemblages or habitat types. A significant number of countries have identified such areas at a national level, but progress on protection and effective management of the sites is unclear. Information has not been consolidated from national site identification to give a global picture of progress.

Target 7 of the GSPC calls for at least 75 percent of known threatened plant species to be conserved *in situ*. This is a difficult target to measure given the partial information on threatened plant species, the fact that relatively few protected areas have plant species inventories and that many threatened plant species occur outside protected areas. GSPC Target 8, which calls at least 75 percent of threatened species to be in *ex situ* collections and at least 20 percent available for recovery and restoration programs, is easier to measure. This is because botanic gardens and similar institutions maintain record systems of their collections. The PlantSearch database maintained by Botanic Gardens Conservation International (BGCI) includes 1.3 million accession names from over a thousand botanic gardens around the world (BGCI 2020).

A recent comparison between PlantSearch and The Plant List (2013) indicates that botanic gardens manage at least 115,787 different species in their living collections—equivalent to 33 percent of all the species listed in The Plant List (Smith and Pence 2017). Estimates of how many threatened plant species are conserved in botanic garden collections vary according to the definition of "threatened," as well as the location and the availability of information (Smith and Pence 2017). In Australia and New Zealand, an estimated 56 percent of threatened plant diversity is found in *ex situ* living collections or seed banks (Hird 2014). In Europe, the figure is 71 percent with 50.5 percent held in seed banks (ENSCONET 2015); however, the conservation status of European plant species has not yet been fully assessed in a consistent way.

Seed banks are a particularly efficient and cost-effective way to conserve plant diversity in an *ex situ* context. Currently over four hundred botanic gardens maintain seed banks. As well as playing a critical conservation role, seed banks with well-documented collections of wild-source material are important in the provision of seed for food security (Castañeda-Álvarez et al. 2016; Smith 2008), and ecological restoration (Bozzano et al. 2014). Conservation progress for CWR plants held in seed banks (generally agricultural seed banks maintained by national or international bodies) has recently been assessed. The distributions of 1,076 taxa related to eighty-one crops of primary importance to food security were analyzed and the extent to which the potential diversity is encompassed in these taxa is conserved in seed banks. The results show that the diversity of crop wild relatives is currently poorly represented in *ex situ* collections with, for example, 313 taxa (29.1 percent of total) associated with 63 crops, not included in seed banks. Over 70 percent of the taxa are considered as high priority for further collecting in order to improve their representation in seed banks. The most critical collecting gaps for CWR occur in the Mediterranean and Near East; western and southern Europe; Southeast and East Asia; and South America (Castañeda-Álvarez et al. 2016).

GSPC Targets that relate to overexploitation of wild plants are primarily Targets 11 and 12. Target 11 calls for "No species of wild flora endangered by international trade." This links the work of the CBD directly with the work of CITES. As noted above, over thirty thousand plant species are subject to CITES controls including ornamental species with all orchids and cacti together with some other succulents, palms, bulbs such as *Cyclamen* and *Galanthus* spp. (snowdrops), and cycads. Some medicinal plants are included in the CITES Appendices of controlled species and more recently there has been a shift towards including commercially traded timbers. This resulted partly from the growing awareness of the level of illegality associated with the timber trade (Oldfield 2013). The first report on illegal wildlife crime from the United Nations Office on Drugs and Crime Unit (UNDOC

FIGURE 3.4 South West China Germplasm Bank of Wild Species, Kunming, Yunnan Province China in 2017. Courtesy of Visual China Group via Getty Images.

2016) highlighted that the illegal trade in one group of timbers, rosewoods, represents about 35 percent of the total illegal trade in wild plants and animals and their products.

The concern for rosewood timbers, particularly those used in the manufacture of Chinese furniture, has led to inclusion of the main rosewood genus, *Dalbergia*, with about 280 species on Appendix II of CITES in 2016. Previously a smaller selection of *Dalbergia* spp. were subject to CITES controls. Additional rosewood species including the African species, *Pterocarpus erinaceus, Guibourtia tessmanni, G. demeusei*, and *G. pellegriniana* were also added to CITES in 2016.

CITES has definitely had an impact in raising awareness of the impact of international trade on wild harvested plant species and has provided effective mechanisms for controlling trade. Implementation remains challenging, however, and less emphasis is given to botanical aspects of the Convention than to animals. It is also clear that many wild plant species impacted by international trade are not yet included in the Appendices of CITES (Margulies et al. 2019).

Ensuring sustainable production of wild-harvested plant products as required by GSPC Target 12 is a huge task given the range of plant species utilized for livelihoods and the paucity of information on many of them. The various certification schemes for timber production and sourcing of medicinal and aromatic plants developed at the end of the twentieth century provide mechanisms to support implementation and monitoring of this target. CITES can also help in a relatively small way through the requirement for sustainable harvest of certain species traded internationally. Sustainable management of forests for timber production does, however, remain a major challenge.

Target 13 of the GSPC calls for "Indigenous and local knowledge innovations and practices associated with plant resources, maintained or increased, as appropriate, to support customary use, sustainable livelihoods, local food security and health care." This target is closely linked to the text of the CBD which states that each contracting Party shall respect, preserve, and maintain knowledge, innovations, and practices of Indigenous and local communities embodying traditional lifestyles relevant for the conservation and sustainable use of biological diversity. In practice, progress towards achievement of Target 13 is difficult to measure.

FUTURE PROSPECTS

By the beginning of the twenty-first century, knowledge and policies were broadly in place to conserve and sustainably utilize plant diversity at the beginning of the new geological era of the Anthropocene. The extent to which humans can transform ecosystems and climatic processes was broadly understood at the scientific level. Stronger political will and public support are, however, needed to ensure progress. Moving forward, exploration of the natural world continues with, for example, new understanding of the genetic diversity and relationships between plants. The scope for finding new solutions to tackle global problems of food, water, and energy security together with addressing poverty and disease is considerable. As stated by the Liaison Group of the Biodiversity Related Conventions in 2015, "Biodiversity and sustainable development are inextricably linked. Biodiversity, at the level of ecosystems, species and genes, forms the foundation of the Earth's life support systems and provides the services that underpin human lives and prosperity. Our social and economic well being depends on biodiversity, as does our future" (Secretariat of the Convention on Biological Diversity 2015).

Agricultural production is anticipated to account for 70 percent of the future loss of terrestrial biodiversity. Addressing trends in food systems with measures such as restoring ecosystem services in agricultural landscapes, reducing waste and losses in supply chains, and addressing shifts in consumption patterns should help to conserve biodiversity in the future (Secretariat of the CBD 2014). At the end of the twentieth century, 47 percent of the human population lived in urban areas. Now cities are home to over half all the people on the planet and account for around three-quarters of the world's natural resource consumption (UNEP n.d.). Planning for sustainable production and consumption will clearly need to recognize the movement of people away from the land, potentially allowing scope for plant diversity to be conserved and restored in natural areas.

There is no technical reason why any plant species should become extinct given the range of conservation techniques that are available. The main challenges in plant conservation are raising the profile of plants at the popular and political levels, explaining their value to livelihoods and trade, and securing sufficient resources for practical conservation action. Collecting the baseline information on plant species taxonomy, nomenclature, distribution, and conservation status remains critically important. The general shortage of taxonomists and field biologists needs to be addressed and botanical capacity increased in many countries. The partnerships between business interests, environmental NGOs, and governments need to the strengthened and the rights and expertise of local and indigenous communities respected and valued. Plant diversity is in decline in the Anthropocene undermining ecological resilience and the continued supply of ecosystem goods and services, but this trend can still be reversed.

CHAPTER FOUR

Plant Technology and Science

Advances in Crop Improvement

PETER LANGRIDGE AND GEOFF FINCHER

INTRODUCTION

In this chapter we will trace the roles of plant science and biotechnology in cereal crop improvement over the last hundred years. For the majority of this period, farmers and breeders achieved enhanced yields, disease resistance, and improved quality largely through the exploitation of natural variation. However, in the last thirty years, rapid advances in breeding technologies and in the identification and characterization of important genes have enabled quantum increases in cereal crop productivity and quality.

Before embarking on a description of these relatively recent advances, it is worth going back several millennia to the early domestication of cereals and the selection pressure placed upon key characteristics by early human cultures. Domestication of wheat and barley from their wild relatives occurred in the Near East during the Neolithic period thirteen thousand to ten thousand years ago and coincided with at least two key evolutionary events, namely an increase in grain size and the loss of the natural process for grain dispersal, seed shattering (Fuller 2007; Harlan et al. 1973; Purugganan and Fuller 2009). Shattering involves natural seed dispersal by the shedding of grain from the spike or ear as each grain matured and dried. Through selection by early farmers, this trait was replaced by the retention of grains on the spike so that the grain could be harvested before it fell to the ground and the grains needed to be collected and resown by farmers (Fuller 2007). Variation between genes and mutations in other important characteristics, such as flowering time, free-threshing so that grains were free of husks making milling and eating easier, an annual growth cycle, and increased grain yield, were also selected by ancient farmers so that the crop matched local production, consumption and cooking cultures (Purugganan and Fuller 2009).

While increased grain size enabled deeper burial during sowing and resulted in enhanced vigor of the young seedlings, the loss of the seed shattering was of particular importance for domestication of cereals. In wild wheats and barley, distinct zones can be seen where the grain is attached to the stem or floral axis. These zones were quite brittle and as individual grains progressively matured and dried from the base to the top of the spike or ear, mechanical forces such as winds and passing animals caused the grains to break off and fall to the ground. This "seed shattering" characteristic was of course a very efficient biological strategy to maximize seed dispersal and hence the survival of the species, but it posed severe harvesting challenges for humans who wished to gather large quantities of the mature grain. The genes responsible for this trait are involved in regulating other genes (Konishi et al. 2006; Simons et al. 2006). In barley the non-shattering trait originated from mutations that occurred independently in two adjacent genes (Pourkheirandish et al. 2015). The functions of the genes remain unknown, but plants with the wild-type genes have different cell structures in the grain attachment zones compared with the non-shattering mutants (Pourkheirandish et al. 2015).

The mutations described above that led to the non-shattering phenotype in barley and hence to more efficient harvesting is believed to have occurred independently some twelve thousand years ago in the southern and northern regions of the Levant, where many early agrarian societies originated (Pourkheirandish et al. 2015). Shortly after the domestication of barley, farmers in the same region selected a mutation in a gene that greatly reduced the period of grain dormancy compared with both the recently domesticated, cultivated barley and wild barleys of the region (Sato et al. 2016).

Dormancy is another trait that is important for wild plants since it allows the seed to sit in the ground for quite long periods and wait until conditions are best to germinate and grow. However, this is not good in cultivation where farmers want all the seed to germinate and grow at about the same rate. The early selection of a mutation in the key dormancy gene by farmers following the domestication of cereal crop species was not only important for farming but allowed the use of barley grain for brewing beer. The initial malting process of brewing involves the controlled germination of the barley grain. A shorter grain dormancy period would be desirable in this context because the grains would require a much shorter period of storage after harvest before the brewing process could be initiated (Sato et al. 2016). The apparent change of barley utilization from a food and source of carbohydrate to an alcoholic beverage soon after domestication can be linked to the debate as to whether the assembly of ancient agrarian societies was driven by the human desire for a source of stored cereal grain for flour and bread production, or by our desire for alcohol and beer. Alcoholic beverages have long attracted human interest, through their mild and generally positive mind-altering characteristics, their catalysis of social and cultural behaviors, and their benefits to human health through the partial purification of water during the brewing process (Dietrich et al. 2012; Katz and Voigt 1986).

It is likely that both "bread and beer" cultural drivers were in place in ancient agrarian cultures and that the domestication and mutant selection events described above represent some of humans' earliest forays into plant breeding, science, and biotechnology. We will focus this brief review on the role of science and technology in the improvement of quality and productivity in wheat, and on the scientific and technological advances that have enhanced the quality of barley, specifically for the malting and brewing industries. Many of these scientific and technological advances can be applied to other cereal crops and to crops more generally.

Let us now leap forward from Neolithic times to the twentieth century and examine how breeding technologies have developed in the last hundred years. A key milestone in crop improvement in the early twentieth century was the mission of Nikolai Vavilov to use his broad knowledge of the origins of cultivated plants and their diversity to end famines in Russia (Vavilov 1922; Zakharov 2005). Thus, Vavilov initiated crop improvement programs through the identification and exploitation of natural variation and crop species diversity. Indeed, the exploitation of natural variation, involving crossing of compatible species and breeder selection of improved lines, remained the central approach to crop improvement until the advent of molecular biology and functional genomics in the 1990s and 2000s significantly broadened the range of technologies available to breeders.

The natural selection approach met with considerable success. For example, Nazareno Strampelli was credited with increasing average yields of bread wheat in Italy by 50 percent by the 1930s (Mugnozza 2005) and his work and that of others led to the birth of the "Green Revolution" that occurred between the 1930s and the late 1900s. As the "Father of the Green Revolution," Norman Borlaug developed semi-dwarf wheat varieties that were high-yielding and disease tolerant (Swaminathan 2009). These varieties were deployed in developing countries and, when combined with the application of fertilizers, agrochemicals, and improved agronomical management, resulted in dramatic increases in yield by the 1960s (Evenson and Gollin 2003). Increases in the average yields for wheat during this period are shown in Figure 4.1, where it is apparent that, on average, food production increased at an approximately linear rate from 1960 to 2006 (Tester and Langridge 2010). While these yield increases have been impressive and have made a massive contribution to food security during the late 1900s, it is also clear from Figure 4.1 that the future needs of an estimated world population of nine billion people by 2050 will not be met at the current rates of crop productivity improvement (FAO 2009; Tester and Langridge 2010). A second Green Revolution will be required, and this will almost certainly depend on emerging technological advances in breeding methodologies and molecular genetics. In tackling these challenges, we need to acknowledge the increasingly important influence of global environmental changes on both productivity and quality of our major cereal crops (Tester and Langridge 2010).

The development and deployment of the new cereal varieties that delivered the Green Revolution was based not only upon the central importance of genetic diversity, but also on our ability to collect, curate, and share potentially useful germplasm. As a result, breeders have established local, national, and international seed banks and developed efficient procedures for maintaining and distributing the germplasm accessions. Landraces (local cultivars that have been improved by traditional agricultural methods) and wild relatives of the common crop species have also been archived in these seed banks and have proved to be an invaluable source of variation in genes that were "left behind" during the domestication of crop species in different parts of the world. For example, wild wheat and barley collections have been successfully mined for disease resistances that are not found in modern varieties (Steffenson et al. 2007). Recently sequenced ancient DNA from barley grown in Israel six thousand years ago might also reveal useful genes that have escaped the attention of breeders, particularly in the area of disease resistance and environmental adaptation (Russell et al. 2016).

In the sections below, we will outline recently developed methods for the identification and isolation of genes that are important in crop productivity and quality, how these have been developed and exploited, and how new molecular technologies have been translated into improved breeding programs for cereals. Examples of key technological breakthroughs are used to demonstrate the translation process, including the impact of techniques such as genetic mapping of key traits and the generation of artificially induced mutants to expand genetic diversity. We will also consider the structures of modern

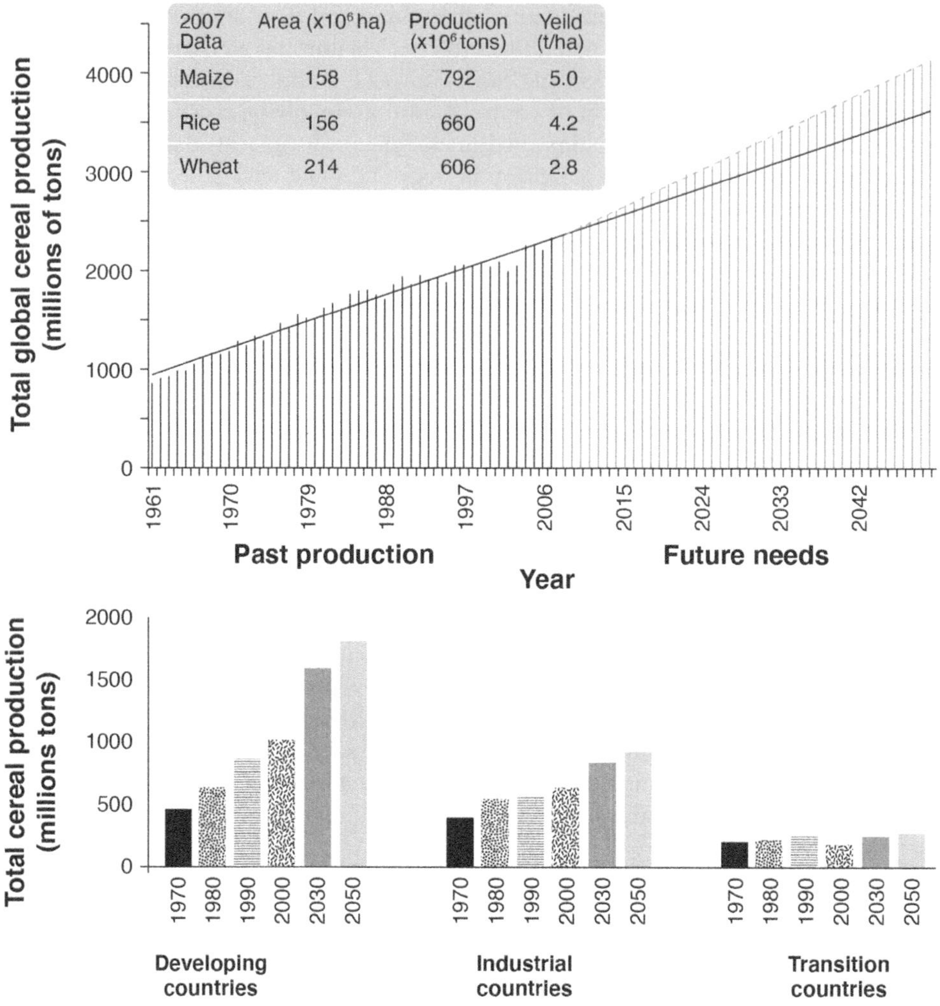

FIGURE 4.1 Cereal production targets and yield improvements over the last forty years. Adapted from Tester and Langridge (2010).

breeding programs and the infrastructure required for their successful operation. Finally, the implications of emerging molecular and agronomic technologies related to precision agriculture are discussed in the context of the crucial advances that are required in crop production over the next thirty years.

ADVANCES IN GENE IDENTIFICATION AND REGULATION

The characteristics of an organism are determined by the complex interaction between the environment and the genetic make-up of the organism. Plants are sessile and, in order to survive, they must cope with a highly variable environment during their growth and development. Developmental patterns and environmental factors will influence the expression of genes and this will lead to changes in the plant's development. Most genes carry the information needed to produce proteins that then drive the biochemistry, physiology, and development of the plant. The study of the relationship between the genes, proteins, and plant development is referred to as molecular biology and has underpinned modern research and breeding of crop plants. In this section we will focus on technological advances that enabled the development of the modern toolkit for molecular genetics and for the evolution of new breeding technologies. Most of these advances have occurred in the last forty years, via successive but overlapping eras of discovery in physiology, biochemistry, and molecular biology. Through these disciplines, we have accumulated greatly improved knowledge and techniques for the identification and characterization of genes that are important for breeders. Not only have these discoveries allowed greater precision for exploiting natural diversity, but they have also allowed us to extend diversity through the identification of useful mutations and genetic modification (GM) and to more quickly and accurately identify genes that are central to crop productivity and quality.

The Biochemical Era

From the 1950s to the 1980s, research in the biological sciences was focused on the definition of biochemical pathways, the proteins that catalyzed individual reactions (enzymes), and the regulation of flow of chemicals through the pathways. During this period, the control of germination of barley grain or seed was used as a model system. Barley germination is controlled by a layer of cells surrounding the bulk of the seed called the aleurone layer. These can be isolated from barley grain used to study the effects of the plant hormones, the phytohormones gibberellic acid (GA) and abscisic acid (ABA), on gene expression and enzyme production. Results from the isolated aleurone layers were subsequently extrapolated back to the intact, germinating grain. Many of the enzymes studied using isolated barley aleurone layers were of commercial importance, especially in the malting and brewing industries (Fincher 1989). At the same time the identification and characterization of different variants of the enzymes provided the first clues to the presence of gene families in the cereals (Fincher 1989). For example, the family of genes that produce the enzymes to degrade cellulose (cellulases), the main constituent of plant cell walls, range in size from twenty-three to twenty-nine members in the cereals and grasses (Buchanan et al. 2012), while another gene family that produces another important cell wall protein, expansins, has between fifty and seventy-five individual members (Mayer et al. 2012). Cellulases mediate the complete or partial degradation of cellulose in the cell walls of growing plants, while expansins have been implicated in the cell wall loosening that is required to allow cell expansion during normal growth and development. These observations were to prove important for future work on the regulation of individual gene family members in different tissues and at different times.

The Molecular Biology Era

In the 1980s and 1990s, the discipline somewhat loosely named "molecular biology" came of age. Associated technologies were focused mostly on the genes and their mechanisms for expression to produce enzymes and other proteins. Isolation or cloning of genes provided an opportunity to look at the fine structure of genes and ask questions about the control of their expression. Most recently, this has led to the definition of the entire genetic make-up of plants through genome sequencing.

Again, isolated barley aleurone layers proved to be useful in defining the effects of GA and ABA on gene expression (Chandler et al. 1984). The improved understanding of gene structure also enabled more detailed analyses of control regions, promoters, of genes, and the components of these promoters that were important for the regulation of gene expression (Gubler et al. 1995; Lanahan et al. 1992; Slakeski et al. 1990). Because promoters are regions of genes that are actively involved in gene expression, they have attracted a good deal of attention as a means of modulating the levels of the gene products in crop species.

Structural Biology

Towards the end of the biochemical era and dependent on the overlap with the early stages of the molecular biology era, there was a surge in interest in the three-dimensional (3D) structures of proteins, and enzymes in particular. Thus, the complete structure of a barley enzyme (1,3;1,4)-β-glucanase was obtained (Varghese et al. 1994). The 3D structures of enzymes pave the way for engineering their properties to make them more efficient, to change compounds they can work with, and to enhance their stability. These enzymes are particularly important in the malting and brewing industries, where an ability to manipulate efficiency and stability would have practical applications in beer and whisky production. As the number of 3D structures in publicly available databases increased, so too did computer modeling programs that predict the 3D structures of proteins from their amino acid sequences.

The Functional Genomics Era

In the 1990s, molecular biology moved from what had essentially been studies of individual or small numbers of genes, to the examination and definition of very large groups of genes and their expression, called transcript profiling. New, high throughput and high-speed technologies, many of which found their origin in the human and Arabidopsis genome sequencing programs, were quickly developed. Linked with the rapid growth in computing power and efficient search programs, these technologies enabled the analysis of multiple and eventually thousands of genes in a single experiment. Many large-scale transcript profiles have now been performed using these technologies and in most cases these profiles are made publicly available in international databases. Research groups around the world can now investigate transcript profiles in their particular area of interest and expertise, whether that be starch or storage protein metabolism, phytohormone synthesis and turnover, pigment formation, plant-pathogen interactions, etc.

The end-point in functional genomics technologies is often considered to be the delivery of a complete genome sequence of the plant of interest. Thus, a physical, genetical and functional sequence assembly of the entire genome is now available for barley (Jayakodi et al. 2020; Mascher et al. 2017), wheat (IWGSC 2018; Walkowiak et al. 2020), and rice (International Rice Genome Sequencing Project and Sasaki 2005), and many other related cereals and grasses. As mentioned above, the assembly of genome sequence information and analyses of transcript profiles have been made possible by recent dramatic increases in computing power and data storage capacity, together with the emergence of software packages that underpin bioinformatics analyses of the sequence data. Just as importantly, the international community has adopted a highly collaborative and cooperative approach in assembling teams of scientists with the wide array of expertise required for the successful completion of these very large genome sequencing programs (Brenchley et al. 2012; IRGS 2005; Mascher et al. 2017; Mayer et al. 2012).

Genetic Modification (GM) and Mutant Technologies

During the molecular biology era it quickly became apparent that genes isolated from a wide variety of plant and non-plant species could be readily transferred into other plant species through a process known as genetic transformation. The introduced genes are expressed in the resultant GM plants or genetically modified organisms (GMOs) and changed their phenotype. The technology provided a new and more precise method of expanding genetic diversity in crop species. Common crop species such as soybean, maize, and canola have been transformed with genes that confer herbicide tolerance, insect resistance, and altered fatty acid composition, to name but a few. These GM crops have been deployed around the world, particularly in North America and in developing countries. It is estimated that global plantings of GM crops in 2017 approached 200 million hectares and were sown by seventeen million farmers in twenty-four countries (ISAAA 2017). This level of uptake of the technology has clearly been driven by observations that the technology has a huge positive influence on crop yields and hence food security, particularly in developing countries. The high adoption rates of major GM crops grown in the USA over the last two decades are shown in Figure 4.2.

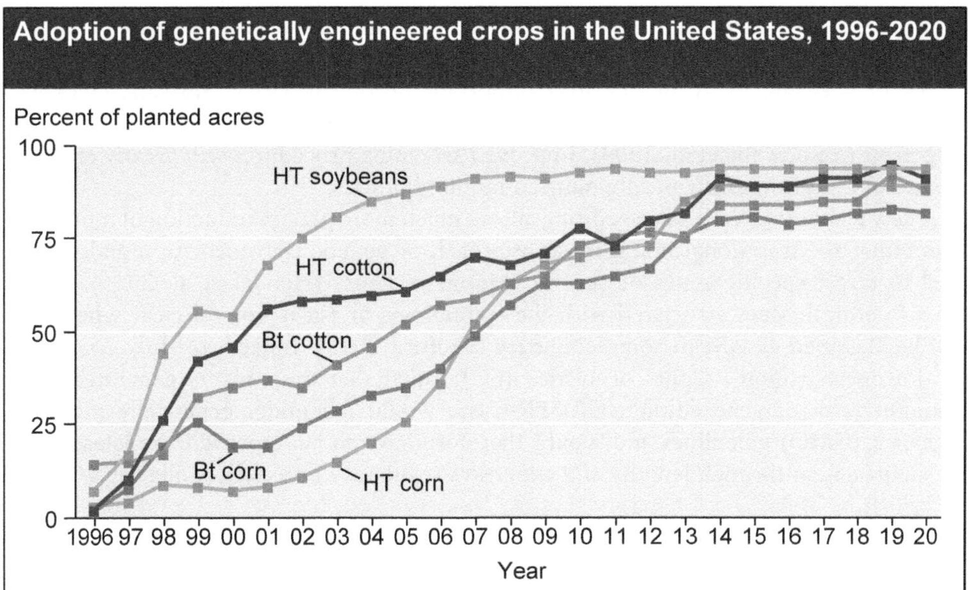

Note: HT indicates herbicide-tolerant varieties; Bt indicates insect-resistant varieties (containing genes from the soil bacterium Bacillus thuringiensis). Data for HT/Bt corn and cotton are not mututally exclusive, as HT and Bt categories include those varieties with overlapping (stacked) HT and Bt traits.
Source: USDA, Economic Research Service using data from the 2002 ERS report, *Adoption of Bioengineered Crops* (AER-810) for the years 1996-99 and National Agricultural Statistics Service, (annual) June Agricultural Survey for the years 2000-20.

FIGURE 4.2 Adoption of broad acre GM crops in the USA over the last twenty-five years. Note that Bt and HT designate plants engineered with insecticidal genes from the bacterium Bacillus thuringiensis (Bt) and those engineered with genes to produce generally "herbicide tolerant" (HT) plants. Adapted from USDA Economic Research Service (2020).

However, GM technology precipitated a massive international debate over perceived dangers associated with GMO crops and over the moral correctness of GM technology. While detrimental effects of GM crops have been widely claimed in the popular media, none have stood up to scientific scrutiny. Nevertheless, although not one single adverse effect of GM crops has been demonstrated with the required level of scientific rigor, the debate rages on. We do not wish to discuss the issues in any detail in this short review, but the ongoing social and hence governmental nervousness about the technology has limited its deployment in many countries (Hartley et al. 2016). Furthermore, the social nervousness has given rise to draconian regulatory hurdles that require tens of millions of dollars to clear if a GM plant is to be certified for broad acre cropping. As a result, scientists have looked back for more traditional, non-GM methods for increasing crop diversity and have also developed new, more precise methods for generating improved crop species that are not classified as GM.

For many years, breeders and their molecular geneticist colleagues have prepared mutated libraries using chemicals, such as ethylmethanesulfonate (EMS) or sodium azide (Caldwell et al. 2004), or ionizing radiation, such as x-rays. The mutant lines carry no external DNA and are therefore classified as non-GM. Screening these libraries for the characteristics of interest and identifying a mutation in a gene of interest, followed by removing the large number of background mutations, has been a time-consuming task. However, well over two thousand varieties of plants have been released that were derived from artificially induced mutations. Among the most important have been rice cultivars widely grown in Thailand, China, India, Vietnam, Myanmar, and Japan, wheat mutants grown in Pakistan, many of the most important European barley cultivars, chickpea, sunflower, and cotton cultivars in many countries and modern varieties of pear and grapefruit (Ahloowalia et al. 2004). Improved screening procedures will greatly enhance the usefulness of these chemically induced mutant libraries.

A new technology called gene editing allows much more precise induction of mutations than either the traditional mutagenesis approach or genetic transformation and can be used to target specific genes or specific regions of genes (Hendel et al. 2015). There are also ethical issues associated with the technology, and it is not yet clear whether it will be classified as GM in some countries (Ledford 2015). Indeed, on July 25, 2018, the European Union's Court of Justice (ECJ) ruled that crop plants generated by a common form of gene editing, CRISPR-Cas9, would fall under genetically modified organism (GMO) guidelines and would therefore have to be approved for release into the environment through lengthy and expensive regulatory processes (Callaway 2018).

ADVANCES IN BREEDING TECHNOLOGIES

In this section we will attempt to tie in with the more general technological advances discussed in the previous sections with the more practical aspects of breeding technologies. Again, we will use examples of where advances in technologies have been translated into advances in breeding technologies associated with cereal crop improvement.

Over the past century plant breeding has achieved great advances in the productivity of our major crops. Since the early 1960s, combined cereal production has increased from less than 1 billion tonnes to almost 3 billion tonnes (in 2014) while only 11 percent more land has been brought into production (FAOSTAT 2017). Although several factors (e.g. mechanization, fertilizers) have underpinned this rapid improvement, only one—the development and release of improved varieties, largely achieved through the development and adoption of new breeding technologies, can realistically be regarded as sustainable. These advances have allowed our farmers to feed the growing world population and reduce the overall cost of food while maintaining the economic viability of our farms.

While the technological innovations that led to production increases achieved over the past fifty years are impressive, they were built on many developments that occurred over a much longer period of time. It is always difficult to go back in time to determine when key events may have occurred. The first breeders were those communities that originally selected for plants suited to farming. In recent years, we have been able to identify several domestication and post-domestication genes (the latter are responsible for adaptation to different environments) and the causal changes or mutations responsible for their impact on the characteristics of the plants. For example, in barley the primary domestication genes have been identified. These include genes that were responsible for shattering or spreading of the grain when ripe. The loss of shattering was critical to allowing farmers to harvest grain before it was spread over the ground. Similarly, the gene responsible for the trait that allows the grain to be separated from the husk was important for milling grain to produce flour (Pourkheirandish et al. 2015; Taketa et al. 2008). The post-domestication traits provide examples of adaptation to particular production environments and include the genes controlling the flowering time network, the seed dormancy, and the green revolution dwarfing genes. Flowering time or maturity is important to allow the crop development to match the growing season, or to grow and put on biomass while the weather is good and then produce the grain for harvest before the climate turns bad for growth, such as cold and snow in winter or a very hot dry summer. The dwarfing genes were important because they helped to keep the crop vertical and easy to harvest after strong winds and also meant that less energy was wasted by the plants on increasing stem length, leaving more energy and carbon for filling the grain, which is a trait called harvest index or the proportion of the total crop biomass that can be harvested as grain. In wheat, in addition to the maturity and the dwarfing genes, another important trait is spike morphology, the shape of the head that carries the grain. This trait is controlled by single gene called *Q*, which regulates several other genes to give the characteristic shape of the wheat head, high grain yield and also makes the grain free-threshing so that the grain can be easily separated from the chaff. A single mutation in *Q* leads to the free-threshing characteristic (Simons et al. 2006). In maize a gene called *Teosinte branched1* (*tb1*) was responsible for the major difference between maize and its wild relative, teosinte. The key differences between the two lies in the expression of this gene, which results in the change from a branched growth pattern seen in teosinte to a single stalk in maize. It appears that human selection targeted lines showing elevated expression of *tb1* (Shi and Lai 2015).

For fruit crops, domestication also appears to have relied on only a few genetic changes. In tomato, most of the variation we see in modern cultivars compared to the wild progenitors can be explained through modifications to only four genes (Monforte et al. 2014). We know relatively little about the domestication of other fruits, but it seems likely that a similarly small number of genes were involved.

A Brief History of Plant Breeding

These basic domestication traits formed the basis for further selection by farmers and led to the diverse set of landraces found for most of our crops. Varieties selected by famers over many thousands of years and grown in specific regions are referred to as landraces to distinguish them from cultivars developed through modern plant breeding. There was a long gap between domestication and systematic breeding. Indeed, it was not until around 1700 that Thomas Fairchild (1667–1729) showed that plants did indulge in sex and he made the first artificial hybrid, between Sweet William and pink carnation. It was then almost another century before Thomas Knight (1759–1838) started breeding a range of horticultural crops including strawberries and peas. In the mid-1800s Louis de Vilmorin (1816–60), with his work on beetroot, initiated the concept of targeted selection for particular traits. This means that he generated large populations of plants from crosses and selected individuals that had characteristics that he believed desirable. De Vilmorin is also credited with starting the seed breeding industry. Progress continued slowly until the rediscovery of Mendel's work by Hugo de Vries, Carl Correns, and Erich von Tschermak-Seysenegg in 1900. From this point, many of the basic plant breeding principles were developed and tested. These included the concepts of developing genetically pure lines to generate uniform crops (Wilhelm Johannsen 1857–1927) and the idea of making specific crosses as a route to generating new genetic combinations (Herman Nilsson-Ehle 1873–1949).

As the importance of genetic variation became clear, concern started to develop around the narrow genetic base used in breeding programs. An important factor in this development was the work of Nikolai Vavilov (1887–1943) showing that there were clear centers of diversity from which our modern crops were derived. He suggested exploring this diversity for new sources of agronomic traits. This approach was linked to practical breeding by Nazareno Strampelli (1916–42), who introduced three innovations to wheat breeding. First, he developed the strategy of using landraces, as sources of new variation to address specific traits that limited productivity; he sought novel sources of rust resistance using this approach. Rust was, and continues to be, one of the most important diseases of wheat. Second, he recognized that modifying crop development could help protect against late season drought; he selected for early flowering wheats that could flower and set seed before water deficit became too limiting for growth. This allowed the wheat plants to escape the summer drought that severely limited yields in the Mediterranean environment. Third, he suggested reducing plant height to protect against lodging and to boost harvest index; the proportion of total biomass that is converted to harvestable grain. This early, largely conceptual, work was later to lead to the shuttle breeding approach used by Norman Borlaug (1914–2009) to produce the Green Revolution varieties in 1964. Shuttle breeding was based on screening lines for high yield in an optimal environment where disease pressure was low; Borlaug used an irrigated site in the desert regions of northwest Mexico. The highest yielding lines were subsequently transferred to a second site at high altitude and humidity near Mexico City where the plants were exposed to diverse diseases. The best lines were taken back to the high yielding site and used in further crosses. The shuttling between sites meant that there was intensive selection for both yield and disease resistance and this resulted in rapid genetic gain for these key traits. Genetic gain refers to the improvements that result from enhanced combination of genes as opposed to the yield gains that come from improvement in agronomic practices.

The rate of genetic gain in a breeding program is determined by several factors: the extent of genetic diversity in the program, the number of crosses and size of the populations generated and available for selection, and the intensity of selection, that is, how effectively can the breeders select for the target trait. In addition, the duration of the breeding cycle will have a major impact. For most annual crops this is one year for each selection cycle, but it can be five or more years for tree crops. These factors are referred to as the "Breeders' equation." These terms and the impact of technological advances are explained below.

Diversity

There is a global imperative to enhance the rates of genetic gain in our crop plants. For many of our major crops, the rates of genetic gain have slowed and will not be able to keep pace with predicted demand. While many approaches have been proposed to address this problem, there is general recognition that the germplasm base in modern breeding programs has become too narrow to allow rapid genetic gain (McCouch et al. 2013). Although extensive germplasm collections are available, these have been poorly utilized. The primary problem is that the effort involved to use even a small number of the hundreds of thousands of available accessions is enormous in terms of both time and effort; for example, it can take over twenty years for a research group to introduce a single novel disease-resistance gene from a wild relative into a modern wheat cultivar. Several major international programs have attempted to address this problem by characterizing germplasm collections at the genotypic and phenotypic levels (http://www.divseek.org, http://seedsofdiscovery.org). While these programs have played an important role in characterizing collections, they have so far had little impact on utilization.

The problem with utilization can be broken down into the following issues. First, characterization of wild germplasm for traits important to modern agriculture is difficult. While it is feasible to screen large wild germplasm collections for some simple traits such as disease resistance, it is very difficult to screen from complex traits like yield under drought. The problem is that landraces and wild relatives are not suited to modern cropping systems and are low yielding. Therefore, how can you tell if there might be some genes that would enhance yield if the plants had the appropriate maturity and growth habit for modern production methods. Consequently, we do not know which of the thousands of available accessions should be used. Second, in most cases, the genetic information of the wild accessions will not mix well with the genetic make-up of modern cultivars. This means that the progeny may be sterile or that very large populations must be used in order to have a reasonable chance of introducing a new gene/allele into elite lines. And, last, if multiple genes are required to transfer a desirable trait, which applies to most yield-related traits, then the populations required are so large as to be completely impractical. These problems have not only severely inhibited the effective exploitation of genetic resources but have also mitigated against the domestication of new crop species.

New Sources of Variation

We generally think of plant breeding as involving crosses between plants that are able to cross readily and produce normal fertile offspring. However, breeders have been far more adventurous in seeking novel sources of variation. Thomas Fairchild provided an early example when he crossed Sweet William and pink carnation. A more dramatic example was provided in 1876 in Scotland when wheat (*Triticum aestivum*) and rye (*Secale cereal*) were crossed to produce Triticale (*Triticosecale*). However, it was not until 1938 that a fully fertile hybrid was generated that went on to form the basis for the modern Triticale industry, which is now grown on over 4 million ha, particularly in regions where its cold and stress tolerance is of greatest benefit and produces almost 17 million tonnes of grain (2014 data, FAOSTAT 2017).

The crosses of the type used to produce Triticale are known as wide crosses and vary in their complexity. In some cases, the crosses produce normal seed that can be grown on. In some cases, the genetic make-up of the progeny is unstable, and the lines die or become sterile over subsequent generations.

While Triticale involved the introduction of the entire rye genome into wheat, smaller regions of rye, and other wheat relatives, have also been important for wheat breeding. For example, a piece of the rye genome was introduced into wheat to produce lines with resistance to several foliar diseases, along with wide adaptation and high yield performance (Dubin and Rajaram 1982). More recently the variety Robigus was released, which carries a genome segment from a wild grass related to wheat. Importantly, 30 percent of all wheat varieties produced by the International Wheat Improvement Program at CIMMYT in Mexico, which delivers improved varieties to breeding programs around the world, are derived from crosses between normal wheat and "synthetic" wheat, where synthetic wheat is derived from crosses between wheat and a wild relative (Dreisigacker et al. 2008).

In some cases, genetic diversity can be sourced from species that cannot be crossed. The process of protoplast fusion involves fusing cells from two different plants and recovering a whole plant from the fused cells. The resultant plants are called somatic hybrids. This technique has been used for a number of crops, including citrus (Grosser and Gmitter 2011), potato (Orczyk et al. 2003), and ornamental plants (Kuligowska et al. 2016).

An important aspect of molecular biology and gene discovery is that it opens up opportunities to access variation from diverse sources. Through genetic engineering, genes can be taken from any organism and inserted into the genome of the crop plant. In most cases the gene will require some modification so that it is appropriately expressed in its new host. This technology has had a major impact in resistance to insect pests and tolerance to herbicides. In both of these cases genes derived from bacteria have been used.

Novel variation can also be generated using chemical mutagens or ionizing radiation to produce mutations. Lewis Stadler (1896–1954) first introduced this technique in 1928. Many modern varieties of crops and fruits, such as some seedless mandarin varieties and many new disease resistances, include these mutations. The advances in sequencing technologies have allowed the rapid screening for lines that carry mutations in specific genes, and these can be passed to breeders for evaluation (Chen et al. 2014).

Advances in molecular biology have greatly expanded our understanding of gene regulation and the processes that operate in organisms to protect themselves against diseases, pests, and herbicides. This information has provided new options to introduce new genes or modify or turn off expression of genes in a highly targeted fashion (Boettcher and McManus 2015). The first GM crop plant, insect resistant cotton, was released by Monsanto in 1987, although large-scale production of GM crops did not start until 1996.

Newer technologies based on a sound knowledge of genes and their regulation allow creation of new traits in plants. Non-browning apples, where the gene responsible for making apples, and other fruit, go brown when cut or damaged, has been silenced, represent the first plant varieties produced through this technology (http://www.arcticapples.com/) but many more are in the pipeline. Genes can also be turned off with genome editing technologies described earlier (Belhaj et al. 2015).

The effects of these techniques for gene silencing are very similar to changes induced by chemical mutagens or ionizing radiation, although the new techniques are more targeted and lead to fewer modifications to the overall genetic make-up of the plants. Several studies suggest that these technologies present no significant risk or hazard to human health or the environment, but regulatory agencies are still unsure on whether these technologies require regulation (see for example Lusser et al. 2011). The regulatory framework that operates within different jurisdictions and regions is likely to determine the extent of application of these new technologies, but the scope for application is extensive.

Selection Intensity and Population Size

Many important traits are quite difficult to assess and effectively select for, since they may be subject to large environmental effects. The genetic control of yield, either the quantity of grain, tubers, fruit, or biomass, is not only the most important trait for most breeding programs, it is also one of the most difficult to measure reliably. For example, wheat yields of over 15 tonnes per hectare have been recorded but the global average yield is only around 2 t/ha and wheat is grown in environments where yields are below 1 t/ha. The yield of any particular variety can vary over tenfold depending on the growth conditions, but most of this variation is usually due to environmental factors, such as rainfall: the genetic component of this variation that is targeted by breeders may vary by only 5–10 percent. Nevertheless, a 5–10 percent increase in yield potential is highly significant. The accuracy of measurements of yield and other traits is critically important to plant breeding.

The reliability of screening is also tied to the scale of screening or the sizes of populations that can be evaluated. For traits such as yield, large field plots at multiple sites and with a high degree of replication will improve the reliability of yield measurements. However, there is clearly a significant cost associated with these screens and many breeding programs need to balance their resources and limit the size of populations they can evaluate. Therefore, development of techniques that allow large-scale screening, at low costs and with high reliability has been extremely important to breeders.

The phenotype of a crop can be described as the observable and measurable characteristics of the crop. For plant breeding, the accuracy of phenotyping is critical since efficient phenotyping is equivalent to efficient selection in the breeding populations. The phenotype of crops can refer to individual plants or the community of plants that form the crop. Phenotype includes developmental traits, morphological, physiological, or biochemical characteristics for single plants, while for crops the phenotype usually refers to the features of a group of plants growing in a defined area such as yield and processing quality of grain. Ultimately, the phenotype can refer to anything we can measure. However, these measurable traits will vary in their importance to the breeding program and in the reliability or accuracy of the measurements. Breeders seek to improve the characteristics of the crop or, more specifically, they work with the genetic potential of the crop. This means that they are primarily interested in traits that are under genetic control. For some traits, such as many disease resistances, this is not difficult to assess—if the plants are exposed to the disease they will either show disease symptoms or not. However, this is not always easy to determine since many traits will vary greatly depending on the environment. Plant height can vary depending on the growth conditions since if there is abundant water, nutrients, and sunlight the plant may be tall but if starved and stressed it may be very short. The interaction between the genetic make-up of the plant (genotype) and the environment is referred to as the genotype by environment or G x E interaction. For breeders to make progress in the genetic potential of the plant, they need to know what proportion of the phenotype (for example plant height) is determined by the genotype relative to the environment. This is referred to as the heritability of the trait—the higher the heritability the more effective will be the screening and selection for improved genetic combinations. We can measure heritability by growing a series of varieties or lines under a variety of conditions and determining the proportion of the variation that changes across sites.

A range of new imaging technologies allows the measurement of diverse crop and plant characteristics. In particular, the cost and weight of cameras used to capture images has come down to a level where cameras can be attached to drones and can take large numbers of high-resolution images of a developing crop. Using normal visible light cameras (RGB cameras), the growth rate of the crop can be measured and plotted. These images also allow detection of abnormal color of the crops due to factors such as nitrogen deficiency or disease outbreaks. Long-wave infrared or thermal cameras can help determine if plants are under stress or suffering from disease. For example, under drought stress, plants are less able to cool themselves via transpiration and they will show an increase in canopy temperature—a bit like the high temperatures we get when sick. It is now possible to use wavelengths of light well outside the visual range (hyperspectral), visible and near-infrared cameras that capture hundreds of images over a broad electromagnetic spectrum (Araus and Cairns 2014; Falhgren et al. 2015). All these data can be fed into models of crop performance.

At the individual plant level, we can also generate detailed information on a plant's behavior under a range of growth conditions. There are now many automated facilities available to scientists and breeders for screening plants. These facilities generally use single plants grown in pots in a greenhouse or growth chamber. The plants can be moved via a conveyer belt to imaging stations where they are photographed using a diverse set of cameras, including visible, infrared, near infrared, and fluorescence (Figure 4.3). At a fine level, we can also generate information on parts of the plant, leaf area and structure, leaf surface, flower size and number, and many other features. Still finer analysis can be made of the composition of plants, such as the levels of metabolites (metabolomics) or proteins (proteomics) and levels of gene expression (transcriptomics). New techniques for root analysis have added further to the long list of plant characteristics that can be measured non-destructively and at increasing throughput (Downie et al. 2015).

Ultimately these measurements are aimed at supporting selection for traits that will be of importance to farmers. Breeders are seeking tools that offer high heritability and consequently rapid genetic gain through selection; are highly correlated to traits of importance to their programs so they can replace difficult, unreliable, or expensive measurements; and are low cost and high throughput, so they can screen large populations at reduced cost compared to conventional screening methods. The information that is now available to plant breeders can be extensive. Therefore, the challenge is often to decide what is useful and what should be ignored. There are several ways the data can be used. One can look for relationships between traits, for example total biomass of the plant is often associated with leaf area and the duration of vegetative growth and is strongly related to yield. Some of the associations between traits can be quite complex, for example canopy temperature can be linked to heat tolerance since a plant that is cool is able to transpire more water than a hot plant and this protects it from high temperatures. However, this trait is also associated with root depth because a plant that can track water down the soil profile late in the growing season is better able to access water (Pinto and Reynolds 2015). Therefore, canopy temperature can be used to select plants that have deep roots.

With detailed sets of phenotypic information, breeders and scientists can undertake extensive correlation analyses. For some of the plant and crop imaging platforms, many different parameters can be measured. These can be correlated with each other

FIGURE 4.3 Automated phenotyping as an aid to capturing genetic variation in large populations of crop plants. Tens of thousands of plants can be grown in randomized locations in a glasshouse and moved by conveyor belt to imaging stations, where phenotypic characteristics such as leaf area, leaf color, photosynthetic efficiency, water status, and thermal distribution patterns can be measured non-destructively throughout the life of the plant. The photograph is from the Plant Accelerator at the University of Adelaide node of the Australian Plant Phenomics Facility. Photo by Geoff Fincher.

and with traits that are important to a breeding program. Examples of how this can work in breeding can be seen if we look at selection for drought tolerance. This is a very complicated trait since the timing and severity of the drought stress can have a large effect on yield. In maize, at moderate drought stress, grain number is associated with the anthesis-silking interval, which is the period between emergence of the male (pollen) and the female flowers (the silk), which in turn is associated with silk growth rate. Therefore, measurement of silk growth can be used to screen for improved drought tolerance (Tardieu et al. 2014).

The Breeding Cycle

The time taken from production of the first cross to release of a new variety can be long. Even for annual crops, such as wheat, this frequently takes between ten and fifteen years and for perennial crops like apples, the period is several decades longer. The breeding cycle is largely determined by the generation time for the plants. For an annual crop, such as wheat, once the first cross is made several rounds of selection are needed, followed by a period of extensive large-scale testing and production of pure seed. For perennial crops, for example apples, the generation time is over five years, making breeding a very slow process. Reducing the cycle or generation time can have a dramatic effect on the rates of genetic gain and is, consequently, an area of intense research.

There are several phases of the breeding cycle where technologies can be applied to accelerate the rate of variety delivery. The techniques vary between species and breeding systems. For example, for some crops modifying the growth temperature, photoperiod, or light intensity or applying mild stress can induce flowering; in chickpea three generations can be produced in one year using these approaches (Sethi 1981). It may also be possible to use tissue culture to bypass maturation or dormancy phases; combining the light and temperature regime with embryo rescue can be used to achieve six generations a year for wheat and barley (Hickey et al. 2017).

A detailed knowledge of the control of flowering time can provide a novel approach to accelerating the reproductive cycle in some plants. Several groups have modified the control of flowering in apples and other fruits so that seedlings will flower within one year of germination rather than requiring five years (Le Roux et al. 2012). This approach has involved genetically engineering the plants by changing the expression of genes controlling flowering time (Yamagishi et al. 2014).

The breeding cycle can also be reduced through the screening techniques that allow the prediction of a particular trait based on an analysis of the genetic make-up of the plant or by assessing surrogate traits. Amongst the most important of these techniques is the use of genetic or molecular markers. These are based around screens for sequence variations, DNA markers, that are associated with specific traits. This is much the same as DNA testing used in screening humans for genetic disorders or in forensic science. A wide range of techniques are available for assessing DNA sequence variants and the technology has moved through rapid innovation with several low-cost and highly reliable screens now available.

The use of molecular marker technology can take a variety of forms. In the simplest case, the DNA marker may be genetically associated to the trait of interest, such as disease resistance. Rather than screening for resistance itself, the breeder can use the marker to predict the presence or absence of the resistance gene. Molecular markers can be particularly important where the target trait is expensive or time-consuming to measure. For example, bread-making quality in wheat is a very important trait for breeders but to measure it, breeders need to grow sufficient grain to mill the wheat and bake a loaf of bread and this takes time. However, we know that a group of grain proteins, glutenins, play a major role in bread-making quality and these can be accurately screened in seedlings using DNA markers, meaning that the breeder will have a good idea of the baking properties of the plants simply by analyzing the grain proteins or looking for specific sequence characteristics in the DNA of seedlings. For long generation crops, such as many fruit or forestry species, the ability to assess traits using DNA extracted from seedlings saves the breeder the time and expense of growing the seedlings to maturity. Other important advantages of using molecular markers include combining multiple disease resistances to increase the durability of resistance, where conventional screening is unable to distinguish a single resistance from multiple resistances.

Over the past decade there has been increasing use of whole genome analysis to predict the performance of lines. In this approach, a model is developed for the relationship between the genetic make-up of the plant and the trait or traits of interest. This information can then be used in the breeding population to assign breeding values to individual lines based on their overall genomic structure.

FUTURE DIRECTIONS

As mentioned earlier in this brief review, we have made impressive gains in cereal crop productivity in the last fifty years. Thus, combined cereal production increased from less than 1 billion tonnes to almost 3 billion tonnes between 1960 and in 2014, yet only 11 percent more land has been used to generate this increased production (FAOSTAT 2017). Nevertheless, if the projected global population exceeds nine billion people by 2050 becomes a reality, which is highly probable, it is also apparent that the future needs will not be met at the current rates of crop productivity improvement (FAO 2009; Tester and Langridge 2010). The Green Revolution delivered a quantum leap in crop yields and it is likely that another major breakthrough in enhancing crop productivity will be required. The requisite improvements will need to be achieved against a backdrop of global climate change, where some areas might become amenable to higher yields while others will see significant decreases in yields. The predictions are that yields of major crops will be severely impacted by climate change, for wheat yields are predicted to decline by 6 percent for every degree increase in temperature (Zhao et al. 2017; Zhu et al. 2016). Farmers and breeders will need to be flexible and nimble-footed with respect to choosing and changing crops and cultivars to be used in ongoing responses to climate change.

We can be confident that increased crop productivity will be partly driven by improvements in agronomic practices, in which precision agriculture, robotics, drones, local sensors, satellite technology, infrastructure, big data collection and analysis will undoubtedly continue to contribute to productivity improvement. There is likely to be an increased emphasis on reducing energy and other inputs during crop production in the future. However, emerging technologies, better access to genetic resources and advances in breeding methodologies and molecular genetics will continue to underpin our efforts to meet the predicted food requirements for nine billion people by 2050. We will need to use all the technologies at our disposal, including GM and gene-editing technologies.

ACKNOWLEDGMENTS

We wish to acknowledge the hard work and dedication of numerous postgraduate students and postdoctoral scientists who have contributed so enthusiastically to the work described here. We also thank the Australian Research Council, the Grains Research and Development Corporation, the State Government of South Australia, and the University of Adelaide for their continued support over many years.

CHAPTER FIVE

Plants and Medicine

From Imperial Divergence to Global Convergence

JEAN-PAUL GAUDILLIÈRE

INTRODUCTION

In 2015, Dr. Tu Youyou, a Chinese phytochemist, was awarded the Nobel Prize in Physiology or Medicine for her contribution to the discovery of artemisinin and the demonstration of its therapeutic activity against the agent responsible for malaria. Artemisinin is a compound isolated from *Artemesia annua,* a plant endemic to many parts of Asia and East Africa and the attribution of the Nobel Prize was widely perceived as a powerful demonstration of the importance medicinal plants continue to play in present-day pharmaceutical innovation. Artemisinin and its derivatives are actually the most important anti-malarial medications at hand. Their access has been for two decades pivotal to major programs in global health including the WHO initiative "Roll Back Malaria."

The 2015 Nobel Prize bore a secondary level of meaning, however. As Dr. Tu Youyou reminded her Nobel audience, species of *Artemisia* have been used for centuries in Chinese medicine to prepare herbal remedies active against fevers. Her reading and reflecting upon the texts from the classical corpus of Chinese medicine played a decisive role in the choices she made when researching the active principles of the plant (Tu 2015). Artemisinin is thus one of the rare success stories that phyto-pharmacology can use in order to illustrate the fecundity of its grounding paradigm, namely the organization of drug research around the prospection of plants with medical uses and the chemical identification of their active compounds followed by their synthesis and/or mass production. The case is all the more interesting as the artificial chemical synthesis of artemisinin, although doable, proved too costly and the drug is still obtained from cultivated *Artemisia annua*.

What the Nobel Prize lectures and celebrations did not reveal is the extent to which this particular research was contentious work. The first reason for the problematic nature of the investigations conducted in China from the late 1960s onward is the fact that medicinal plants did not figure high on the agenda of global pharmacy. It is well known that the decades after the Second World War were a time of major breakthroughs with the emergence of new classes of drugs—from antibiotics to psychoactive substances—and a time of profound reorganization of drug production around major corporations, mounting investments in research and marketing, and a radical shift toward chemistry as the main source of knowledge. As a consequence, medicinal plants definitely lose ground,

FIGURE 5.1 Youyou Tu won the 2015 Nobel Prize for Medicine in 2015 for her contribution to the treatment of malaria. Tu credited "strengths from both Chinese and Western medicine" in her derivation of artemisinin from *Artemisia annua*. She hoped that medical researchers would be able to develop many more novel medicines from the "substantial" nature resources we have available and quoted Chairman Mao to remind us that "Chinese medicine and pharmacology are a great treasure house. We should explore them and raise them to a higher level." Courtesy of Jonathan Nackstrand/AFP via Getty Images.

FIGURE 5.2 A picker harvests *Arnica montana* (wolf's bane) in Le Markstein, eastern France in 2018 to be used as herbal medicine for analgesic and anti-inflammatory purposes. Photo by Jean-Christophe Verhaegen/AFP via Getty Images.

becoming minor sources of bioactive molecules when confronted with the chemists' abilities to synthesize entire families of artificial compounds.

The second reason why researching the antimalarial properties of *Artemisia* species was far from legitimate was the low status of "traditional" medicines even in the case of the highly sophisticated, professionalized, text-based, and elite-oriented systems like those prevailing in Asia. Apart from national initiatives, as in India and China, it is only during the fifteen years period dominated by the 1978 Alma-Ata Declaration's agenda for primary health care that the World Health Organization (WHO) included traditional medical practices among the resources for a viable strategy toward "health for all" and argued for their "integration" in health care policies. That science would provide the backbone for integration was, however, still clear and presented a source of tensions since biomedicine was the reference while non-Western systems provided the practices to be assessed. The Chinese inquiries into classical anti-malarial preparations, for instance, quickly eliminated mixtures of heterogeneous plants and their putative synergies in order to focus on a few species of *Artemisia* and on their chemical composition.

This chapter examines the paradoxical status of medicinal plants and their uses as remedies in a long century starting in the 1870s and ending with the 2000s. It first discusses the origins of what may be called the "big divergence." While medicinal plants had been at the core of European practices of medicine and pharmacy, their uses became less and less important and visible in the late-nineteenth-century and early-twentieth-century world of biomedicine. This process in turn reinforced the disjunction between the European and Asian medical worlds, whose difficult but significant previous dialogue had deeply relied on shared interests in a plant-based *materia medica*. Secondly, the chapter interrogates the recent and unexpected shift of fate associated with the return of medicinal plants in the global world of medicine and the rise of "alternatives," whose roots are not only to be found in the increasing recourse to heterodox Western systems, beginning with homeopathy, but also—and more importantly—in the globalization of re-invented industrialized Asian medicines.

FROM COMMON GROUND TO HEGEMONY, FROM TRADE TO EMPIRE: THE *MATERIA MEDICA* OF THE OTHERS

As western European trade with Asia radically expanded from the early sixteenth century onward, it rested not only on the consumption of spice and luxury commodities but also on the circulation and use of medicinal plants. A detailed knowledge of natural things, notably the flora of the East, was then deemed critical for two related reasons: the first one was the simple fact that such knowledge could nurture trade with the circulation and commercialization of new and/or rare therapeutic material; the second was the high mortality of the thousands of sailors and traders who found themselves in the hostile climates of the tropics and the necessity of maintaining their health with local resources since travelers could rarely access European remedies (Cook 2007).

But how could such knowledge of plants and their properties be collected? The historiography of botany has insisted upon the massive enterprise of collection and cultivation, which took place in the modern era. It resulted in the creation of large botanical gardens in Europe with their numerous outposts on the road to Asia or America. This was the infrastructure for a universalist classificatory project that culminated in the late eighteenth century with the debates about the Linnaean system (Drayton 2000). That this enterprise resulted in major epistemic changes, turning botany into an exemplary science, the model of comparative natural history is well acknowledged (Foucault 1963; Pickstone 2000). What has been less emphasized is the fact that the "golden age" of plant collection and classification was as much a question of botany as a question of pharmacy and medicine and, even more importantly, the fact that the expansion of Western mastery of plants simply could not have existed without an equally massive inventory and appropriation of local, non-Western knowledge.

A famous episode in the Western combination of botany, trade, and medicine that characterized the modern voyages to Asia may illustrate the point. In the 1670s, in response to a request for an inventory of local natural resources from his superiors in the Dutch East India Company (VOC), the civil and military Commander of Dutch possessions on the Malabar coast in southwestern India, Hendrik Adriaan Van Rheede tot Drakenstein (1636–91), commissioned a treatise on the flora of this region. Its drawings of some 720 species were accompanied by a detailed description, which did not only tell the reader about the plant morphology, its cycle of life, or its requirements for soil or climate, but also delved at length into its properties and usefulness, mixing the therapeutic and the culinary. The book was published in Latin, partly posthumously, under the title of *Hortus Indicus Malabaricus* (Garden of Indian Malabar) in Amsterdam between 1678 and 1693 and was soon to become a major reference. Van Rheede's work was Linnaeus' main sources for the flora of Asia.

How was Van Rheede's herbal assembled? Van Rheede himself was a VOC man who neither possessed training in botany nor good command of languages, not even Latin. In the Preface to the third volume of the *Hortus Malabaricus* (1682), dedicated to the Raja of Cochin, he described the construction of this work:

> By my orders, ... Brahmin and other physicians made lists of the best known and most frequently occurring plants in their language. On this basis, others classified the plants according to the season in which they attracted notice for their leaves or flowers or fruit. This "seasonal" catalogue was then given to a certain number of experts, who were entrusted with the collection of the plants with their leaves, flowers, and fruit, for which they even climbed the highest tops of trees. These experts went out in groups of three to designated forests. Three or four draftsmen, who stayed with me in a convenient place, would accurately depict the living plants as the collectors brought them. To these pictures a description was added, nearly always in my presence.
>
> (Van Rheede 2003: 3:viii)

Historians have argued that multiple levels of translation and intermediation were actually necessary behind this highly simplified account. As historian Kapil Raj, for instance, recalled, the working language was neither Dutch nor Latin but Portuguese—then the intermediary language between Europeans and Indians; moreover "it was the Konkani Brahmin and physicians who sent the men out to collect the flowers, fruits and seeds of the desired plants, and it was they who compiled the descriptions from written sources as well from their own experience over a period of two years" (2016).

The making of the *Hortus* thus exemplifies the critical role of "go-betweens"—these Europeans and Indians who acted as intermediaries and knowledge-holders—in what was not a simple translation but a process of appropriation resting on the (then fragile) asymmetry of power between the VOC commissioner and his informants, who became as invisible as the collectors, healers, and gardeners associated with professional botanists in Europe. But how could Portuguese and Dutch listeners make sense of the knowledge Brahmin pandits (scholars) were passing to them and commenting upon? Van Rheede's account solved the problem in a quick and easy manner: he witnessed the whole work of description and drawing, thus testifying to its accuracy. When the botanical and morphological dimensions of the work were considered, a reference to direct observation from a noble witness could suffice in the world of seventeenth-century natural philosophy and its aristocratic base. But how could therapeutic claims be legitimate if the Indian physicians hired by Van Rheede seemed to operate in a non-Western—radically different—medical system that had its own understanding of disease and *materia medica*? The mere fact that indications of use originating in Sanskrit treatises or in the decades-long practice of Brahmin physicians were included in the *Hortus* without mentioning any further inquiry suggest that there was no such divide, that the pandits' statements linking plants and diseases were all the more trustable that they referred to a common ground, i.e. a shared world of medical belief and practices.

This common ground was inextricably economic, social, and cognitive. As the Europeans sought to enlarge their trade in Asia, they inserted themselves by force in a very dense network of commerce involving Chinese, Javanese, Japanese, Malaysian, Indian, and Arab merchants dealing with silk, porcelain, tea, and silver as well as spices and medicinal plants. For instance, in the late eighteenth century the *materia medica* that was used by the English East India Company and could be bought locally included roots, leaves, and barks or resin imported from all parts of South and East Asia or East Africa (Chakrabati 2010; Harrison 2010).

The cognitive common ground originated in a rather straightforward manner from the nature of therapeutic intervention in the neo-Hippocratic paradigm of modern European medicine. If much has been said in the history of medicine about the "birth of the clinic" in the early nineteenth century, one major paradox of this revolution was its limited impact beyond the walls of teaching hospitals and medical faculties. The clinic was certainly a radical departure, as it redefined pathologies on the basis of anatomy, lesions, and pathological changes of tissues (Foucault 1963). However, the clinic was an academic phenomenon that produced new classifications and etiological scenarios without much consequence for the practices of treatment and care (Ackerknecht 1976; Léonard 1981; Pickstone 2000; Rosenberg 1987). When the latter are considered, continuities with the neo-Hippocratic paradigm prevail until the late nineteenth century. For most physicians, health remained perceived as a matter of individual balance between the various functions of the body, even if the pivotal status of the four basic humors was challenged by the accumulation of anatomical knowledge.

Two critical elements of convergence with the medicines of Asia were thus left intact: 1) the idea that health and disease are a matter of individual equilibrium and adaptation to the changing influences of climate, living conditions, and age; and 2) the notion that the struggle against disease is made possible by the body's natural ability to restore this equilibrium through processes of crisis that are reflected in the visible symptoms. As a consequence, therapy was not only highly individualized, but had to be comprehensive—that is, based on a complete regimen taking into account the nature of the patient's food, exercise, and housing conditions as well as the medications whose role was to help the body mitigate symptoms and survive. This provided for significant commonalities with the medicines of Asia, especially Ayurveda or Unani medicines in India, as these were also based on the existence of an individualized healthy balance of humors: an integrative approach of the body and its conditions of life. Climate, food, and plants were accordingly shared targets as well as basic tools of intervention.

A good example of how this common ground determined practices not only in the eighteenth century but until the mid-nineteenth century is the case of cholera, famous for being at the center of European debates on infection and miasmas (Delaporte 1990). Europeans approached the post-1830s multiple epidemics of cholera as a product of India, where it was endemic, and did so with major uncertainties regarding its path of transmission and its etiology (Hamlin 2009). Beyond the schematic divide between theories of contagion targeting the bodily contacts between individuals and theories of miasmatic infection focusing on the unhealthy environment, the role of water, vicious air, and bad climate, explanations were overlapping and diverse. The deep commitment of British experts to miasmatic theories—well reflected in the positions they defended during the nineteenth-century international sanitary conferences—was therefore not only a reflection of commercial interests opposing quarantines and other measures limiting the circulation of boats, merchants, and goods but an effect of the emergence of public health as field of inquiry linking diseases, the urban milieu, and poverty and of the centrality of India as site of medical knowledge production.

India was central in providing evidence for the endemic and local nature of the disease, which was therefore understood as a problem of climate and environment. Moreover, the colonial experience of cholera was that of double-sided exchanges with British physicians asserting the specificity—the fragility—of white bodies under tropical conditions and borrowing from local preventive and treatment habits while their Indian interlocutors adopted the symptomatic definition of the disease and forms of intervention like homeopathy. The impact was the formation of a hybrid medical culture especially visible in the work of "daktari," Bengali physicians practicing European medicine (Mukharji 2011). For them cholera was a distinct disease but rarely an epidemic problem. Case narratives focused on the person, his or her family, and their ways of living. Treatment was a combination of food prescription and of a shared set of remedies including mercury, opium, camphor, and various combinations of plants originating in Ayurvedic recipes, which started to become commercially available in ready-made forms.

By the turn of the twentieth century, this common world of cholera had recessed under the influence of three factors: 1) the mounting distance between colonizers and colonized following the 1857 uprising and the ensuing reorganization of British rule; 2) the production of new knowledge in the metropolis, which turned cholera into a transmissible, often water-borne disease, thus reinforcing the part played by sanitation and spatial isolation in its prevention; and 3) the laboratory revolution in European medicine, which placed the cholera germ center stage with the development and mass production of vaccines, and mass inoculation becoming the main form of collective intervention (Harrison 1994).

The "big divergence" between European and Asian medicines thus appears as the end of a form of medical "orientalism" that permitted the circulation of medicinal plants in both directions as well as their hybrid uses in the period between the 1750s and the 1890s. Historians of colonial medicine have explored the social and cognitive changes that brought this encounter to an end. They have on the one hand insisted on the changing nature of empires from the establishment of commercial routes and trading posts to the systematic occupation, exploitation, and administration of territories deemed essential for their financial and natural resources. By the end of the nineteenth century the colonial order relied on patterns of structural violence that required natives and their cultures to be radically different from the civilized and industrious white men. This was directly reflected in the "racialization" of ethnic differences, which took place during the second half of the nineteenth century (there is an extensive literature in this area—in relation to India, see for example Dirk 2001). Race as major factor in understanding the bodies of non-Europeans decisively contributed to marginalizing the environmental reading of diseases and rendered the use of local remedies meaningless. "Native" remedies were considered products of a non-scientific culture bordering on magic and were therefore devoid of efficacy or, at best, they were pragmatic discoveries adapted to radically different bodies and therefore of doubtful value. On the other hand, the divergence was a consequence of the transformation of European medicine itself as the rise of public health and of laboratory medicine did not only create spaces of knowledge production without equivalent in Eastern medicines but established—in parallel—modes of intervention focusing on populations and biological tools, which bore a distant relationship to the previously shared practice of biographical medicine.

One should add to this that the "hybridization" of treatment and the import of herbal remedies from the East were marginalized on the Western scene (be it metropolitan or colonial) under the influence of a fourth trend, namely the transformation of European pharmaceutics (Bürgi 2011; Chauveau 2009; Gaudillière and Hes 2013; Wimmer 1994). During most of the nineteenth century European pharmaceutics was a full-fledged professional discipline distinct from medicine. Pharmacists were legally recognized practitioners trained in specific academic institutions, working in small workshops with a few apprentices or employees. This discipline resulted from the delegation of expertise, which granted pharmacy graduates a monopoly over the sale and preparation of therapeutic agents. Such regulation was justified as a means to avoid unnecessary competition among pharmacists, as well as to bar the entry of supposedly untrained and unskilled practitioners into the market. Under this regulation, professional governance emanated from corporations, pharmaceutical societies, and their medical counterparts, possibly supplemented by special committees of experts set up by academic journals or public-health authorities. Within this framework, judgment regarding the value of drugs rested on the acquisition of pharmacological knowledge focusing on the links between dosage, patterns of elimination, and the balance between toxic and therapeutic responses. The pharmacist was therefore an expert in *materia medica*, which he knew how to prepare; that is, he was an expert in all the substances and remedies deemed legitimate by the profession. The basic corpus of knowledge that he mastered consisted in protocols for the preparation of each type of drug, the definition of appropriate dosage, and recommendations regarding indications and contraindications.

This was directly reflected in the structure of national pharmacopoeias (compendia of drug information), which became the reference books for the practice of pharmacy. National pharmacopoeias performed several functions. First, they were tools in professional life: they were written by recognized pharmacists (and physicians)—most of them academics. All trained and/or established pharmacists had to keep one exemplar of the reference book in their practice and were supposed to know its contents, which defined a corpus of collective knowledge without links to therapeutic uses (dosage, indications, prescriptions mode). Second, the pharmacopoeia was a legal instrument. It was a list of accepted methods and authorized drugs—with or without enforcement power by the state—and it opposed the "commercial" world of secret remedies sold on the fringes of legality and whose composition was not known. The pharmacopoeia was therefore a means of collective appropriation alternative to the development of intellectual property rights, i.e. trademarks or patents on production processes. Third, the national pharmacopeia was a means of standardization. Pharmacopeias defined the accepted names and raw materials, detailed the protocols for preparation and isolation as well as the formulas of combined preparations associations, and listed the investigations of, and ways of controlling, identity/quality.

Up to the First World War the various European pharmacopoeias first and foremost included herbal preparations, which could be either extracts of one single plant or poly-herbals and constituted the core of European *materia medica*. The French Codex, for instance, was organized as a dictionary of products whose description included the following headings: a) physical properties (morphology, color, odor, taste plus eventually density, fusion point, solubility, optical properties); b) chemical properties (acidic or basic, reactions to light and heat); c) falsification and contamination; d) control of identity and assays; e) toxicity. This started to change by the turn of the century. In contrast to its earlier versions, the edition of 1908, for example, introduced three innovations. On the one hand, it eliminated all reference to basic pure chemicals, which were considered easily available industrial products outside the realm of professional pharmacy while reinforcing the reference to botany with the introduction of extended morphological description of medicinal plants. In parallel it gathered various preparative modes under one single heading corresponding to the isolated active principle which was then considered to be present in the preparation and responsible for its potency, thus favoring a chemical understanding of therapeutic efficacy.

These changes illustrate the tensions originating in the mounting industrialization of pharmacy. The late nineteenth and early twentieth centuries were actually a time of rapid growth in the production and distribution of ready-made therapeutic specialties, many of them produced by large companies like Bayer, Rhône-Poulenc, or ICI, whose initial expertise was in the production of chemicals. These firms had developed drug production facilities focusing on purified substances and active principles, initially using plants as raw materials but selling active principles. This mounting competition of industrial drug producers was a direct threat to the professional order of pharmacy as it tended to transform the apothecary into a simple retailer, storing and dispatching substances invented, produced, and tested outside his premises. The threat was also cognitive. Large chemically oriented firms did not build on the knowledge of *materia medica*; they favored an alternative vision of animal or plant preparations as a source of specific, highly potent, physiologically tested “natural” entities. This is well illustrated with the trajectory of the German firm Schering, which proved highly successful in the development of hormones as therapeutic products, focusing first on their purification and later on their mass production through (partial) artificial synthesis (Gaudillière 2005). *Materia medica* was thus increasingly challenged by the rise of “biologicals,” that is, standardized industrial compounds of biological origin whose main exemplars were the recently invented preparations of hormones, vitamins, enzymes, sera, and vaccines.

THE BIG DIVERGENCE AT HOME: THE FAILED INDUSTRIALIZATION OF HERBAL MEDICINE IN TWENTIETH-CENTURY EUROPE

This process of industrialization and molecularization did not proceed without raising oppositions, resistance, and open conflicts. Protests focused not only on the economic rewards of local pharmacists, whose competencies and revenues were reduced. They also revolved around the very definition of pharmaceutical science, the identification of effective remedies, as well as the investigation, characterization, production, and marketing of drugs. Biologicals then surfaced as a major category of drug, characterized by particular methods of invention, production, sale, and regulation during the first half of the twentieth century. Biologicals were not only therapeutics "made" out of things biological. They were technical objects, socially and culturally associated with nature at the same time as they were elaborated through industrial mass preparation. Their importance was certainly rooted in the relation that sera, hormones, or enzymes—these "new" products of nature bearing the aura of frontier science—maintained to physiology and laboratory experiments with the living (hence the strong consensus regarding their therapeutic potency). But the visibility of biologicals also resulted from the fact that the category included old and new plant extracts, thus creating a porous boundary between the passing professional order of pharmacy, with its emphasis on local preparations, and the industrial order centered on standardized specialties. Up to the Second World War the European pharmaceutical scene was accordingly not simply that of an increasing reliance on chemistry through which plant extracts were replaced by pure *active principles*,[1] it was also the locus of various attempts to maintain and expand the uses of medicinal *plants* through the industrialization of their preparation.

One fascinating example of such attempts is that of the German Madaus pharmaceutical company. Madaus operated at the boundary between academic and popular medicine, and participated in a movement for more natural and holistic therapeutic practices, which became increasingly visible and influential during the 1920s and 1930s (Dinges 1996; Jütte 1996; Rothschuh 1983). Madaus was founded in 1919 by the three sons of a church minister's wife who practiced medicine herself as a non-licensed healer. Benefiting from the then widely debated "crisis of medicine," the firm's production combined herbal and homoeopathic remedies. Success was quick to come. In 1925, Madaus had already opened branches in Amsterdam, Barcelona, and in several German cities in addition to its headquarters in Dresden. On the eve of the Second World War, its employees had reached eight hundred in number. Its production included a wide range of "biological" therapeutic agents, from classical homeopathic preparations of metals to hormones, enzymes, or combinations of bacterial antigens. The bulk of the firm's ready-made specialties were, however, composed of various types of plant derivatives (Timmerman 2001).

Madaus maintained a complex and contradictory relationship with German movements in favor of alternative healing practices and their strong interest in "natural" and "biological" therapeutic agents. On the one hand, the firm and its main figure, Gerhard Madaus, were strongly associated with the critique of "academic medicine," writing articles and pamphlets, sponsoring events, publishing journals, leaflets, and books promoting the discussion among *Heilpraktiker* (medical professionals) and spreading their views on diseases or remedies to the public at large. On the other hand, Gerhard Madaus pushed the company on the path of industrial and mass production of homeopathic and plant remedies, arguing for the modernization of practices as well as for more experimentation and evaluation. The aim was to revisit, modify, and improve the corpus of known recipes and traditional preparations in order to optimize and generalize their uses, bringing them back into the arsenal and purview of local practitioners as well as academic physicians and pharmacists.

The broad palette of documents and publications on the firm's products and activities left in its archives is associated with two discourses (Gaudillière 2010). The first is a

FIGURE 5.3 Packaged herbs in a herbalist's shop in Fouesnant, western France. Since the removal of the herbalist profession and diploma in Vichy France in 1941, only pharmacists are authorized to sell medicinal plants registered with the pharmacopoeia, except for a list of 148 set in a 2008 decree. Without a legal frame for their trade, herbalists are limited to the 148 authorized plants, but are hoping for a renewal of the profession to extend their practice. Photo by Fred Tanneau/AFP via Getty Images.

medical discourse based on a global vision of the body that recommended attention to the systemic and multiple dimensions of diseases, pled for an ecological understanding of the relationship between people and their environment, and argued for the need of more natural or biological means of intervention. The luxurious *Jahrbücher* (Yearbooks), published by the firm for its clients and, more strategically, for local doctors, featured eloquent discussions of this holistic and natural perspective, which strongly resonate with the nineteenth-century medical orientalism discussed above.[2] The second discourse is that of industrialization. The stated goal was to produce more for less money in order to enlarge the market and to facilitate access to remedies. Homogeneity, standardization, and quality control were key concepts opposing essential ideas in the culture of alternative practitioners such as the extreme variability of diseases or the adaptation of formulations and preparations to the needs of individuals.

The two discourses met in the strong critique of purity that G. Madaus and its company associated with the promotion of complex extracts. The firm gradually elaborated a local understanding of "scientific preparations" stressing biological mixtures and combinations. Several arguments were at play, namely that plant extracts are composed of many ingredients, that efficacy is not to be attributed to a single active principle, that clinical efficacy may be linked to the synergies among these constituents, that the potency of a preparation is not to be attributed to the isolated effect on a single organ, that pathologies have to be seen in a global life context including the influence of environment and habits, and that the power of plant extracts is related to the life of the plants themselves as well as to their interactions and to the dynamics of ecological systems. As a consequence of these claims, the industrialization of plant preparations at Madaus focused on three types of practices, which were rarely combined in other companies: a) the mechanization of processing; b) the standardization of products and the expansion of quality testing; and c) the domestication and mass culture of medicinal plants. These corpuses of practices were rooted in three forms of obtaining knowledge that the firm sought to develop in its experimental sites, which included a chemical laboratory, a biological institute, and a farm.

Madaus' innovation strategy was primarily linked to the mechanization of processes (Kuhn 1936). The most important changes, those for which Madaus submitted several patent applications (Deutsche Reich Patent 129588 and 131837), had to do with the machine-based treatment of fresh plants (Gaudillière 2010). The company continued to sell dried extracts of entire plant organs, but gradually transformed their presentation into sugar-coated pills. In the eyes of Madaus, the so-called "*Teep*" (for *Tee-Pulver*, "tea-powder") preparations had the immense advantage of providing such pills without any isolation or purification steps. As proclaimed in the Teep advertisements, the Teep kept all the plant constituents, "they did not eliminate proteins or plant hormones and are therefore of a much higher quality than classical homeopathic decoctions and tinctures." Teep were obtained by mincing and grinding freshly collected plant parts with sugar, followed by drying under warm air. This raw "0" preparation was stored in dry storage rooms for later mixing (or "dilution") with additional sugar in order to get the dosage of plant material sought for the Teep pills, which were produced with exactly the same type of machinery used in all pharmaceutical companies: grinders, mixers, dryers, and, most importantly, pill-compounding machines. Madaus considered that Teep preparations were better, not only because they were less expensive and more stable than the classic "soft" plant preparations, but more importantly because the Teep process preserved the mixture of substances found in the fresh plants. Numerous experiments were conducted at the company's chemical and biological laboratories in order to document this quality of Teep preparations.

The second and dominant aspect of industrialization was standardization. The sensory experience of plant connoisseurs with their knowledge of forms, odors, texture, and tastes was not given up, but rather systematized and complemented with laboratory testing. Industrial standardization was thus connected with multiple activities including: a) writing and implementing formal production protocols; b) systematized control of the quality of incoming raw materials; c) measurement of the potency of the final plant preparations; and d) occasional evaluation of their clinical side-effects. The chemical laboratory under A. Kuhn's guidance was central to this effort. Kuhn and his colleagues spent considerable amounts of time investigating the composition of Teeps made of various plants such as *Mentha, Valeriana, Viscum, Digitalis, Lycopodium, Oleander, Aloe,* and *Arnica*, focusing their studies on extract composition rather than on isolated substances but including broad categories of interacting substances such as alkaloids, oils, proteins, and carbohydrates. Much of these investigations consisted in straightforward quality control. For instance, in December 1939, following complaints that a given lot had acquired a suspicious color, Kuhn and his colleagues determined the content of various *Arnica* Teeps. Although the quantity of oils that could be extracted with ether was highly variable, all the preparations analyzed presented a normal arnica smell and the same yellow deposit (Madaus 1939). The grinding protocol followed in the production department was in this case not changed.

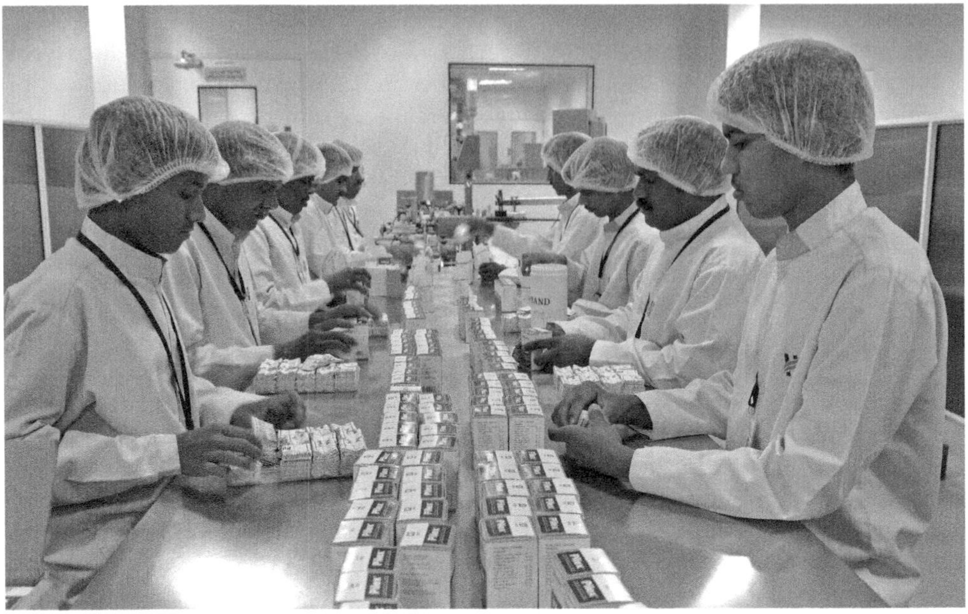

FIGURE 5.4 Workers at the secondary packing unit of Himalaya Drug Company in 2013. The raw materials are plants and its products respond to Ayurvedic texts dating back millennia and contemporary knowledge of the impact of plant active principles on symptoms resulting in novel formulations. Photo by Manjunath Kiran/AFP via Getty Images.

Chemical testing was important, but the local quality control measurements relied even more strongly on the practices of biological standardization. The latter is often viewed as a pragmatic and incomplete substitute to molecular analysis; something indispensable when the composition and structure of substances is not known; something to be later replaced by more specific and accurate chemical tests. Historically, physiological and biological controls were, however, not necessarily preliminary steps toward a molecular redefinition of pharmaceutical activity; they were valued for their own sake. Madaus' biological standardization worked this way. Biological testing was accordingly a way to "mine" the corpus of popular and alternative medicine in order to make new mixtures and find new indications. Standardized animal models were, for instance, used in three different ways to deal with the complexity of plant extracts: a) to evaluate and control the variability of collected plants and adapt mixtures to ensure some stability of effects based on the global composition; b) to survey the implementation of production protocols (although the issue was less the surveillance of operators and the detection of errors than the adaptation of practices to highly variable materials; c) to provide experimental information on new specialties and/or indications. The example of garlic may illustrate this combination of old and new—of experimentalized and traditional—herbal pharmacy. The first one is that of garlic (*Allium sativum*), a plant widely sold by Madaus in the form of *Alliocaps* pills to treat rheuma, inflammation, and atherosclerosis in conformity with the German pharmacopoeia. The role of the laboratory was not to find new indications but a new model, as in the case where mice became atherosclerotic following the administration of high dosage compositions of vitamin D, which could be cured with parallel ingestion of *Alliocaps* (Koch 1936).

The third dimension of industrialization at Madaus was the long-term work devoted to the cultivation of key medicinal plants. The annual harvest was a recurring source of worry regarding the quantity and the quality of the material collected in the wild, which usually stopped at summer's end with bitter complaints about shortages (Madaus 1937–42). The dream solution was cultivation as a means to minimize collection in fields and forests. A few species like *Digitalis* were actually domesticated, but they remained exceptional and they raised numerous questions regarding the consequences of cultivation on plant physiology and on the composition of extracts. Research on *Belladona* (*Atropa belladonna*), for instance, confirmed the idea that its agricultural production, coupled with the use of fertilizers, had increased the toxicity of dried preparations (Madaus 1938: 675–90).

A specific line of research thus focused on the ecology of medicinal plants. Relations between all living entities were central to the vision of plants as containing active elements that affected life processes in animal and human bodies. Madaus' botanical science was holistic, viewing plants not as isolated and self-sufficient organisms but as integral members of stable communities maintained through relations of synergies and antagonisms among species. Although G. Madaus never used the concept of ecosystem, he nonetheless insisted on associations among plants as characteristics of a given environment, dependent on, and influenced by, poorly understood physiological exchanges such as the circulation and exchange of nutrients through the plant roots (Madaus 1938: 78–9). Theoretically, a good system of cultivation would replicate the most important relations between the targeted medicinal plant and its accompanying community of species. Experiments were therefore conducted to document and select meaningful combinations. Only those performed in the biological laboratory and the greenhouse have left archival traces. In the case of *Viola odorata* (the extracts were used against skin diseases, eczema in particular), the association with wheat and barley gave opposite results as the latter stimulated growth and germination while the former inhibited them (Madaus 1934: 14–23). The cultivation scheme of *Digitalis* was modified following similar experimentation showing that co-cultivation with *Galega officinalis* (a plant used against diabetes) increased the amount of alkaloids (Madaus 1935, 1938b: 1407). It seems, however, that most cultivation processes, either in the field or in greenhouses—as was the case for plants of tropical origin like *Cactus grandifolia*, which was needed for a very successful cardio-tonic called *Goldtropfen*—were organized without any reference to this ecological approach of plant physiology.

A final component of the regulatory activities of Madaus was the definition of the proper medical uses of the firm's products, and more broadly, the standardization of herbal treatments. Madaus followed the path of many companies that sought to "discipline" the practice of physicians by disseminating their own scientific information. Scientific marketing through the simultaneous organization of research and promotional activities became an important tool in building pharmaceutical markets in the 1930s (Thoms and Gaudillière 2014). At Madaus, this development resulted in two different forms of "propaganda." The oldest and most classic of these used advertisements in professional and lay journals that focused on the name of the firm and built its image through slogans and trademarks. The second approach focused on the mobilization of laboratory and clinical knowledge and took the form of numerous articles, leaflets, exposés, and textbooks written for regular physicians and biological healers. A typical publication was the above-mentioned *Biologische Heilkunst* (Biological Medicine), a journal of academic format circulated among therapists, supported by the firm through many advertisements, edited by a board including G. Madaus, with himself and his collaborators signing many articles.

Among the new tools of scientific marketing, Madaus' *Lehrbuch der Biologischen Heilmittel* (Manual of Biological Medicines) is worth considering as it shows a structure similar to that of a national pharmacopoea, including more than four hundred plants and counting more than three thousand pages. Each plant had a specific chapter listing its names, locations (with maps), morphology, composition, points of interest, physiological effects, toxicological symptoms, therapeutic indications, and modes of preparation and conservation. Differences were nonetheless significant: a) indications were rooted in long-term history with systematic references to old medical treatises dating back to the sixteenth century; b) lay therapeutic experience was given a significant place; c) toxicology did not mean animal experimentation for modeling the dose-response relationship but the reporting of clinical cases; d) homoeopathic conditions were central to the definition of proper uses. The chapter on *Atropa belladonna* thus included not only everything that could be found in a pharmacopoeia but also all sorts of facts about the ecology of the plant and a lengthy discussion of its therapeutic history, with references to treatises written by Dioscorides (*c.* 40–*c.* 90 CE), Conrad Gesner (1516–65), Albert the Great (*c.* 1200–80), and Herman Boerhaave (1668–1738). Madaus, for instance, discussed a "Bulgarian cure" against epilepsy, shaking, or muscular seizures advertised in German newspapers by a certain Iwan Raeff as a secret remedy well known in Italy, where it benefited from the Royal Court's promotion. G. Madaus criticized the "cure" severely on the basis both of its putative composition as defined by the academic pharmacists who had tried to analyze the mixture and of what he thought about the interactions between *Atropa belladonna*'s alkaloids and the rest of the components including animal charcoal, roots of *Acorus calamus,* and nutmeg.

Madaus represents just one of numerous attempts to adjust the remedies of plant-based *materia medica* to the new world of industrial pharmacy while preserving traditional cultural elements that linked biographical medicine and herbal pharmacy. Although they were often less systematic and less documented than in the German context, where herbal preparations were in significant demand and backed by a strong movement opposing academic and popular medicine, parallel developments occurred in France and in the United Kingdom.

The postwar years decisively weakened the position of such firms and further marginalized the uses of medicinal plants. The so-called "therapeutic revolution" of the 1940s–1970s, which originated in a radical phase of concentration and reorganization of industrial pharmacy, resulted in the adoption by all large players in the pharmaceutical industry of the "screening" model of drug development, initially invented at Bayer in the 1930s (Lesch 2007). This model was entirely based on the mobilization of chemistry and molecular pharmacology in order to fabricate hundreds of variants of a given molecular structure and test them in animal models and mass produce the most potent among them. The screening model did industrialize discovery with the organization of a pipeline of systematic trials from the chemist's bench to trials in humans that made extremely difficult the importation into the system of the botanical and clinical knowledge associated with the use of medicinal plants. In order to become valuable, this knowledge had to be molecularized: translated into the idiom of chemical bonds and groups, radicals and cycles, ligands and receptors. The task became the province of a reinvented discipline, namely *materia medica* turned into phyto-chemistry or phyto-pharmacology. Even if they did not altogether abandon the preparation and sales of plant extracts, firms like Madaus engaged in the process and from the 1950s onward they increasingly relied on the identification of purified active principles as the main focus of innovation. The irony of this generalization of the screening model is, however, that many of the compounds viewed as emblematic of the fecundity of the model, from the antibiotics to the steroids, had first been "biological," complex preparations obtained from plants, animals, or bacteria, rather than inventions of the chemical laboratory.

GLOBALIZATION AND REFORMULATION OF ASIAN MEDICINES: EPISTEMIC ALTERNATIVE AND MARKET-BASED CONVERGENCE

In 2002, the Indian Himalaya Drug Company started to market a poly-herbal medication called Menosan.[3] This was a preparation made of eight different plants well known in Ayurvedic medicine. Menosan was then introduced as a safe response to the problems recently identified with hormonal replacement therapy (HRT) of menopause. In the mid-1990s the US women's health movement had obtained funding for a very large epidemiological study of HRT—significantly called the Women's Health Initiative (WHI) study—from the US National Institutes of Health. In 2002, when the first results were made public, they clearly opposed the claims (and previous studies) made by the promoters of HRT, showing that women under medication evidenced not only an increasing risk of breast cancer but also an increased risk of cardiovascular disorders. The WHI resulted in a major crisis of confidence with women abandoning treatment. By 2003–4 prescriptions had dropped by 30–50 percent in the US and in most European countries. The problem became even worse when epidemiologists started to correlate this change in drug consumption with a drop in the incidence of breast cancer; a major finding in the recent history of the disease. Given that a majority of women in Europe and the US were not as keen as their grandmothers simply to wait for the end of the menopausal transition and the disappearance of painful symptoms, the HRT crisis opened a vast potential market for alternative treatments, a niche Himalaya was trying to seize.

According to Himalaya's promotional material, Menosan is an industrial poly-herbal whose ingredients have been selected on the basis of their (phyto) pharmacological properties, ideally defined in terms of chemical composition and specific classes of ingredients. However, Menosan is a formulation—a *yogam*—whose action remains hard to explain even in Ayurvedic terms. The valorization of Menosan targets a biomedical category with no existence in the classical Ayurvedic corpus, defines the problem with the same symptoms as biomedicine, and insists on the need for drugs as the appropriate response. In the eyes of an observer with some knowledge of the Indian system of medicine, Menosan may easily appear as a typical example of the biomedicalization of Ayruvedic medicine (Banerjee 2009; Bode 2008). The view associated with this reading is not only that "traditional" Ayurveda and "modern" biomedicine are two mutually incompatible medical systems but also that the former is gradually disappearing through its alignment with the latter. Such a view, however, misses the ways in which the reformulation practices involved in the current industrialization of Ayurvedic remedies borrow from heterogeneous medical knowledge; they combine bits and pieces of various corpuses (more than two) in order to invent previously unknown preparations and to provide for modes of intervention irreducible to biomedicine.

The origins of Menosan and its "multiple ontology" are not, however, reducible to Himalaya's pharmacological discourse. When discussing the reformulation practices involved in its invention, the Ayurvedic practitioners (*vaidyas*) at the Himalaya Drug Company actually reveal another world—a world strongly committed to the mining of Ayurvedic knowledge. As Dr. Rangesh, the head of Himalaya's formulation department explained:

> The initial formula for Menosan is none of the published ones. Actually we envisioned dozens of combinations. There were two complementary ways of selecting the ingredients to be included. The first one was to come back to the classical texts and reflect upon the formulations mentioned in the context of women life and health. The second mode of selection was to start with specific symptoms such as vascular, sweating, depression and mood problems, and target plants used to alleviate such symptoms either in the classical texts or in novel formulations.
>
> (Pordié personal comment)

What is at stake here is much more than the formation of hybrid species combining the traits from two varieties. Inventing such a formulation blends forms of knowledge and registers of legitimacy but does not provide for an undifferentiated mix. A major resource for this reformulation work was the multiplicity of combinations, which may be imagined on the basis of the four hundred core formulations of the Ayruveda corpus, which often include several dozens of different ingredients. Given such variability and flexibility of preparations, it is not surprising that the Himalaya Drug Company scientists explored the properties of more than forty formulations in the early stages of Menosan research and that the trajectory of a poly-herbal with menopause as the major indication was actually several Menosan, which appeared or disappeared through the contingencies of reformulation.

As Dr. Rangesh's quote indicates, the firm's mining of Ayurveda rested on moving back and forth between the classical texts and the laboratory. These two resources obeyed different logic. The first was a logic of translation and *materia medica*, focusing on the supposedly stable composition of plants and a series of equivalences between botanical denominations on the one hand and vernacular and Sanskrit names on the other—equivalences worked out along two centuries of interactions between pharmacists, botanists, and *vaidyas* (physicians). This logic makes it possible to examine the classical formulations, looking for combinations including plants whose active *ingredients* (not necessarily molecules) are linked both by the biomedical and the Ayurvedic pharmacologies with processes responsible for menopausal symptoms (for instance neurological regulation of the vascular system). The second logic was one of commonalities and clinical knowledge. It focused on the supposedly stable symptoms of a "real" pathology, mobilizing another series of equivalences, this time between the symptoms used in both the Ayurvedic and the biomedical clinical descriptions. This logic makes it possible to look for classical formulations targeting the components of the menopausal syndrome Himalaya imported from biomedicine.

Reformulation thus appears as a complex mode of drug innovation that mobilizes medicinal plants in the form of poly-herbals, insists on the benefit of synergies originating in the interactions between ingredients, chooses them through a complex assessment of Ayurveda shastric formulations but breaks with decisive aspects of this knowledge basis, for instance with the understanding of disease as an imbalance between the three fundamental humors of the body. Two terms have been used to delineate the dynamics involved in this regime: industrialization and pharmaceuticalization.

One essential dimension of reformulation is its industrial nature. Industrialization means that the main actors in the supply chain of remedies are no longer *vaidyas*, local collectors, and merchants, or household members, but Indian Ayurvedic drug-producing companies, some of them large enough to operate as global players seeking consumers all over Asia and possibly Europe or the United States. Industrialization also refers to the manufacturing process and the production technologies. Even if the introduction of mechanical, grinding, pill-making machinery into Indian traditional medicine can be traced back to the first half of the twentieth century, these processes remained peripheral and were often associated with the trajectory of producers crossing the boundaries between allopathic (Western, evidence-based) and alternative medicine; between drugs, food, and cosmetics. In contrast, in the past two decades, the search for productivity and large-scale output by using mechanized processing and automated machinery as well as the quest for standardization through quantitative, laboratory-based, quality control have become pervasive and provide Indian Ayurvedic firms with their main tools to occupy what they perceive as fast-growing urban and global markets for herbal remedies (Banerjee 2009; Bode 2008; Farquhar and Rajan 2014).

Contemporary drug-innovation practices in traditional Indian medicine, however, are not restricted to this mass production and mass distribution logic. The term "pharmaceuticalization" has been used to stress the importation into the new "techno-Ayurveda" of a whole body of knowledge and practices associated with late-twentieth-century "global" pharmacy, with its ways of testing, producing, and selling molecularly defined therapeutic agents. However, one major reason why the reformulation strategies of traditional preparations promoted by Indian firms and researchers are in their essence foreign to the chemical-screening model discussed above is the strong emphasis most actors involved in the new regime place on complex poly-herbal formulations and their opposition to strategies of isolation, purification, and in vitro synthesis (Pordié and Gaudillière 2014). Like bioprospection, which was also revived by the rapid growth of biotechnology in the 1980s and 1990s, Indian firms favor the use of medicinal plants. However, unlike bioprospection, the purpose is less to control a small set of active principles than to exploit the synergetic properties of poly-herbal compositions—as it was the case in the early European attempts to industrialize *materia medica*.

The dynamics of reformulation thus entail a deep change of Ayurveda because they create a world of pharmaceutical practitioners focusing on the collection and the manipulation of medicinal plants in a sphere that had thus far been basically clinical, claiming a "holistic" and individual approach to illnesses and remedies. In other words, reformulating and simplifying Ayurvedic medicinal compositions in order to create new poly-herbal drugs relies more on the ability to identify, collect, manipulate, and combine the plants than on any form of clinical work or encounter with patients.

The emergence of actors, sites, and practices focusing solely on the management of *materia medica* is not completely new. Already in 1948, the Chopra report on the future of indigenous systems of medicine under the national health system stated that "everywhere the professions of medicine and pharmacy have separated or are in the process of becoming so" and suggested that specific teaching and registration should also take place in Indian medicine (Ministry of Health 1948). The institutionalization of Ayurveda in post-independence India accordingly brought with it the writing of a national pharmacopoeia (the first committee for this purpose was set up in 1963 by the Central Council for Research in Ayurvedic Sciences), the teaching of "pharmacology" courses in Ayurveda colleges, and the institution of a federal laboratory with the mission to establish reference assays and standards to control the raw materials used in Ayurvedic formulas. The current reformulation regime, however, introduces different and deeper boundaries between medical practitioners, plant collectors/merchants, and drug-makers, and between the sites where they operate or the forms of knowledge and expertise they command. The most visible actors of pharmaceuticalization are accordingly not only Ayurvedic drug companies, which integrate research, formulation, fabrication, and distribution but also new state institutions seeking to promote the growth of the sector and to regulate its activities.

One major example is a new central body under the Ministry of Health Department of Ayurveda, Yoga & Naturopathy, Unani, Siddha and Homoeopathy (AYUSH) in New Delhi called the National Medicinal Plants Board (NMPB). The creation of this board in 2000 was a major sign that a new kind of federal policy was emerging, targeting the management of the plants used in Indian traditional therapeutic formulations, and claiming to integrate health, industrial pharmacy, and agriculture. The NMPB was established to coordinate initiatives of the relevant ministries and foster an all-embracing strategy to increase the supply of medicinal plants. As defined by the National Planning Commission (NPC) Task Force, which had recommended the formation of this board, the increasing difficulty in finding adequate quantities of medicinal plants was a mounting problem and a major bottleneck in any attempt to increase the production of drugs, facilitate the diversification of formulas, and improve the nation's export capacity. Given its broad mandate and mediating role, the NMPB could easily turn into a typical bureaucratic structure, whose means, operational role, and influence would scarcely go beyond the preservation of its precarious existence. Evaluated on the simple basis of medicinal-plant output, its achievements during its first ten years of work are currently highly contested, as the supply crisis is far from being eased. Besides stimulating plant cultivation, the board has, however, accomplished other more lasting regulatory functions, thus confirming its pivotal role in the new reformulation regime.

This may be illustrated by the status of a list of prioritized plants the NMPB adopted in 2002. The first significance of the list was to deliver a reference sheet for what initially was the main function of the board: to provide subsidies in the form of contracts with medicinal-plant producers or collectors. Priority also meant that the selected plants should become targets of initiatives for their conservation (*in situ* and *ex situ*), for the standardization of supply protocols, and for the definition and implementation of quality-control procedures. These thirty-two plants were deemed important and rare enough to require a significant influx of public money. But how were these thirty-two items selected out of the presumed thousands of species used by the industry and sold on the local or regional markets? Choice could be a matter of medical benefits, taking as proxy for such benefits the frequency of use in the corpus of classical formulas, or the importance of the associated clinical targets, or the consumption of specific remedies. The logic of the NMPB was different. The list did not originate in a medical but in an industrial and agricultural hierarchy predicated upon the autonomy of plant management. Rather than collecting health-related data, NMPB experts focused on market indicators as main proxy for demand and utility. Moreover, a critical feature of the final list was the fact that twenty-five out of the thirty-two prioritized plants were already being cultivated in 2002. This does not necessarily imply that consumption was exclusively based on field production but that protocols for this production had been developed and were to some extent being used routinely.

Between 2002 and 2010, NMPB prioritization policy increasingly focused on this agricultural management of medicinal plants. Accordingly, the inclusion of a species in the priority list implied that it became eligible for financing cultivation projects. The basic scheme negotiated between the three ministries (health, agriculture, and industry) was that individual farmers, cooperatives, or clusters committing to the cultivation of prioritized plants on a minimal acreage would receive 30 percent of their costs as a subsidy. To avoid useless production, signing a contract was conditioned to the existence of a supply agreement with a drug producer, usually in the form of a payback contract. Priority was also a matter of "promotional activities," which were less a matter of education or awareness than research and development initiatives ranging from the establishment of nurseries to provide enough planting material, to the study of cultivation protocols, or the definition of morphological and biochemical quality-control procedures.

The world of *herbalized* Ayurveda thus appears as a form of alternative modernization of both the Indian system of medicine and *materia medica*. It does not only represent a third reinvention of traditional medicine (after the early-twentieth-century professionalization and the post-independence institutionalization) but echoes the kind of industrialization encountered in early-twentieth-century Europe in reaction to the molecularization of pharmacy. A convergence with the ways in which the uses of medicinal plants have been rejuvenated in the West in the past three decades is therefore visible at several levels. The first one is that of the market: Ayurvedic herbal remedies are produced, sold, and consumed on a segmented but global market with products circulating far beyond Asia and firms appropriating the practices of scientific marketing developed by Western firms. The second one is that of production with the use of machinery, standardization, and quality control as well as an enlarged mobilization of cultivation in order to solve the supply problem. The third one is that of research with the invention of new combinations, eventually patented, which embody at the same time claims of science, naturalness, and holism.

CONCLUSION

In a recent book on the history of medicinal plants in the Spanish modern colonial world, the French historian Samir Boumediene powerfully showed how the status of medicinal plants in the sixteenth- through eighteenth-century colonial Americas differed from the situation we have encountered in Asia (Boumediene 2016). American medicinal plants were definitely a major target of investigations, imports, and uses. However, there was no "divergence" because from the beginning of the colonization process Europeans focused on the inventory, collection, and testing of plants without entering anything like the uneven but significant conversation Europeans conducted in Asia with native experts up to the mid-nineteenth century. In the eyes of the Spanish investigators of plants, the information native healers, shamans, or tribe leaders could provide was always incommensurable to European medicine: it was rarely recorded in writing, highly variable, and, more importantly, overladen with rituals and religious practices that the colonial power wanted to eradicate for the benefit of Christianization. As a consequence, medicinal plants were not the aim of an asymmetrical encounter but things to be extracted from the wilderness just like silver or gold.[4]

The history of medicinal plants in Europe and Asia, by contrast, brings to light a long period of what may be called "medical orientalism." Medical orientalism was a discourse of hegemony, but hegemony based on the recognition that if "oriental" systems of medicine were less advanced and less rational than the knowledge Europeans had accumulated they were part of a common history, providing a legitimate source of knowledge in order to identify new remedies and to adjust to the specificities of disease in the tropical worlds. As argued in this chapter, medical orientalism died—just like its cultural equivalent—in the wake of imperialism with the shift from commerce to territorial control and large-scale systematic extraction of resources. Its death had, however, several roots, including the radical transformation of European medicine and pharmacy, which took place in the late nineteenth century with the "laboratory revolution," the rise of public health, and the industrialization of drug making. The big divergence was not only a divide between the East and the West. It took place both in the East and at home, marginalizing Asian medicinal plants and formulations as well as the European *materia medica*, both becoming traditional and unscientific in comparison with pure, chemically defined and industrially produced therapeutic agents.

Divergence, however, does not mean eradication. The hegemony of biomedicine did not eliminate uses of medicinal plants, though their use was radically different from the chemical paradigm that argued for the identification of active substances, their isolation, and their industrial synthesis. Forms of "alternative" modernization were pursued both in the East and in the West unevenly following the parallel but connected paths of institutionalization, industrialization, and standardization of herbal medical practices. The irony of the most recent phase of health globalization is therefore that it opens new spaces of legitimacy, circulation, and convergence to the agents and to the products of these alternative modernizations. The cognitive and social landscapes associated with the contemporary uses of medicinal plants are therefore much more interesting, diverse, and laden with conflicts than any simple theory of a global and all-potent biomedical culture allows for.

CHAPTER SIX

Plants in Culture

MARK TREDINNICK AND STEPHEN FORBES

INTRODUCTION

This chapter focuses on the symbolic or metaphorical role of plants since 1920; the representation of plants in art and design and the use of plants as ornaments in gardens and landscapes are addressed elsewhere in this volume. Plants are admired for their beauty, and they are pressed into service to articulate love and grief and virtue. Plants are enlisted for political, sporting, and charitable causes. Plants represent spiritual values and the divine—the importance of sacred lotus in Buddhism and Hinduism powerfully illustrates this aspect. Plants make places; they decorate, elaborate, and stand as synecdoches for the landscapes they inhabit. Plants, especially trees, endure as witnesses to historical events and are seen to embody memory transcending place and time. We harvest plants for food; and exploit their legion biochemical pathways to utilize plants for medicine, cosmetics, adhesives and cord; for textiles and paper; for dyes and inks; for fuel and construction, and for ornament. In short, plants provision human well-being and culture; humankind has been intimately connected to plants from the start of human history.

The modern era has seen accelerating human population growth. The better the internal combustion engine got the further road networks pushed across continents, the more vegetation was cleared for agriculture—and the further urbanism sprawled. Increasing demand for energy for domestic use, transport, and industrialization has resulted in vast reserves of fossil (plant) fuels extracted and burned and has led to increasing atmospheric CO_2 and exacerbated climate change. These changes, particularly evident since the second half of the twentieth century, have resulted in scientists viewing human impact as the equivalent of the geological epochs that previously defined the planet with the modern era being labeled the Anthropocene. Such changes have profoundly damaged the ancient, direct connection between plants and people.

Our relationship with plants extends beyond provisioning—we make use of them as symbols, and we deploy them as metaphors. At the symbolic level, we remain as connected to plants as ever. The domestication of plants is a reciprocal relationship; plants domesticate and cultivate us, as we cultivate and domesticate them: "The offspring of the ancient marriage of plants and people are far stranger and more marvelous than we realize. There is a natural history of the human imagination, of beauty, religion, and possibly philosophy too" (Pollan 2001: xviii). In this chapter, we make a start on the project of restoring plants to a more honorable place in our imaginations; and of recalling just how deeply plants have made our cultural articulations and practices what they are.

PLANTS AND THE DIVINE

The Search for Enlightenment

Humans in every culture we know of, everywhere, have used plants—as offerings, as symbols, as objects of contemplation, intoxicants, and scents—to take them closer to the notion and experience of the Divine. Many ancient relationships between plants and people retain significance in spiritual life into the modern era: celebration of the lotus in liturgy and iconography as a symbol of divine perfection rooted in the profane; pilgrimage to particular bodhi trees connected with the one under which Gautama Buddha received enlightenment; offerings of flowers at temples; and the widespread ritual and religious use of plant-derived smoke as incense (e.g. Bruland 2013: 426–7; Goody 1993: 328–33; Staub, Geck, and Weckerle 2011: 1–3). Beyond symbolism and metaphor, plants and plant products are consumed to negotiate a direct connection with the Divine. In the Catholic and other Christian churches, the bread and wine used in the communion ritual are viewed as undergoing transubstantiation—transformed by faith and ritual into the body and blood of Christ. The direct ingestion of various psychoactive plants and fungi has continued to provide communion with the Divine in some traditional societies; it has been appropriated in pagan rituals and the psychedelic movement.

The sacred lotus (*Nelumbo nucifera*) is perhaps the most venerated plant in world religions. In Hinduism, it is fundamental in cosmogonic myths, where "the lotus unfolds the universe" and is "associated with Brahma the demiurge Creator and the pristine embodiment of the universal spirit who was born of the lotus" (Gupta 1971: 46; Mitra and Kapoor 1976: 126).

Celebrated astronomer Carl Sagan (1934–96) viewed the Hindu Creation story's cosmic dream as "a premonition of modern astronomical ideas" (2013: 273–4). In his 1980 television series, *Cosmos*, Sagan narrated these words:

> There is the deep and appealing notion that the universe is but the dream of the God, who, after a hundred Brahma years, dissolves himself into a dreamless sleep and the universe dissolves with him until after another Brahma century he stirs, recomposes himself and begins again to dream the great cosmic lotus dream. Meanwhile, elsewhere there are an infinite number of other universes each with its own God dreaming the cosmic dream. These great ideas are tempered by another, perhaps still greater. It is said that men may not be the dreams of the Gods but rather that the Gods are the dreams of men.

His book accompanying the television series was included in the US Library of Congress' 2012 exhibition of *88 Books that Shaped America*.

Perhaps the most striking example of an enduring spiritual relationship with a single instantiation of a plant is the veneration accorded—by Buddhists, in particular, those of the Theravada tradition—to the Jaya Sri Maha Bodhi tree in Sri Lanka. This *Ficus religiosa* (known as peepal in Hindi or esathu in Sinhalese) is the oldest cultivated tree in the world with a written history. This tree was propagated from the southern branch of the tree under which the Buddha achieved Bodhi (enlightenment) by the river Neranjana in the Buddhagaya, India (see Forbes 2016a)—trees with a direct lineage to this particular tree are referred to as Bodhi (from the Sanskrit word for wisdom).

The tree that the Buddha chose to sit under was already a sacred tree. In India, Hindus and Jain holy men choose them for meditation, and many Hindus still practice *pradakshina*, which includes circumambulation, or meditative pacing, around peepal trees

FIGURE 6.1 The Jaya Sri Maha Bodhi at Anuradhapura, Sri Lanka—the oldest historical tree in the world—established from the southern branch of the peepal tree (*Ficus religiosa*) in whose shade the Buddha achieved enlightenment. This branch was brought to Sri Lanka from India in 236 BCE. Photo by Stephen J. Forbes.

with an accompanying chant of "vriksha rajaya namah" ("salutation to the king of trees") (Krishna and Amirthalingam 2014). Sal (*Shorea robusta*) is also shared by Buddhists and Hindus. The Buddha, Siddhartha Gautama (b. *c.* 567 BCE), is supposed to have been born and have died under a sal, which is also held sacred by Hindus for its association with Vishnu in his incarnation as Ramachandra. In the twentieth century the Guyanan cannon-ball tree (*Couroupita guiannesis*) has been mistaken for the sal and planted widely at temples in Sri Lanka, Myanmar, and Thailand (Thann 2015). In a commentary on the confusion between the sal and the cannon-ball tree, Lakshman Ranasinghe (2015) described the complex iconography adopted for the flower:

> The beauty of its flower defies description. 6 pink lotus-like petals (but more circular) surround a central male unit (stigma). The ring-like arrangement resembles the padmasana of the Samadi Buddha-image. The central complex is a masterpiece of nature. A ring of creamy filaments surround a single central drop-shaped white replica of a Stupa—surrounded by a ring of filaments comparable to the grassy moat seen around the Yatala Stupa in Debaraweva. A pink snake-head hood-like fleshy structure curves and overhangs the "dagoba". Yellow and pink-tipped stamens provide a brush-border to the hood, in close proximity, but above the yellowish filaments of the stigma. The entire flower is a miracle of nature, and ideally suited for religious offerings.
>
> (Ranasinghe 2015)

Ginkgo trees (*Ginkgo biloba*), regarded as auspicious and even godly, were cultivated at shrines and temples in China, Korea, and Japan from around the eleventh century (Chen and Fan 2012; Nesbitt and Prance 2005: 223). An enduring tradition of attaching red perfumed pouches, containing prayers or wishes, to a ginkgo tree on the last day of the Mid-Autumn Festival continues on Jinli Ancient Street in Chengdu, in Sichuan Province, China. Prayer and wishing trees are widespread globally illustrating a common belief in a connection between such trees' qualities and human needs and wants (e.g. China News Service 2014; Dafni 2002; Nugteren 2005). Reverence for ginkgo's longevity, beauty, and utility is a common response to the tree, and ginkgo plantings may reflect these values as much as any clear links to religious history or iconography.

Sacred gardens, such as Japanese Zen gardens, have a continuing history to the modern era, while Christianity, Islam, and Judaism have established Biblical, Qur'anic, and Talmudic gardens as a form of devotion and religious education during the modern era. The Neot Keumim Biblical Landscape Reserve in Israel is illustrative (Alon-Mozes 2013). In 1924, Ephraim and Hannah Hareuveni proposed establishing a "Garden of the Prophets and Sages." As Ephraim Hareuveni later wrote,

> If we can no longer hear the ringing words of the prophets, but we can see what they saw and smell what they smelled, let us therefore arrange a plant garden of our ancient literature, and learn to write in it and to read there from the "Book of Books" and the "Song of Songs" in the letters and the colors which were used for writing the Book of Books (the Torah) and was the source for all of the psalms.
>
> (1927, quoted in Neot Kedumim 2011)

FIGURE 6.2 The ground is covered with golden fallen leaves from the ancient ginkgo tree at the Ancient Kwanyin Zen Buddhist Temple (Gu Guanyin Buddhist Temple) in the Zhongnan Mountains in Xi'an city, northwest China's Shaanxi province. Photo by Imaginechina Limited. Courtesy of Alamy Stock.

The search for enlightenment through plants extends beyond contemplation of their form or their association with cosmology and religion. The term "entheogen" is sometimes applied to psychoactive plants employed to induce a spiritual experience intended for spiritual development (rather than for the recreational use of the same plant). Psychotropic plants have been used in most religions of the world to induce spiritual experience and a sense of the divine. However, they have not had much history in Christian practice. Various North American entheogenic churches have been established (and disestablished) in the modern era (Stuart 2002). The continuing use of peyote (*Lophophora williamsii*) among some American Indians resulted in conflict with United States drug enforcement law, which was later resolved by the passing of the American Indian Religious Freedom Act 1978, providing legal protection for members of the Native American Church (Maroukis 2010). Aldous Huxley's *The Doors of Perception* (1954), documenting his experiences with mescaline, was influential in promoting psychedelic experience. Although mescaline was originally derived from the peyote, it had been synthesized in the laboratory as early as 1919 and the mescaline ingested by Huxley was manufactured.

The psychedelic movement, which reached its high point in the 1960s and which was marked by artistic and musical production reflecting the experience of altered consciousness, is thought to have begun when Swiss chemist Albert Hofmann (1906–2008) synthesized LSD in a laboratory in 1938; he ingested LSD-25 in 1943, thus becoming the first person to experience its effects first-hand. While a semisynthetic compound, LSD-25 was derived from ergot (typically *Claviceps purpurea*), a psychoactive fungus (Schiff 2006). Ergot-infected cereal has long been identified with ancient ritual and religion. Hofmann suggested that *kykeon* consumed during the Eleusinian Mysteries ceremonies in Ancient Greece was a flour mixture contaminated with ergot (Wasson, Hofmann, and Ruck 1978). The value of the laboratory is in controlling dosage—high levels of ergot fungus results in ergot poisoning (ergotism) manifested as St. Anthony's fire (and perhaps responsible for the infamous Salem witchcraft accusations and trials). In addition to Hofmann and Huxley, writers such as Carlos Castaneda, ethnobotanist Richard Schultes, Timothy Leary, and Wade Davis popularized plant-based hallucinogens and the roles of shamans to provide pathways to spirituality and the Divine (Castaneda 1968; Davis 1998; Leary 1964; Schultes, Hofmann, and Ratsch 1979). Michael Pollan's *Changing Your Mind* (2018) contends that psychedelics help humans by relaxing the ego's hold and opening an awareness that one shares a mind, a consciousness, with much more than merely one's self. Biological-psychology researcher David Kennedy has reviewed such plant secondary metabolites (those that are secondary to standard biosynthetic and metabolic pathways)—including alkaloids, phenolics, and terpenes, and their effect on the human brain (2014: 111–13).

TREES AND CULTURE

The Harmonious Landscape and how Forests Think

It is not surprising, given their size and beauty and how many benefits trees provide humans—canopies for shelter from rain and sun, fruits to eat, wood to burn, timber to build with, limbs to climb, bark for boats and shields and roofs—that trees have been held sacred. Both the Epic of Gilgamesh (*c.* 1800 BCE) and the Bible record the planting and destruction of sacred groves. But the scale of their destruction in the modern era is unprecedented. Nevertheless, cultural traditions and beliefs associated with sacred groves and trees endure where both the trees and the people remain (Laird 1999). Some examples here illustrate the continuing contests for sacred groves in the modern era.

First, the "church forests" associated with Ethiopian Orthodox Tewahedo churches and monasteries protect some of the last native forest in South Gondar and support the highest diversity of tree species in the region, providing a range of social benefits to community members as well as containing many of Ethiopia's endangered plant and invertebrate taxa (Klepeis et al. 2016). These sacred groves are considered to have been appropriated to Christianity from ancient Agao paganism, and noted twentieth-century scholar of Islam John Spencer Trimingham (1952) observed that certain groves had also been subsequently appropriated to Ethiopian Islam. Second, in Central Australia isolated stands of the waddy-wood (*Acacia peuce*)—an exceptional feature in an otherwise treeless desert, remain significant to the Arrernte people. Their spiritual beliefs have, in recent years, led to the trees' protection following years of cutting for fence posts (Hercus 2012). Third, in Cameroon, moabi trees (*Baillonella toxisperma*) are used as cemeteries and valued by the Baka people for their contribution to traditional pharmacopeia, for their fruits, and their oil-rich seeds. Here, the individual moabi trees are viewed as sacred. The contest between the defence of the Baka's livelihood and of their customary institutions against monetary exploitation is fierce if manifestly one-sided in the context of the power imbalance in relationships and in modern law (Veuthey and Gerber 2011). Finally, in the Hong River catchment of southern Yunnan Province in China, the traditional Hani people pursue "the harmonious landscape," where various forest types have instrumental values servicing water conservation and agricultural production as well as spiritual values. Of special importance to the Hani are the Zhai Shen Lin (sacred woods of the village), which represent for them the numerous gods living in the nearby forest (Zhan and Jin 2015).

In *How Forests Think* anthropologist Eduardo Kohn (b. 1968) discussed the nature of a language for communicating with forests, based on his experience with the Ávila Runa people of the Ecuadorian Amazon and their interaction with the forest. Kohn considered that the Runa "... come to think with the forest's thoughts, and, at times, they even experience themselves thinking with the forest's thoughts in ways that reveal some of the sylvan properties of thought itself" (2013: 100). French anthropologist (and student of Claude Lévi-Strauss) Philippe Descola (2014: 268) held that Kohn's goal was to "... (repopulate) the social sciences with nonhuman beings, and thus shift the focus away from the internal analysis of social conventions and institutions toward the interactions of humans with (and between) animals, plants, physical processes, artifacts, images, and other forms of beings" and that Kohn's thesis was one where "Human and nonhuma become aware of each other and develop modes of relating prior to the usual processes of categorization and communication embedded in historically and linguistically contingent frameworks."

Concern about the impacts people are having on the ecosystems that supply us with food, water, and climate security has seen the emergence of many political and philosophical movements connected to ecosystem conservation. These "green" groups and the green movement as a whole reflect both spiritual and scientific concerns about the exploitive ethos humans have employed in relation to the environment in the modern era, as discussed below.

The Representation of Trees in Modern Culture

The representation of trees in modern culture is rife with anthropomorphism in various guises. Our projection perhaps begins with the morphological similarities between humans and trees—both of us generally standing upright and having skin, trunks, limbs, and crowns in comparable locations. In the modern era, J. R. R. Tolkien (1892–1973) saw this projection deriving from humanity's wish to associate or communicate with other living things. In this context, trees may be enchanted by others, may be inhabited by spirits or souls, or may have their own souls. Writing in *Tolkien Studies*, Cynthia Cohen discussed literary trees in a number of traditions and suggested four categories: trees that do nothing unusual, appearing essentially as Primary World trees; trees that remain rooted, but are able to talk, think, and feel; trees that remain rooted but can move their branches or trunks; and trees that can uproot themselves and move from place to place. She suggested that Tolkien's conception of forests and trees is singular for its diverse characterization of trees in all of these categories and for his knowledge of tree mythology (Cohen 2009: 91). His trees can sometimes even metamorphose, as Treebeard explained to Pippin and Merry, protagonists in Tolkien's work.

As a linguist and philologist, Tolkien was acutely aware of the importance of language in shaping relationships and particularly those with plant life. Treebeard admonished Pippin and Merry for divulging their own names used for their race, the hobbits, and for themselves: "Real names tell you the story of the things they belong to in my language" (Tolkien 1954b: 464). Treebeard might well be a critique on the poverty of Linnaean plant classification. Indeed, the stories that can be read from both common and scientific plant names are one-dimensional in comparison with the folk taxonomies of indigenous people and seen against the narratives of any individual tree: Tolkien had 141 plant species in Middle Earth (Judd and Judd 2017) and wrote to his publisher, "I am (obviously) much in love with plants and above all trees, and always have been; and I find human maltreatment of them as hard to bear as some find ill-treatment of animals" (Tolkien in litt. June 30, 1955; see Tolkien 2013). Tolkien's forests and trees are important in presenting a perspective on trees beyond the anthropomorphic view of them only embodying qualities that we value, such as endurance, timelessness, grandeur, beauty or terror and seasonality. The vegetative soul may be well beyond our understanding. But Tolkien did masterfully create an alternative vegetal world and, through it and the relationships he depicted between trees and beings, he allowed readers to think more deeply about their relationships with trees.

While Tolkien's trees represent the most complete portrayal of anthropomorphic trees in literature, popular culture in the modern era includes a variety of trees in books and film. The presence of trees in film, includes Peter Jackson's adaptation of Tolkien's *The Lord of the Rings* (2001–3) and ranges from the talking apple tree in *The Wizard of Oz* (1939), the whomping willow in J. K. Rowling's 1999 book *Harry Potter and the Prisoner of Azkaban* book and subsequently in Warner Bros' Harry Potter movie *Chamber of Secrets* (2002), to James Cameron's movie *Avatar* (2009) with spiritually important trees (Home Trees, the Tree of Voices, and the Tree of Souls) and Marvel Studio's *The Guardians of the Galaxy* (2014) where Groot is an extra-terrestrial, supernatural, tree-like character.

PLANTS AND THE COMMUNICATION OF IDEOLOGY AND POWER

While the representation of trees in culture may ascribe power to trees, trees are also used to ascribe power to humans and to articulate particular ideologies. The tradition continues in the modern era as in the landscape gardens of the mansions in Newport, Rhode Island, and the Philadelphia Main Line in the United States, and of the Côte d'Azur in France. Such gardens are experienced directly or vicariously through private garden tours and open gardens schemes, and in books, magazines, and film developed by specialist garden writers, photographers, film makers, and presenters. Parks and gardens of a certain scale and style stand for the wealth and power of nobles, rulers, of men and women of means.

Beyond wealth and power, there is, as garden historian John Dixon Hunt (2015: 173–4) argued, a nationalist text implicit in the development of botanical and other gardens focusing on native species in modern landscape architecture. Native-plant selection may be intended to strengthen local landscape character, improve plant persistence in the face of reducing maintenance inputs, and contribute to local biodiversity conservation values. While there is no explicit nationalist intent here, the fervor with which locally native plantings may be promoted deserves further scrutiny.

For example, the infiltration of a Nationalist Socialist philosophy into landscape architecture in Germany has been well documented by landscape theorists Gert Gröning and Joachim Wolschke-Bulmahn's research on the history of native plants' popularity in Germany. As they noted, landscape architect Albert Krämer argued in 1936 that Germans "still lack gardens that are race-specific, that have their origins in nationality and landscape, in blood and soil. Only our knowledge of the laws of the blood, and the spiritually inherited property, and our knowledge of the conditions of the home soil and its plant world (plant sociology) enable and oblige us to design blood-and-soil-rooted gardens" (their translation), and a team of Saxonian botanists who saw the invading *Impatiens parviflora* as replacing *Impatiens noli-tangere* in light shaded areas of forest and sought "a war of extermination" contending, "As with the fight against Bolshevism, our entire occidental culture is at stake, so with the fight against this Mongolian invader, an essential element of this culture, namely, the beauty of our home forest [is at stake]" (both parties quoted in Gröning and Wolschke-Bulmahn 1992: 124; 2003: 81). Perhaps one of the most overt expressions of Nationalist Socialism in this context was the planting of trees to form swastikas in German forests (Boyes 2000; Imort 2000).

The story is really more about the woods than the trees. But still, it is an instance of a State conscripting plants to an ideology—a planting used to symbolize a State, to represent its values and speak, with special power because nature and landscape were doing the speaking, of the power and persistence of the regime.

The ideological perspective of the Nationalist Socialist state in Germany was extreme, although similar less strident views are evident in other nationalist garden projects. American landscape architect Jens Jensen (1860–1951) is viewed as a major influence in the development of the Prairie School style in landscape architecture in the United States (where Frank Lloyd Wright is credited as the major architect of this style). The nationalism inherent in Jensen's style has been drawn attention to by Joachim Wolschke-Bulmahn in quoting an article for the German journal *Gartenkunst* (Jensen 1937, quoted in Gröning and Wolschke-Bulmahn 2003: 85). Jensen's views have been discussed by Dave Egan and William Tishler (1999), who endeavored to contextualize his nationalism with contemporary advocacy preferencing of native plants. Renowned evolutionary biologist and science-historian Stephen Jay Gould (1941–2002) rejected both evolutionary and nativist arguments for preferencing native plants (Gould 1998).

Gardens, across the political spectrum, were, as John Dixon Hunt argued, put to the service of the modern state, where gardens remain potent forces of nation-building, powerful signifiers of authority and ideology (2015: 173–4); as Hunt observed, "Some modern botanical gardens privilege a nationalist agenda, in part as a means of focusing on indigenous materials ..." together with "... the need to see that a nation acquires its own botanical garden as an emblem of its scientific capabilities as well as national prestige" (and "... usually a concern to reveal a distinctly modern design"). Hunt's exemplars include the Australian Garden in Melbourne, the National Botanic Garden of Wales, Jardín Etnobotánico de Oaxaca, Mexico, and the Roberto Burle Marx-designed Maracaibo botanical garden, Venezuela.

The distinction between political and religious influences on a such a nationalist agenda can be difficult to separate. Landscape scholar Tal Alon-Mozes (2013) explored history of the Neot Kedumim Biblical Landscape Reserve (see above) and read the project as principally a nationalist text. She considered it as an example where gardens and landscape plantings have manifest political and doctrinal implications. While on a smaller scale, the Gluck Talmudic Garden at the Jewish settlement of Beit El is explicit in this vein: Eugen Gluck, the garden's founder, observed this garden is "... about reminding ourselves and the rest of the world that our presence in this land goes back long before any of those European countries existed. The Gluck Talmudic Courtyard proves that all the towns like Beit El are not here because Jews needed an asylum. Rather, the Jews are the rightful owners of the land" (Sheva 2011).

Critics such as Joanna Long consider the Jewish National Fund's history of landscape afforestation projects in Israel "assisted the material dispossession of the Palestinians and the deletion of the Palestinian narrative from the land and the history" (2009: 75). The Al Aqsa Foundation for Waqf & Heritage criticized a "nine-garden project" they believed had been proposed by the Israeli government to surround the Al Aqsa mosque. The Foundation described "these [as] biblical gardens ... counterfeiting of geography, history and archeology," and suggested that "Israel is striving to steer the situation in Jerusalem into its ambition for an illusory history by having the original Islamic and Arab landmarks blurred and labeled with Hebrew fake names." The Foundation's reaction underscores the ideological and political uses to which gardens have been put, and not just in distant times and places (Occupiedpalestine 2012).

The connection of politics and ideology to plants extends beyond the theater of gardens and landscape to influence exercised by political regimes, driven by ideology, upon the plant sciences. The outstanding example here was the rejection of Darwinism and mainstream genetics in the Soviet Union from the 1920s to the 1960s. In this context Soviet plant breeder Trofim Lysenko's (1898–1976) rhetoric asserted the priority of a Marxist Darwinism doctrine in the sciences, and with the support of Josef Stalin, oversaw a purge of biologists with dissenting views. Darwin's "natural selection" was supplanted by a notion of "natural co-operation" that had profound consequences for plant breeding and agriculture. His opponent, Nikolai Vavilov (1887–1943), President of the All-Union Academy of Agricultural Sciences of the Soviet Union from 1929–35 and a leading researcher on the origins of crop plants and their geographical centers of diversity, was arrested in 1940 and subsequently died of starvation in prison in 1943. Lysenko, President of the Academy from 1938–56 (and subsequently from 1961–2), with Josef Stalin's support, used his 1948 speech "The Situation in the Science of Biology" to the Academy to impose what became known as Lysenkoism, which was only formally ended in 1964 (Krementsov 2010).

The Stagecraft of Power

In the political realm, flowers have long been an element of what American artist Taryn Simon (b. 1975) described as "the stagecraft of power." Simon's work illustrates in microcosm the symbolic and metaphorical roles of plants in politics and stagecraft for the communication of power. Her interest began with examination of a floral arrangement in a photograph of the Munich conference of Hitler, Mussolini, and Chamberlain's meeting in 1938 and led to her exhibition "Paperwork, and the Will of Capital" that began at the 2015 Venice Biennale and has subsequently been shown internationally. Here she recreated floral arrangements associated with the signing of peace treaties and other international accords among countries represented at the 1944 United Nations Monetary and Financial Conference in Bretton in New Hampshire that led to the establishment of the International Monetary Fund and the World Bank. Using historic images, and working with botanist Daniel Atha from New York Botanic Garden, she shipped thousands of plant specimens from the Aalsmeer flower auctions in the Netherlands to her New York studio, where she recreated and photographed the original "impossible" bouquets that combine flowers from different places and seasons (La Force 2016; Jobey 2017).

ANTHOLOGY OF PLANTS IN LIFE, LOVE, AND DEATH

Life and Love as well as Purity, Innocence, Beauty, and Truth

The way humans use plants to connect with the Divine is discussed above; but plants are also employed to articulate secular values and in the quest to acquire qualities attributed to plants. Certain plants are ascribed qualities such as beauty, purity, and agelessness; we turn to those plants, symbolically or actually, to invest our own lives with those same qualities. In a largely secular world plants are still viewed as talismans and even as imbued with magical agency. Humans seek to attain these qualities by association with those plants (or their representations) and even by consuming their extracts (regardless of any scientific evidence for their efficacy). Our search for our needs and wants within the plant world has been remarkably effective in shaping human history; our talismanic association with plants may simply be another manifestation of this exploration.

Ginkgo, for instance, is widely seen to embody continuity, resilience, and the survival of the ancient or eternal (Nesbitt and Prance 2005: 223–4). Because of those qualities it is seen to embody, it has been, over the years, used in perfumes, cosmetics, medicine, and design. The qualities of ginkgo imbue products ranging from perfume, to Daum's ginkgo leaf crystal perfume bottles and vases (the hype on luxury store Gearys' Beverly Hills website exhorts gift buyers to "Show her she is unique as the Ginkgo plant with this perfume bottle from Daum"), and even to a housing complex skinned with prints of ginkgo leaves: Housing Beekbergen in Apeldoorn, the Netherlands by architects Casanova–Hernandez (Casanova–Hernandez 2007–11; Daum 2019; Gearys 2018).

The talismanic nature of plant extracts is exemplified in the promotional hype for Comme de Garcons' "Comme 3" perfume—the ingredients include Ginkgo extract, and much more, "An imaginary flower: the flower you want it to be, the flower in your dreams: A new rose, incandescent, electric, opalescent" and in the list of twenty-one botanicals *viz* "Rose Oxide, Mandarin, Magnolia Leaf, Basil, Gingko Leaf, Blackcurrant Flower, Red Pepper, Angelica Root, Cardamon, Mace, Black Rose Leaf, Paradise Grains, Jasmine Sambac, Lebanese Cedar, Sandalwood, Gaiac Wood, Javanese Vetiver, Patchouli,

Birch Essence, Olibanum Incense, Amber." It seems the story is as important as the scent here (Comme des Garçons 2017).

Plants as Life—Plant Obsessions

While the relationship between food plants and humans might be viewed as a reciprocal relationship, collectors of ornamentals may well be in their plants' thrall. In her book *The Orchid Thief*, Susan Orlean (1998: 94) quoted from Norman Macdonald's 1939 book *The Orchid Hunters*: "When a man falls in love with orchids, he'll do anything to possess the one he wants. It's like chasing a green-eyed woman or taking cocaine ... it's a sort of madness." Orchids might be the most celebrated subjects for plant collectors' obsessions, but other striking plants such as cacti and succulents, bromeliads, carnivorous plants, ferns, cycads and palms, and showy flowers such as roses, fuchsias, begonias, and heliconias fire individual obsessions and drive devotees into specialist societies. Garden writer Noel Kingsbury's perspective from researching daffodils and their collectors is illustrative:

> Many cult plant enthusiasts will grow only the objects of their obsession and never let a cabbage or a rose enter their gardens. Cult plants attract a wide range of people, but (until recently) nearly all were men, with a strong tendency for growers to be part of what could broadly be called the skilled working class. Some of the cult plant growers are true obsessives, and given to a clannish secretiveness ...
>
> (2013: 13)

These obsessions are evident in modern plant societies and are explored widely in popular and specialist books, blogs, and magazines focused on both plant collectors and the plants that are the objects of their desire. Two examples illustrate the summits that privately established and held specialist plant collections might reach: Les Cèdres on the Saint-Jean-Cap-Ferrat in southern France and Nong Nooch Tropical Garden at Pattaya, southeast of Bangkok in Thailand. Les Cèdres' plant collections were established by Alexandre Marnier Lapostolle in 1924 and substantially developed by his son Julien Marnier Lapostolle (1902–76). Julien's passion for cacti and succulents saw a number of species named in honor of him, and of his wife Suzanne. The plant collections have seen Les Cèdres described as the "Louvre of botany." The property was considered the most expensive residence to be offered for sale on the global market in 2018. The future of the plant collections remains unclear, however (Glass and Foster 1975; Sage 2018). Nong Nooch Tropical Garden was established by foreign-film distributor Pisit Tansacha and his wife, Nongnooch, in 1954 and has continued to evolve as both a display garden and as a specialist plant collection under the guidance of their son Kampon Tansacha. While the garden has exceptionally rich plant collections, the cycad collections are perhaps the most globally renowned for Nong Nooch's active involvement in cycad conservation and research programs globally (Mazza 2012; Nong Nooch Tropical Garden 2020).

War and Peace

Plants have been recruited for war and mobilized for peace in the modern era.

In times of war, plants are viewed as *materiel* and conscripted for the war effort—as food, medicine, fiber, latex, wood for construction, and so forth. But there is more to the role of plants in war than that. The multifaceted role of botany in war has been pursued by a number of contemporary researchers. Ecologist and military historian James Wearn has described the intersection of war, plants, and culture as "polemobotany"; his work examines the impact of war on plant communities and the intricacy of the relationship between people and plants in times of conflict, well beyond plants as *materiel*. His work considers landscapes shaped by war—by the physical effects of battle (shell holes, trenches, etc.), chemical defoliation, the introduction of new species on plant communities, and post-war recolonization and revegetation programs. In some cases disputed territories in conflict zones provide a botanical refuge for rare species and plant communities: Wearn noted the incidental role of conflict zones, including minefields and the Korean DMZ, as de facto plant conservation reserves (2016: 280). Conflict also, generally, presents an impediment to botanical exploration: botanist David J. Mabberley (2009) noted recent plant discoveries in Ogaden, a region of Ethiopia where unrest has long dissuaded botanists from botanizing. According to Wearn, polemobotany includes cultural perspectives on plants and warfare, such as the psychology of soldiers' engagement in botanical activity in battle zones (such as plant-collecting and illustration) and the production of botanically inspired trench art.

Military concealment represents another intersection of culture and plants. While concealment has long been part of warfare, the First World War saw the first dedicated camouflage unit established by France. Disruptive patterns were used to hide lookouts, equipment, and troops in plain sight. Such designs illustrate a response to landscapes, plant communities, and plants. Camouflage in the First World War relied on hand painting or the application of actual plant material—the technology to print camouflage designs onto fabric was only available after the First World War, in the 1920s (Imperial War Museum 2018). The construction of imitation "spy trees" that could replace actual trees and provide lookouts on the battlefield of the First World War presents a curious subset in camouflage history (Billock 2017; Imperial War Museum 2015).

The importance of plant *materiel* in the Second World War is vividly illustrated by the Japanese occupation of Malaya and the Dutch East Indies from January to March of 1942. That occupation captured the larger part of the world's sources of quinine and rubber plantations and led to sudden and significant shortages of both, with major implications for the treatment of malaria for Allied forces in tropical theaters of war, and for manufacturing and transport, dependent on rubber for tires, drive belts, seals, and much more (Sundin 2016). As a result, the US Foreign Economic Administration (FEA) responsible for accumulating strategic *materiel* commissioned botanists and foresters to find new sources of quinine and natural rubber (Howard 1994, 2000; Sumner 2019).

After wars, nations and local communities have historically been moved to honor those who died and were injured with memorials; war memorials also celebrate valor and victory and assert the value of peace. Such memorials are built on battlefields, in town squares, and in specially designated war cemeteries. And most often, of course, they include gardens and plantings, and they often incorporate floral elements in their design.

Plants are, of course, also directly associated with peace. The olive branch continues as a symbol of peace—as, for example, on the United Nations flag and in the form of a gold replica left on the surface of the moon during the Apollo 11 landing in 1969 by astronaut Neil Armstrong. The connection between flowers and peace is powerfully illustrated by two of the most famous anti-war photographs from the modern era: "Flower Power" by Bernie Boston and "The Ultimate Confrontation: The Flower and the Bayonet by Marc Riboud" were both taken during a demonstration against the Vietnam War in 1967 in Washington, DC. Both powerfully capture 1960s "flower power" (a term coined by Allan Ginsberg) and peace movements (Gottschalk 2016). Anthropologist Jack Goody outlined

FIGURE 6.3 "Flower Power" by Bernie Boston—one of two iconic images of the 1960s flower power movement taken on the same day during an anti-war protest at the Pentagon in 1967. Here George Harris places carnations into gun barrels. Photo by Bernie Boston/The Washington Post via Getty Images.

FIGURE 6.4 "The Ultimate Confrontation: The Flower and the Bayonet" with the photographer Marc Riboud and the subject Jan Rose in 2005—thirty-eight years after the original photo taken in 1967. Photo by Michael Loccisano/Patrick McMullan via Getty Images.

more recent struggles where "the gift of flowers [has] stood opposed to the blood sacrifice, flower power against fire power, and love against war and authority" (1993: 311–12).

The most celebrated flower of the twentieth century is quite possibly an exceptional rose, known as Peace (the commercially renamed cultivar, 'Madame A. Meilland') bred just prior to the Second World War. Peace roses were given to delegates at the first United Nations meeting in San Francisco in 1945 with a note reading, "We hope the 'Peace' rose will influence men's thoughts for everlasting world peace." Rosarian and writer Peter Beales argued, "'Peace,' without doubt, is the finest Hybrid Tea ever raised and it will remain a standard variety forever" (Meilland and Lambert 1984; Stehkämper 2020).

Victory and Laurels

The bay laurel, olive, myrtle, date palm, and apple are all plants that have had an association with victory in varying contexts since antiquity. Floral bouquets have been presented to medalists in the modern Olympics, and at the 2004 Athens Olympics, medalists also donned, sometimes with difficulty, olive wreaths, which served as the official symbol of those games (IOC 1999; Rhizopoulou 2004; Sachs 2004). However, while the 2016 Rio Olympics maintained flowers as part of the stagecraft for the medal presentations, bouquets were dispensed with—nor were bouquets presented at the 2018 Pyongyang Winter Olympics (Mather 2016).

Love, Death, and Grieving

Plants have also been a way of symbolizing love, marking grief, and paying tribute to lives ended since ancient times (Iriarte-Chiapussoab et al. 2015; Nadel et al. 2013). The Western lexicon retains vestiges, in the modern era, of ancient practices. Plants, especially their flowers, provide symbols of life, the gift and beauty of life, and its transience. The language of flowers, or floriography, is locally inflected, constrained by local flora, custom and practice, but flowers help us humans deal with love, death, and grieving in all cultures.

Cypresses and yews have been a symbol for death in Europe since at least Greco-Roman times (see Chapter 6 of Volume 1 in this series). Perhaps it is the funereal aspects of their dark foliage that recruit them to this solemn role; and the wood of cypress has long been favored for coffins, perhaps because its strong scent warded off evil spirits and the odors of decay. The foliage and seeds of the yew are poisonous, too, which can't have helped their reputation. And still we wear rosemary to remember the dead in Western countries at funerals.

And then there is the poppy. Authors Jennifer Iles (2008) and Nicholas Saunders (2013) explored the trajectory of the Flanders poppy as a symbol of remembrance following the First World War and cite the influence of Col. John McCrae's poem *In Flanders Fields* written during the second Battle of Ypres in 1915. The British Legion adopted the poppy as part of its Appeal Fund and ordered 1.5 million artificial poppies for the first Poppy Day on November 11, 1921. Although it was aware that the poppy was known as the flower of oblivion, the Legion intended "'to change its significance to the flower of 'remembrance'" (*The Times*, 1921 quoted in Iles 2008: 206).

Significantly, as art historian James Fox (2014) noted, "The poppy never found its way into the cultural practices of the war's defeated nations, and that may be because the only men whose sacrifice was believed to deserve such a symbol were those who had fought on the 'right' side. Poppies, in other words, had been converted into victory medals."

In response to a view that the red poppy may be construed as celebrating war, the Women's Co-operative Guild produced white poppies that first went on sale on Armistice Day 1933 "as a pledge to peace that war must not happen again." It has been noted that the white poppy is thus a secondary symbol as it exists only in opposition to the Flanders poppy (Iles 2008: 209). The Peace Pledge Union (PPU), the successor to the white poppy campaign, saw them as "represent[ing] remembrance for all victims of war, a commitment to peace and a challenge to attempts to glamorise or celebrate war" (Peace Pledge Union 2020). In 1986 Prime Minister Margaret Thatcher expressed distaste for the white poppy campaign in the British Parliament.

The poppy has continued to court controversy. In 2006, in response to criticism, British Channel 4 television newsreader Jon Snow outlined his concern with favoring any cause on air. "Additionally," he wrote, "there is a rather unpleasant breed of poppy fascism out there ...," his critics stating that "he damned well must wear a poppy!" (Snow 2006). In 2015 came #Poppygate: this hashtag appeared on Twitter when Downing Street added a digital poppy to a photograph of British Prime Minister David Cameron on its official Facebook page. A flood of memes followed (*The Independent* 2015). It was observed in this context that "The poppy is no longer a delicate flower; its powerful petals have become a parliamentary instrument of fear. The poppy has become as ruinous as its opiates to the citizens of a free democracy" (Braidwood 2015). In 2016, England, Northern Ireland, Scotland, and Wales wore poppies in their World Cup qualifying matches and were sanctioned by the football organization FIFA, whose rules forbid teams from wearing political or religious symbols in FIFA-sanctioned football matches. The incident earned the ire of British PM Theresa May who criticized FIFA in Parliament (de Menezes 2016).

Flowers have always been good for articulating profound sentiment. But they also lend themselves, perhaps especially in a visual age, an era of mass media, to sentimentality. In the modern era, the death of Diana, Princess of Wales (1961–97) of the British Royal Family was a striking example of use of flowers in love and in public grieving. Over 10,000 tons of tributes, some fifty million blooms, were estimated to have been placed outside Kensington and Buckingham Palaces in London. Rosemary was placed for remembrance, wreathes for mourning, bouquets and red roses for love. Singular messages accompanied many of the flowers. Some golden chrysanthemums bore a note: "The colour of these flowers reflect the sunshine you brought into people's lives." A withered rose was accompanied by the explanation that this was the "last rose cut from our garden in memory of you Diana / Love always / Allison, John and Persephone" (Monger and Chandler 1998). Then there were the floral offerings by members of Diana's family: son Harry laid small white roses, for innocence (his and hers and for grace) on his mother's coffin; William laid white tulips; and her brother, Earl Spencer, laid her favorite flowers there, white lilies (Hoge 1997). Individual blooms and bouquets offered to Diana at her death carried the intentions of the bearer, and meanings the state wished conveyed; but en masse the millions of blooms continued an ancient tradition, but also signified what has been called a "floral revolution": Diana's death had a "floral architecture" (Greenhalgh 1999: 40, 43). Those whose response to her death was more measured felt marginalized. To give voice to these other voices, journalist Ian Jack (1997) wrote "Those Who Felt Differently" for the Winter issue of *Granta*, the London literary journal he edited: to some, "the flowers were quite repulsive. I felt scared when I saw all those flowers. It seemed a kind of floral fascism ... a country patrolled by the grief police." While the scale of public floral tributes for Diana was remarkable, there were precedents, ancient and modern. For example, the 1966 Aberfan disaster in Wales, when a school was destroyed

FIGURE 6.5 Floral tributes and balloons laid in the gardens of Kensington Palace after the death of Princess Diana, Princess of Wales, August 31, 1997. Photo by Jayne Fincher/Getty Images.

by an avalanche of coal waste, killing 144 people including 116 children, saw wreaths sent from all over the world to make a cross a hundred feet tall (Monger 1997: 113).

SIGNS, SYMBOLS, AND METAPHORS

Beyond love, death, and grieving, plants, particularly trees, are often where people turn for metaphors of moments or movements of consequence. In his 2016 Toni Morrison Lecture at Princeton University, "In Praise—and Dread—of Trees," Nobel Laureate Wole Soyinka (b. 1934) proposed that the most iconic image of the twentieth century may have been the birth of a child, Rosita Pedro, to her young mother Sofia, in the crown of a tree in the midst of floodwaters in Mozambique in 2000 (Soyinka 2016; Wang 2016). To Soyinka this may be the perfect image of hope in the face of impossible adversity—specifically the possibility of a resurgence of humanism against abstraction and terror—in Africa and in the world. And what stands for that hope, what birthed one life and saved two, is a tree (actually a Natal mahogany, *Trichilia emetica*) (Agence France-Presse in Chibuto 2017; Paul Smith, Secretary-General, Botanic Gardens Conservation International pers. comm. 2019).

Plant Symbolism and Floriography

Plants—trees and flowers—are a language. They are a complex international language made of metaphor, in which plants are codified with meanings and employed to say, symbolically, what we may find it hard to say, but wish to say so that it stays said. Forms of communication involving plants, particularly flowers, abound. We give flowers as gifts in reciprocity for hospitality, as affirmations of affection; we employ them in rituals such as weddings and funerals; we decorate churches and temples with them, especially on Holy days (such as Christmas and Easter in the Christian Church); we use them in secular stagecraft. While floriography as a codified form of floral communication has declined since its zenith in the Victorian era, elements of this language are still manifest in contemporary culture. The slogan "Say it with flowers" captures the modern era's simplified lexicon but still attests to the continuity of flowers as a means of communicating feelings and emotions. Said to have arisen from a casual conversation between ad-man Patrick O'Keefe and the former president of the Society of American Florists in Boston in 1917, the London-based commercial creativity magazine *Creative Review* ranked "Say it with flowers" in their top twenty advertising slogans ever (*Creative Review* 2018).

FIGURE 6.6 "Say it with flowers," viewed as one of the most effective advertising slogans of the twentieth century, used here in a 1954 Christmas advertisment for Interflora. Photo by Neil Baylis. Courtesy of Alamy Stock.

Plant Emblems

Plants, usually flowers and trees, are adopted, formally and informally, by nations, states, and cities and by organizations as signifiers of natural or cultural heritage or as metaphors for economic, social, or cultural aspirations. The plant becomes the emblem—emblematic of the qualities the state or team, or organization or cause wishes to be identified by. Plants are sometimes emblems, somewhat surprisingly, of sporting teams: the South African Proteas (the national cricket team), the New South Wales Waratahs (*Telopea speciossima*, Australian State rugby union team), and the Irish Shamrock Rovers (*Trifolium dubium* and *T. pratense*, Irish Premier Division football team), and the Fighting Artichokes (*Cynara cardunculus var. scolymus*, Arizona's Scottsdale Community College American football team, which illustrates a wonderful idiosyncrasy). In England the English oak, *Quercus robur*, has assumed the status of national tree, and is viewed as embodying English strength and endurance. In the United States the oak (including native *Quercus* species), seen to illustrate much the same qualities, was formally designated as an official symbol in 2004 through a national tree voting process and a Congressional Bill. Israel and Palestine share the olive (*Olea europea*) as their national tree.

As has already been demonstrated, flowers are variously appropriated and interpreted from different perspectives. An example is the daffodil. Although they were not native to many of the parts of the world where they are cultivated or read about, daffodil flowers (*Narcissus* species, hybrids, and cultivars) are memorable due to their distinctive "trumpet" corona, borne above the tepals (Waters et al. 2013). They are easily cultivated in temperate regions, they flower brightly at the end of the cold months, and in this context seem inherently cheerful. William Wordsworth's famous poem, "Daffodils" was inspired by "wild" *Narcissus pseudonarcissus* at Glencoyne Bay on Ullswater in the Lake District of England. While the poem was written in 1802 and has come to be regarded as emblematic of the Romantic movement, the flower the poem made famous across the world has caused environmental and cultural problems beyond its native settings. Antiguan-American writer Jamaica Kincaid (b. 1945) has written about the trial of being forced to memorize a poem about daffodils, when none were to be found where she grew up. This gap between the lived experience and the colonizing English literature is more general but has been referred to by Commonwealth post-colonial writers and critics as "the daffodil gap" and raises broader questions about the symbolism of the daffodils in particular and perhaps of flowers more generally beyond their native realms (Bergren 2015: 304; Smith 2002: 801).

The daffodil's cheerfulness has been appropriated for raising awareness of cancer, and raising funds to fight it. Daffodils have been employed in this role in Canada since the 1950s, and have been subsequently by the American Cancer Society, the Irish Cancer Society, the Cancer Council Australia, and Marie Curie's Great Daffodil Appeal in the United Kingdom. In 2004, to celebrate the 200th anniversary of the publication of William Wordsworth's "Daffodils," the poem was read aloud by 150,000 British schoolchildren to promote poetry and to support cancer awareness and the Marie Curie charity (Lam 2004). In a similar vein, The Daffodil Project was founded by the group New Yorkers for Parks in 2001 as a living memorial to September 11 "as a powerful memorial to the victims of the September 11 attacks … its spirit is defined by the thousands of New Yorkers who join together to make their neighborhoods, and their city, a more beautiful place to live"; by 2018, over 7.5 million daffodil bulbs had been planted across New York (New Yorkers for Parks n.d.).

TREES AS WITNESSES—OUT OF TIME

Nature writer Robert Macfarlane (b. 1976) has defined a "witness-tree" as "a tree that stood as a record of property boundaries, marked as such by scores in its bark. Now broadened to mean a tree that has seen remarkable things, that stands as 'a repository for the past'" (2018). The phrase, revived and made popular by Robert Frost (1874–1963), who named his 1942 collection *A Witness Tree,* has indeed come to signify mostly trees that have stood, and in human imagination, observed the passage of long periods of time or the occurrence of remarkable events such as battles, massacres, invasions, peace treaties, and marriages. But it has this older usage that Macfarlane chronicled: in the US, land transfers traditionally followed survey methods called metes and bounds. Metes and bounds surveys consist of a series of bearings and distances, in which trees, posts, rock piles, or natural features describe corners where bearings changed. The trees used as the parcel corner or located close to the corner are called "witness trees." Deeds or grants documenting transfer of ownership of a parcel of land also document tree species existing at the time of transfer through these witness trees.

Frost's "Beech," with which *A Witness Tree* begins, employs the phrase in this way, but also put the phrase to a wider metaphorical use:

> One tree, by being deeply wounded,
> Has been impressed as Witness Tree
> And made commit to memory
> My proof of being not unbounded.

Today, witness trees have become an anchor for the ecological analysis of changing species composition in contemporary forests (e.g. Van Gundy and Strager 2012).

Projects cataloging remarkable trees have proven a publishing success in recent decades. Historian and arborist Thomas Pakenham's *Meeting with Remarkable Trees* and *Remarkable Trees of the World* are perhaps the most famous (Pakenham 1996, 2002). In *Tree Stories* Australian photographer Peter Solness portrayed Australians' relationships with trees. He sought out witness trees such as those marked ritually or scarred through bark removal for the production of canoes, the Dig Tree (a river red gum, *Eucalyptus camaldulensis*) associated with the ill-fated Australian explorers Burke and Wills continental crossing, and the Tree of Knowledge (a ghost gum, *Corymbia aparrerinja*) under which a meeting took place in 1898 that led to the foundation of the Australian Labor Party. However, his interviews with Australians witnessing their relationships with particular trees are perhaps even more powerful in exploring notions of identity (Solness 1999). Trees as witnesses and participants in human endeavor have been also been a significant theme in modern literature. Examples include British author Roger Deakin's *Wildwood: A Journey Through Trees*, American authors Annie Prouix's *Barkskins*, and Richard Powers' *The Overstory: A Novel*, and Australian author Inga Simpson's *Where the Trees Were* (Deakin 2007; Powers 2018; Prouix 2016; Simpson 2017).

PLANT BLINDNESS, PLANT RIGHTS, AND PLANT PHILOSOPHY

While the qualities of plants—resilience, beauty, longevity—are celebrated and appropriated daily, the utility of plants, our complete everyday dependence on them goes largely unnoticed. Plants provision human lives and sustain the ecosystems that sustain us. While studies establish the value of everyday contact with plants, and particularly trees, for our cognitive development and psychological well-being, such studies so far barely touch our urban planning, and our approaches to education, health, and welfare. Mainstream cultures in the modern era are profoundly disconnected from the plants that keep them alive. "Plant blindness" is a term that has been coined to encapsulate this disconnection (Balding and Williams 2016; Wandersee and Schussler 2001).

Nevertheless, in the twenty-first century there has been renewed interest in the human relationship with plants reflecting environmental and well-being concerns. The role of plants and plant communities for food, water, and climate security, and the links between interaction with plants and improving mental health have been highlighted. Recent scientific research in plant communication and philosophical studies on plant-"being" have also informed the nature of our relationship with plants. Daniel Chamovitz's *What a Plant Knows* (2012) and Peter Wohlleben's *The Hidden Life of Trees* (2015) have effectively promoted the scientific basis (in plant physiology and biochemistry) for plant communication, while Michael Marder has revisited the nature of plants and our relationship with them through rigorous philosophical inquiry (Marder 2013; Pettman 2013). These works have revitalized interest in plants in their own right rather than for their instrumental value to us. Michael Pollan began to explain the reciprocal nature of our relationship with plants in his own garden in *Second Nature: A Gardener's Education* (1991), and his subsequent writing in relation to food and psychotropic plants has made significant contributions too. Indeed, Pollan is commonly cited as both a writer and an activist (Pollan 2020). The above-mentioned works by Pollan, Chamovitz, and Wohlleben contrast strongly with the pseudoscience and paranormal expositions exemplified by Peter Tompkins and Christopher Bird's best-selling 1973 book *The Secret Life of Plants*.

Nevertheless, alternative perspectives of plants as agents in their own right such as those of the Ávila Runa people presented by Eduardo Kohn (see above) and those of today's Australian Aboriginal people deserve attention. Peter Yates, a bush food wholesaler, described the sight of "bush tomatoes growing in neat horticultural rows" as "disturbing" to Aboriginal people, and explained why:

> Bush tomatoes, as with everything else in the world, are supposed to be made through ceremony, not grown by people .... These captive plants may have seemed to the women to challenge the proper order of the world; in short, to be sacrilegious .... These are not just foods: they are bound up in stories of creation, in kinship, and in multiple layers of personal and collective memory .... Bush foods are an inseparable part of themselves.
>
> (Yates 2009)

The challenges for integrating a deep cultural respect for plants as fundamental to the impact of people on the global environment remains a profound one, especially while such respect remains outside of any enduring cosmology or powerful new philosophy.

A NEW FLORIOGRAPHY? PLANTS AND PLACE

Various approaches to a cosmology or philosophy that encompass the importance of plant life to our future have developed in the modern era. Among them the Gaia Hypothesis of James Lovelock's (b. 1919) formulated in journal articles in the early 1970s and popularized in his 1979 book *Gaia: A New Look at Life on Earth* has been widely discussed (Lovelock 1979). Another is Deep Ecology, most often credited to Arne Næss (1912–2009), which attempts to establish a secular philosophy seeking a reconciliation with the planet (rather than directly to plants) (see Keller 2008; Naess 2012: 15). Further, endeavors to recalibrate religion to address the issues of environmental loss and to climate change are at the heart of Pope Francis' 2015 encyclical on the environment and human ecology—*Laudato Si*.

The difficulties here are legion. American author and environmental activist Wendell Berry (b. 1934) has traversed a path straddling Christianity, nature, and humanity. He has witnessed and mourned "… the triumph of industrialism and industrial values over the lives of living creatures, and over the life of the living world" in the modern era, and he has drawn out the implications of these values for the sustainability of human communities and the land that supports them (Berry, quoted in Olmstead 2018). His roots in farming inform the "agrarian values" he has posited as the necessary basis for a future for humanity (Berry 2018).

CHAPTER SEVEN

Plants as Natural Ornaments

KATE CULLITY, STEPHEN FORBES, JEN LYNCH, AND MIKE MAUNDER

INTRODUCTION

Reclaiming the traditional connotations of "ornament" is productive and relevant to twenty-first-century discussions of sustainability, stewardship, and connection to place. Ornament raises questions of how we "craft" in a contemporary context. For us as landscape architects, horticulturists, garden designers, and environmental stewards, ornament is a device that links the artifacts, practices, and materials of our work to natural systems; to the local, authentic, and communal; to new technologies and traditional cultural practices.

There is an ancient lineage of the relationship between people, ornament, and plants. Authors such as Richard Mabey (2015) have explored elements of this connection, from the Paleolithic to the present. However, this chapter explores the historical role of ornament and the key relationships it has traditionally sustained, as well as the changing connotations of ornament in the modern era, here defined as the twentieth and twenty-first centuries. Specifically, it explores the relationship between ornament and plants—how plants have traditionally inspired ornament, their status as "ornamental" in the twentieth century, and the role of technology and modern horticultural production in shaping the relationship between ornament and plants. Ultimately it also argues for the reclamation of "ornament" and its original meanings. Ornament has inspired a lineage of manifestos, and their principles are illustrated through contemporary horticultural, landscape architectural, and landscape and floral art projects.

This chapter first outlines the traditional and modern meanings of "ornament," including its historical link to plants and gardens. It then describes the status of plants as ornamental in the context of modern horticultural and gardening practices. Finally, it outlines a manifesto, a case for reclaiming ornament's traditional connotations and reimagining them in a contemporary horticultural and landscape design context.

ORNAMENT: A LINEAGE

Ornament's connotations have evolved historically, and the term's meaning shifts significantly between a traditional and modern context. "Ornament's" etymology is Latin, from *ornare*, "to adorn." The *Oxford English Dictionary*'s dual definitions—one connoting the superfluous, the other the sacred—reveal the term's conflicting associations (*OED* n.d.-a). These definitions are historically situated: while twentieth-century connotations of "ornament" as superficial inflect its contemporary usage, the traditional resonances of the term contain important potentials.[1] In its traditional meanings, ornament is defined by the key relationships it embodies—between pattern and form, nature and aesthetics, abstraction and symbolism, authenticity and adaptation, craft and connection. Notions of ornament have been fundamental to design with plants in the modern era. Influential reviews before this period, including those by Owen Jones (1856) and Adolf Loos (1910), have informed the dialogue and trajectory of ornament in the modern era.[2]

PLANTS AS ORNAMENT: MODERN ERA ORNAMENTAL HORTICULTURE

Contemporaneous with the dialogue exploring the value of ornament in architecture and landscape architecture in the early twentieth century, the cultural status of plantings as "ornament" for buildings and gardens was also shifting in response to changing patterns of production and consumption. Over the course of the modern era plant production has been transformed from family-run, and often specialist, nurseries focused on values driven by the quality and qualities of their nursery stock, into industrial plant factories producing disposable consumer artifacts, valued primarily for their visual and superficial qualities.

Before the First World War (1914–18), ornamental plant selections developed in Europe from a global palette led plant fashion and production. Japanese gardening traditions were also influential, with the Yokohama Nursery, founded in 1890, opening offices in New York (1898) and London (1907) (Yokohama Nursery 2019). At this time the ornamental horticulture industry depended on cheap labor and a complex hierarchy of practically trained professionals, and it was commonly delivered largely in defiance of the local ecology and culture. At that time, ornamental horticulture operated in a pre-plastic world and was largely fueled by coal power alongside human and animal labor.

The shattering of the Edwardian horticultural legacy after the First World War reflected profound social and technological change. The enduring appetite for novel ornamental plants could often be satisfied locally or globally at accessible prices through global shipping, and later, air freight. The ornamental horticultural business had changed from a locally resourced profession, supplemented by an international trade in specialty products and exotics, into a global business network. Historically, major cities were surrounded by a zone of agriculture and horticulture that fed those urban areas (examples include the Lee Valley for London and the Redland for Miami, Florida). These agricultural zones supplied the nursery plants, house plants, and cut flowers, with some crops brought by train from more distant areas (such as Cornish cut flowers for London markets). Horticultural materials, such as plants, containers, and composts, that today are supplied in bulk by brand or specification from largely unknown and distant sources, were produced and obtained locally (Higginbotham 1990).

Since the 1920s there has been the rise, and perhaps the beginning of a fall, in the use of plastic as a key resource and material for growing, transporting, and marketing plants. The increased use of plastic has revolutionized the production and retailing of plants and thereby driven the ability of the horticultural trade to supply a vast range of new plants. Previously, trees and shrubs would be lifted either bare-rooted or burlapped from nursery beds, a complicated process that risked damage to the plant. Containerized specimens were grown in wood, metal, or clay containers. Today, the majority of plants are sold in plastic containers. The increased demand for landscape plants in the 1940s through to the 1960s, as a result of house-building booms, promoted the use of easily transportable potted plants. Plastic containers extended the life of the plant, made shipping logistics easier, and were cheaper to transport (Conneway 2013: 9–10; Higginbotham 1990). A visit to a garden center or a "Big Box"-store garden section, both manifestations of a late-twentieth-century business model, reveals annual plants sold in Philadelphia to have been grown in Florida, orchids sold in Miami to have been grown in China, and, in Britain, ornamental plants are imported from all over Europe. This extraordinary globalization of horticulture has resulted from large commercial nursery companies able to take advantage of cheaper labor in different parts of the world and being able to move container-grown plants with relative ease and speed.[3] The development of post-Second World War highways, air cargo, and the global network of containerized shipping has facilitated these changes (see Eschen et al. 2015: 228; 2017: table 1, 3246). This is particularly true with cut flowers, where roses sold in Moscow or London are likely to have been grown in Ethiopia or Kenya, while in the US roses and tropical cut flowers such as species of *Heliconia* are imported from Colombia (Xia et al. 2006).

The centers of plant breeding have also shifted over the twentieth century. Florida grew and faded as a world center for plant breeding; California emerged as a global center, while, for tropical ornamentals, the Far East became the major hub.[4] Such developments have not only transformed some crops, such as orchids, from luxuries available only to the elite into universally available supermarket disposables. It has also created completely new ranges of ornamentals, an example being the desert rose, *Adenium obesum*, until recently cultivated in botanical garden glasshouses and held by a few passionate hobbyists.

FIGURE 7.1 Roses at Paloquemao flower market in Bogota, Colombia. Photo by Jeremy Pembrey via Alamy Stock Photo.

After investment by nurseries in Thailand, this African succulent is propagated by the tens of millions and has become a widely traded and loved ornamental shrub for the tropics and sub-tropics.

The globalization of horticulture to supply new mass markets for ornamental plants has had many unforeseen consequences. The international trade has shrunk the world and facilitated the spread of plant diseases and pests, redistributed jobs for horticulturists from the north to the south, and linked the beauty of cut flowers with social and environmental issues such as the pesticide poisoning of floriculture workers in Colombia and water-depletion and pollution in Lake Naivasha, Kenya (McQuaid 2011; Styles 2019).

At the same time as globalization, one of the most important developments in ornamental horticulture in the modern era is the regionalization of design, where gardens reflect local or regional culture and ecology. One such example is "tropicalization" of gardens, the shedding of European colonial identity in the tropical world (Maunder 2015). Author Christian Larsen aligned the rise of American interest in Latin American tropical flora with America's cultivation of political interests in Latin America during the Second World War ([1939–]1941/5) and with the directions of modern design (Larsen 2015). Hawaiian horticulturist Lorraine Kuck and landscape architect Richard Tongg used the term "tropicalia" to describe the characteristics of a true tropical garden in *The Modern Tropical Garden* (1960): "In the minds of many people there flashes, when the words tropical garden are mentioned, a vision of lush, big leafed exotics, creepers clinging to the trunks of trees, cool dim shade from a high canopy of palms, and close at hand, strange exotic flowers, probably orchids" (Kuck and Tongg 1936, 1960).

A similar process has occurred in many other regions. A glimpse of this process can be viewed by comparing the first edition of *Gardening in East Africa* (Jex-Blake 1934) with the latest edition (Cameron 2017). The earlier versions emphasized the use of European plants, including an astonishingly comprehensive section on roses for East Africa (120 cultivars described) such that the impression is of horticulturists bringing the Edwardian, English, Romantic style of Gertrude Jekyll's garden designs into the tropics. Today the verandah as focal point has been replaced by chapters on water management, soils and composting, pollinators, and, importantly, expanded promotion of native plants and a recognition of the local edible plant heritage.

The reception of locally native flora began to shift substantially in the modern era. While precursors for ecological landscape styles may be found, for example, in the eighteenth century's English Landscape Movement and in William Robinson's *The Wild Garden* first published in 1870 (see Forbes, Cooper and Kendle 1997), reconciliation with localness as a criterion for plants as natural ornaments has required a substantial change in perspectives on plant beauty.

In the United States, Wilhelm Miller (1869–1938) chronicled the formation of the influential Chicago Prairie School of landscape architecture (and architecture) in *The Prairie Spirit of Landscape Gardening*. He attributed the movement to Ossian Cole Simonds (1855–1931), because of the latter's early use of native plants and credited the Prairie School's development to Jens Jensen (1860–1951)—and, to a lesser extent, to American architect and landscape architect Walter Burley Griffin (1876–1937) (Grese 2000; Vernon 2002); clear links to Germany's Jugendstil gardens have also been noted (Vernon 1995). Griffin's 1911 commission to design Canberra, Australia's national capital, reinforced the growing native plant movement in Australia (Vernon 1997). There, garden designer Edna Walling (1895–1973) is viewed as one of the pioneers of native plant gardening (Dyson 2016: 54; Watts 1981). Walling's recognition of the value of

Australian natives came in the 1930s, her epiphany apparent in a 1938 magazine article: "this will be a lovely native garden, and once and for all let us hope it will help to lay [to rest] the ghost that frightens people and makes them look so depressed when one mentions native plants. You really cannot wonder" (Holmes 2019: 126). Ironically, the modern era, which has seen an accelerating destruction of native flora globally, has also seen the adoption of a stronger local conservation ethic in landscape design.

The trajectory of plant selection in vernacular or suburban gardens during the modern era is, overall, poorly documented when compared with celebrity gardens. In a chapter entitled "The Stylish House: Fashions as an Ecological Factor," which appeared in *Reading the Landscape*—the landmark work undertaken while she was at the Morton Arboretum in Illinois—American naturalist May Theilgaard Watts (1893–1975) traced the course of one such garden's plantings from 1856 to about 1956, through the house's five owners during that time. She viewed the ecological forces shifting the garden's composition as "style," reflecting the social and cultural currents characterizing the period. She wryly noted the survival of five plant species through these changes, "Only the white pine tree [*Pinus strobus*], the bleeding heart [*Dicentra formosa*], the apple tree, the fern-leaved peony [*Paeonia tenuifolia*] and the Christmas cactus [*Schlumbergera* x *buckleyi*] have seen all the ways in which the house has been stylish" (Watts 1957: 219). Sociologists Eammon Slater and Michel Peillon (2009: 101) continued Watts' exploration in fifty suburban front yards in Dublin, Ireland, where they (also) found "neighbourly 'works of art'," "determined by diverse social forms of visuality ... metabolized with the natural processes of the garden plant ecosystem" (101).

Plant societies established for particular ornamental plant groups have risen and declined since the late nineteenth century and into the modern era. Examples of specialist plant groups and societies include the Royal National Rose Society in Britain (est. 1876), the American Fern Society (est. 1893), the North of England Orchid Society (est. 1897 as the Manchester Orchid Society), the British Daffodil Society (est. 1898), the Rhododendron Society in Britain (est. 1916 and later, in 1945, constituted as a Specialist Group within the Royal Horticultural Society), the Cactus and Succulent Society of Australia (est. 1927), the Cactus and Succulent Society of America (est. 1929), and the American Rhododendron Society (est. 1945). The fortunes of such societies have waxed and waned, reflecting both fashion and the passion and capabilities of committee members. While the membership of general gardening societies remains solid, membership of specialist societies has been more volatile. The generalist Royal Horticultural Society (2019: 11) reached 502,666 members in 2019, while the Royal National Rose Society membership declined from a hundred thousand in the 1970s to around one thousand by 2017—indeed, the prestigious Society became insolvent (Appleby 2017).

The question of what makes a plant an ornamental one, then, is essentially a cultural one. The selection of a plant as worthwhile to ornament a garden or a room illustrates a range of considerations including availability, durability, and fashion. However, in a globalized world focused on markets, the success of an ornamental plant is largely realized through commercial success. British garden designer and author Noel Kingsbury suggested that ornamental plant selection, breeding, and development essentially constitute "an endless search for novelty," and novelty is certainly a key driver in marketing and sales for ornamental plants (2009: 329). However, while the market illustrates fashion trends, there is little consensus beyond commercial success for the ornamental values of a plant. Cultural factors, including economic, environmental, and social factors, profoundly influence perceptions of ornament in plants.

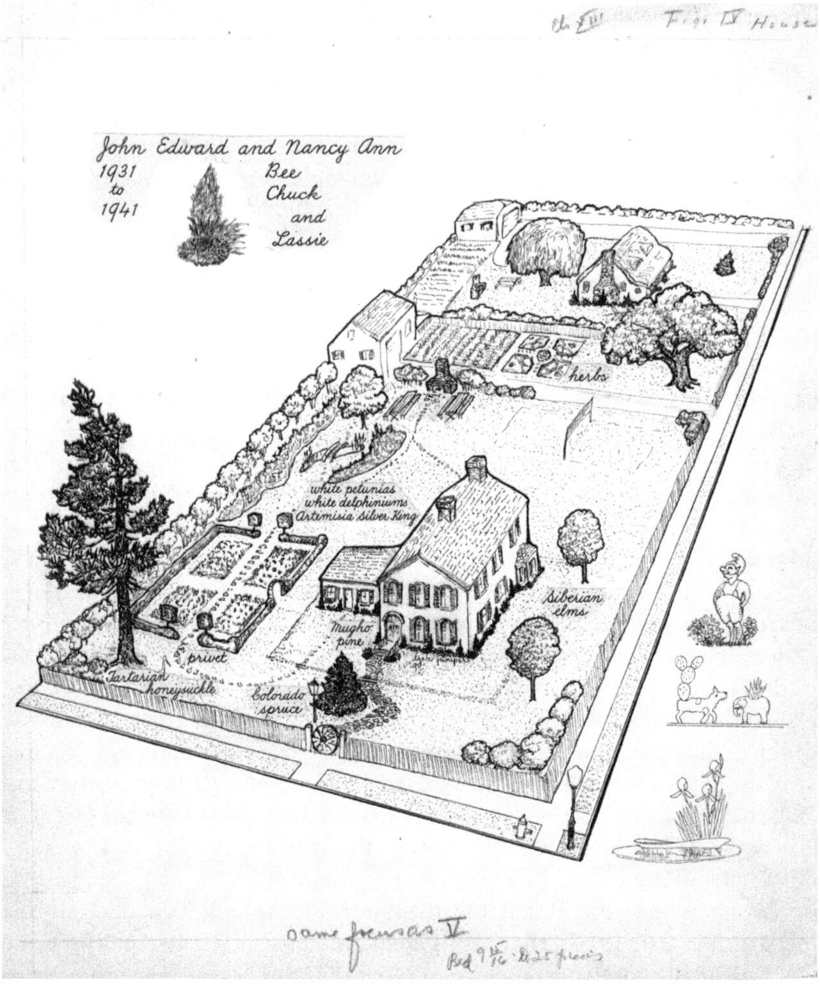

FIGURE 7.2 The Stylish House 1931–41. Illustration from May Theilgaard Watts, *Reading the Landscape: An Adventure in Ecology*. © 1957, 1985, reprinted with the permission of Bridget Watts.

Artist, writer, and iris breeder George Gessert (b. 1944) has explored the goals of plant breeders and flower show judges (Gesset 2010: 53). He observed that "No plant breeder has been trained by historians of ornamental plant breeding or has suffered the judgment of a plant-breeding critic, because those specialties do not exist." He recalled Shakespeare's *A Winter's Tale* in which Perdita rejects streaked carnations as "nature's bastards" while Polixenes defends plant breeding as, "… this is an art / Which does mend Nature, change it rather, but / The art itself is Nature" (Act 4 Scene 4). This observation addresses an issue that has continued to occupy philosophers since the Classical era—the relationship between art and nature. However, Gessert (1996: 297) concluded that while "Polixenes rejected the notion that ornamental plant breeding betrays nature, … he left unanswered a more difficult question: are streaked carnations good art? Three hundred and seventy years have passed, and we still cannot answer that question."

ORNAMENTS FOR THE ANTHROPOCENE

The modern era represents the "Great Acceleration"—a period of human-driven, massive, and catastrophic global change (McNeill and Engelke 2016). Post-Rachel Carson's warning, and post-recognition of the Anthropocene, the era's gardeners are the first gardeners to translate those legitimate anxieties into the design of gardens. Contemporary gardens and the choice of design and ornament operate in this context. Nursery plant production is commonly dependent on either fossil fuel for heating nurseries or for global transport (or both), as well as for the production of plastic containers. Gardeners consciously choose whether or not to douse gardens with petrochemicals in the form of herbicides and pesticides. Indeed, gardeners now decide whether to create a garden that is an ecological cost or one that delivers ecological regeneration.

The ornaments of a contemporary garden comprise not only the plants, the structures, and the decorative ornaments (whether a mass-produced garden gnome or the work of a renowned Modernist sculptor like Barbara Hepworth (1903–75), but also the biodiversity associated with a garden—butterflies, bees, birds, and other animals. This is a remarkable transformation in the value of a garden and is a recognition that gardens are part of global ecology rather than merely private enclaves of ornamentation. Increasingly, the traditional ornamental value of a garden is now supplemented by an appreciation of the garden's role in providing habitat for birds, invertebrates, and even mammals, while contributing to improving surface and subsurface water management, and carbon sequestration.

However, these concerns became mainstream only in the latter half of the twentieth century. Beforehand, scientific and technical solutions in ornamental horticulture were largely viewed as progressive, with specialization of endeavor an essential element of progress. Holistic views of the environment were rare, and atomization seen as the basis of progress. Horticultural writer Walter Wright (1913: 51) observed:

> It is an age of specialization. The times demand specialists. The development of science has been so great that it is impossible for any man to become a master of several great subjects. He must either be a specialist or a dilettante. Gardening, which is held so lightly by many unreflecting people, and looked upon as more or less of an amusement is a great subject.

One of the consequences of this perspective may be the ascent of the lawn in this period (see below).

Public dialogue in contemporary landscape and garden design largely focuses on elite gardens that express professional design, "plantsmanship," and "prestige of place" aligned with significant investments of time and money. Such gardens fuel a rich garden literature for academics, professionals, dilettantes, and amateurs, while visits to celebrity gardens have expanded to become a global garden tourism industry. Symposia, books, and television series reach into this space with titles such as Dumbarton Oaks' *Contemporary Garden Aesthetics, Creations and Interpretations* (Conan 2007), John Patrick's *Contemporary Australian Garden Design: Secrets of Leading Garden Designers Revealed* (Patrick and Wade 2012), Monty Don's television series *Around the World in 80 Gardens* (BBC Two, UK, 2008), and Rae Spencer-Jones' *1001 Gardens to See Before You Die* (2007).

Garden historian (and former director of Garden and Landscape Studies at Dumbarton Oaks) Michel Conan reflected on John Dixon Hunt's introduction to the Dumbarton Oaks' 1993 colloquium on the vernacular garden:

> the word "vernacular" ... calling attention to works of architecture without architects, designs without designers, and cultural differences between custom-bound people and their sophisticated rulers. The simple question was: What can architects learn from folk design that designers ignore? The same perspective could be adopted with regard to gardens. A large crop of new ideas could be expected since gardens studies had been mostly given to the study of design in *self-consciously designed* gardens.
>
> (1999: 181–2)

The vernacular gardens of today's suburbs in some measure reflect the aesthetic established by professional designers and plantspeople. However, the gardens of the homeless in New York, documented by landscape designer Diana Balmori and photographer Margaret Morton (1993), powerfully illustrate the diversity of economic, social, and cultural issues influencing garden design. Between "self-consciously designed gardens" and vernacular gardens there are gardens, as there are individuals, of every stripe.

LAWNS: GRASS AS ORNAMENT

Perhaps the most remarkable shared element in celebrity and vernacular gardens is the ubiquity of the lawn. Indeed, lawn grass should be viewed as the plant ornament par excellence defining the modern era. Already in the nineteenth century, the enviable lawns of Oxford and Cambridge college gardens, and of wealthy estates in Europe as well as the United States, demonstrated what was possible (Lamson-Scribner 1897: 367):

> Among the finest lawns in this country are some of those at Newport, R. I. The best of these are composed almost entirely of either creeping bent [*Agrostis stolonifera*] or Rhode Island bent [*A. capillaris*]. There is ample moisture, and no labor is spared in keeping the surface in perfect order by frequent cuttings and rolling and by removal of all weeds. Nothing can be more beautiful than these broad, unbroken stretches of velvet-like sward.

The twentieth century brought such beauty within reach of the middle and working classes. Indeed, the democratization of the lawn has been a remarkable social, cultural, and commercial achievement. Lawnscapes illustrate the highest ideal for plant beauty during much of this period. As Frank Lamson-Scribner (1851–1938), United States Department of Agriculture (USDA) agrostologist declared: "Nothing can be more beautiful."

The sheer scale of the endeavor needed for lawn establishment and maintenance dwarfs all other ornamental plant culture. It has been calculated that over one hundred thousand square miles (163,800 km^2, ± 35,850 km^2) of land in the continental United States are occupied by turf grasses, an area three times larger than that of any irrigated crop (Milesi et al. 2005: 431–2). This includes residential, commercial, and institutional lawns, parks, golf-courses, and playing fields and covers 1.9 percent of the total area of the continental United States. This compares with 3.5–4.9 percent of the total surface estimated to be devoted to urban development across the country in 2005. Although other studies have been undertaken in specific urban areas (Hedblom et al. 2017: 1110), this one appears to be unique in estimating the total cover of lawns in an entire country. The lawn, then, is fetishized beyond any other ornamental plants in gardens.

The expansion of lawn into vernacular gardens required technological solutions for applying water and mowing the sward. These included a reliable water source, reticulated piping, sprinklers, and rubber hoses, and, of course, a lawn mower. These technologies had evolved to become accessible to suburban home owners by the end of the nineteenth century. The supply of lawn seeds suited to particular edaphic (that is, relating to the water content, acidity, aeration, and the availability of nutrients in soil) and climatic situations was also a critical factor for the expansion of lawns. The American enthusiasm for golf saw the establishment of the first golf course in the United States in 1888, with a thousand established by 1902. By 1915 the United States Golf Association and the United States Department of Agriculture had established a collaboration on turf culture that facilitated effective turf establishment and management (Jenkins 1994). The USDA's standard for turf was an influential ideal presented in its 1897 Yearbook: "A perfect lawn consists of the growth of a single variety of grass with a smooth, even surface, uniform color, and an elastic turf which has become, through constant care, so fine and so close. texture as to exclude weeds which, appearing, should be at once removed" (Lamson-Scribner 1897: 367).

While the rapid trajectory of technology for lawn cultivation is clear, the desire for home owners to cultivate lawn is rather less so. In her comprehensive history of the American lawn, Virginia Jenkins (1994) explored the invasion of lawns in the United States, concluding that suburban lawnscapes only became ubiquitous after the 1930s and, in particular, after the Second World War. The eight-hour day, five-day work week and rising affluence evidenced in increasing car and home ownership provided a "seedbed" for suburban lawns. However, the transfer of the ideals of lawn culture between upper-class and middle- and working-class gardens is a more complex issue (Gaynor 2020: 70–2; Groth 1994: 298). Lawns were not only an aesthetic ideal but also a moral one. Lamson-Scribner, cited above as praising the lawn's aesthetics, had more to say in praise of the lawn's virtues:

> Nothing is more beautiful than a well-kept lawn, whether it be of large or small extent. Even the small plots fronting city dwellings are points of attraction when covered with a soft, even turf. Lawns are the most fascinating and delightful features in landscape gardening, and there is nothing which more strongly bespeaks the character of the owner than the treatment and adornment of the lawns upon his place.
>
> (Lamson-Scribner 1897: 355)

The growth of a gardening culture in the United States is illustrated by the success of *Better Homes and Gardens* magazine, which reached a circulation of over a million in 1928, within four years of its launch (Ainsworth 1930: 262). Garden clubs also played a significant role in promoting lawns. By 1936 the Junior Garden Clubs of America, established by *Better Homes and Gardens* in 1929, grew to include 59,876 clubs with 370,000 active members. Garden clubs had already played a significant role with a heritage including nineteenth-century town beautification projects, yard improvement campaigns in company towns, and industrial-welfare work programs, and small-lot gardening contests in the 1920s and 1930s (Jenkins 1994). The lawn was a new frontier.

Today the lawn is under assault. The lawn, of all garden features, best illustrates the profound changes in horticulture during the last century. A traditional component of the European garden and landscape, it became a key twentieth-century garden feature almost worldwide. The uniform green sward, free of weeds, was indicative of an ordered and decent household, a moral measure. The lawn is now viewed with moral suspicion. The lawn is increasingly seen as an ecological liability, a sterile monoculture that consumes scarce water and needs regular dosing with polluting pesticides and fertilizers. There is a desire for a more biologically valuable sward, in part a nostalgia for species-rich meadows

FIGURE 7.3 Eastern Airline pilot enjoying mowing his lawn with his children on his day off from flying, 1949. Photo by Ralph Morse/The LIFE Picture Collection via Getty Images.

and grasslands, as well as a recognition that such habitats are vital for many animal and plant species (Smith et al. 2015). In some parts of the world (such as southwestern US, the Middle East, and Australia) the lawnscape is increasingly described as an alien, thirsty, and demanding landscape feature unsuited to arid or Mediterranean environments. Author Michael Pollan aptly captured concerns over global lawn culture in an essay entitled "Why Mow?":

> Lawns, I am convinced, are a symptom of, and a metaphor for, our skewed relationship to the land. They teach us that, with the help of petrochemicals and technology, we can bend nature to our will. Lawns stoke our hubris with regard to the land.
>
> (1991: 64)

Despite these currents, some have suggested mitigating value for lawns as a carbon sink, especially if clippings are retained on the sward (Milesi et al. 2005: 64), and others have identified the specific goals of lawn management associated with community, family, and a "green" environment, viewing these as a triumvirate of "alienated desires":

> the lawn-chemical industry ... implemented new and innovative styles of marketing that help to produce an association of community, family and environmental health

> with intensive turf-grass aesthetics and reflect an increasing local demand by consumers for authentic experiences of community, family, and connection to the nonhuman biological world through meaningful work.
>
> (Robbins and Sharp 2003: 425, 443)

While alternatives to a monocultural lawn sward such as Corsican mint (*Mentha requienii*), creeping thyme (*Thymus serpyllum*), clover (e.g. white clover, *Trifolium repens*), evergreen moss (e.g. Irish moss, *Chondrus crispus*), ornamental grasses, and native perennial beds have been advocated with various claims for reduced maintenance and a reduced environmental footprint, their adoption has, to date, had a limited impact on public and private greenspace (see Jenkins 1994: 186).

EXPLOSIVE CREATIVITY

The garden has long illustrated the link between plants used or viewed as ornaments and design. Commencement of the period under discussion in this chapter overlapped with the Arts and Craft Movement, an international trend in the decorative and fine arts that arose in Britain, spreading from there to Europe and America, where it flourished from the last decades of the nineteenth century through the first two decades of the twentieth. This movement was a celebration of hand-craftmanship in a reaction against a perceived artificiality and lack of sensitivity for and understanding of materials resulting from factory versus manual production; ornament, it was argued, should not exist for its own sake but complement, "be secondary to" and "fit" that which was being ornamented (Pevsner 2005). In terms of garden design, the Arts and Crafts Movement saw a close collaboration across the landscape, architecture, and interior design alongside traditions of expansive ornamental horticulture, the integration of a home and an effusively planted garden, extensive use of herbaceous borders, and the creation of "rooms" or series of discreet garden spaces within the larger whole. In Britain, architect Edward Lutyens (1869–1944) and garden designer Gertrude Jekyll (1843–1932) are celebrated as exponents of the Arts and Crafts style (Tankard 2018). Indeed, Jekyll's partnership with Lutyens, for whose projects she created numerous landscapes and who designed her home Munstead Wood, near Godalming in Surrey, proved to be the most influential and historical partnerships of the Arts and Crafts Movement. Jekyll's attention to textures and to color in the garden, using of flowers grouped by "temperature" of hue (groupings of flowers with blossoms in warm colors and others with blossoms in colors deemed cool), created the exquisite painterly effects for which her designs are known.

Among the numerous gardens spawned by the Arts and Crafts Movement, and perhaps the most influential English garden of all, is Sissinghurst in Kent, which was established by Vita Sackville-West (1892–1962) and Harold Nicholson (1886–1968) after 1930 and bequeathed to the National Trust following Sackville-West's death. The garden consists of a series of distinct garden rooms, each of which offers a unique, intimate space linked via openings in hedges or paths to another room: the Top Courtyard; the White Garden, the Yew Walk, the Lime Walk, the Nuttery, Delos, the Orchard, the Moat Walk, the Cottage Garden, and the Rose Garden number among the discreet gardens. The notion of the wild garden, including wildflower gardens, promoted by William Robinson (1838–1935) is also a significant influence during this period and continues to resonate with ecological approaches to contemporary gardens (see Robinson and Darke 2009). In the United States, Beatrix Farrand (1872–1959), the only founding woman member of the American

Society of Landscape Architects in 1899, drew on the work of Jekyll and Robinson, as well as the classical designs of Andre Le Notre (1613–1700), landscape architect and the principal gardener of King Louis XIV of France, in her work as a landscape gardener, her preferred title. Her design for Dumbarton Oaks in Washington, DC, one of over two hundred designs she completed, is recognized for its terraced garden rooms including a rose garden, wisteria arbors, herbaceous flower beds, and orchards intricately woven into an increasingly natural landscape of meadows, woods, and water (Way 2020).

While design is a critical element of gardens, plant collectors' gardens are perhaps better characterized by their passion for the adventure of rarity, novelty, and botanical or horticultural distinction in their plantings. Here the plants themselves are the key to ornament, although garden design (usually) remains highly valued. A few examples here are illustrative. Bodnant in North Wales, established in 1874 but developed by the same family over five generations, continues to be held in high esteem, reflecting a constancy of purpose in the garden and the owners' impeccable credentials in plant collecting, plant collections, and standing with the Royal Horticultural Society—the garden is now managed in partnership with the National Trust (England, Wales and Northern Ireland). In the French Provence-Alpes-Côte d'Azur, two remarkable plant collector's gardens are especially noteworthy. Les Cèdres was established by Alexandre Marnier Lapostolle (1857–1930) in 1924, and the plant collections developed by his son and plant explorer Julien (1902–76). The botanically and horticulturally significant cactus, succulent, and bromeliad collections are globally renowned (Glass and Foster 1975). Les Cèdres has recently been sold, and the future of the plant collections is unclear. Serre de la Madone was developed by Lawrence Johnston (1871–1958) from

FIGURE 7.4 View of Sissinghurst Garden established from 1930 and illustrating "garden rooms." Photo by Jeff Overs/BBC News and Current Affairs via Getty Images.

1921 to 1939 and acquired to secure conservation of the gardens and plant collections by the Conservatoire du littoral in 1999. (Lawrence had previously established Hidcote Manor Garden in Gloucestershire—the first garden accepted by the National Trust.) In the United States, Longwood Gardens in Pennsylvania was developed by Pierre S. du Pont (1870–1954). He purchased the site in 1906 and actively led the garden's design, technological innovation, and plant exploration prior to establishing the Longwood Foundation in 1946 to secure the garden's future. The garden remains renowned in the arenas established by du Pont and has also established renown in horticultural education and ornamental plant research.

A shift in garden design accompanied the advent of Modernism, or the International Style, in architecture. Architectural Modernism was identified and formalized as a distinct new style or movement by American architect Philip Johnson (1906–2005) via the 1932 "Modern Architecture: International Exhibition" that he curated at the Museum of Modern Art, New York. "Characterized by an analytical approach to the function of buildings, a strictly rational use of (often new) materials, structural innovation and the elimination of ornament" (Royal Institute of British Architects 2019), Modernism appeared to privilege a "form-follows-function," minimalist, technologically driven architecture—which, importantly, eschewed "useless" ornament—over the natural environment. While formerly prevailing tastes for painterly, impressionistic floral displays and classicizing, symmetrical formalism had now been eclipsed, it would be wrong to conclude that gardens and landscape played no part—wrong, too, to believe that gardens could no longer be ornamental. Rather, architectural Modernism, which philosophically and physically dictated the necessity of establishing a dialogue or responsion between architecture and nature, issued in a trend in garden design marked by abstraction, asymmetrical formality, a focus on light and space, and the juxtaposition of contrasting forms, textures, and colors, all entirely in harmony with the Modernist architectural aesthetic and its underlying philosophy (Betsky 2002; McHarg 1969). Modernist gardens' numerous noteworthy pioneering designers include Lawrence Halprin (1916–2009), Roberto Burle Marx (1909–94), Dan Kiley (1912–2004), Garrett Eckbo (1910–2000), James C. Rose (1913–91), Thomas Church (1902–78), Emmet Wemple (1954–95), and Laurie Olin (b. 1938), Modernist gardens and their plantings were and are still ornamental, just radically less "ornate" than those of their classicizing predecessors, which looked back to the Renaissance, and their design principles persist in myriad variations through the present day, meshing with and driven by new concerns about sustainability, the planting of natives, and ecosystem management. As a result, contemporary gardens and their plantings defy strict stylistic definition (Hunt 2015: 18–23), ranging as they do from Spanish gardener Fernando Caruncho's lyrical, abstracted homages to Islam, Zen, and European classicism to New York City's High Line, an elevated, linear park created on a former railroad spur on the west side of Manhattan. The latter project constitutes a synthesis of landscape architecture, urban design, and ecology achieved through the collaboration of James Corner Field Operations, design studio Diller Scofidio + Renfro, and garden designer Piet Oudolf (see also below).

The Modernist-driven abstraction of garden design brought with it a union of gardening and contemporary art to create extraordinary new spectacles and a new sense of what comprises ornamentation. That most ossified of plant displays, flower arrangements, have gone from the marquee of the flower show to join the world of contemporary art, whether through the dried-flower arrangements of Ignacio Canales Aracil (b. 1984), the astonishing floral art of Azuma Makoto (b. 1976), the application of preserved flowers by

FIGURE 7.5 Thomas Church and the Modernist Garden; the Donnell Garden, Sonoma, California (completed 1948): minimalism, sculpture, and abstraction. Photo by Peter Anderson. Courtesy of Alamy Stock Photo.

FIGURE 7.6 Brazilian artist and landscape architect Roberto Burle Marx, painting in the garden of his Brazilian home in 1943. Photo by Thomas D. Mcavoy/The LIFE Picture Collection via Getty Images.

Rebecca Louise Law (b. 1980), the public spectacle of "Puppy," a monumental "living" sculpture by Jeff Koons (b. 1955), or the giant flowers, fruits, and leaves sculpted by artists such as Marc Quinn (b. 1964) and Yayoi Kusama (b. 1929) that grace sculpture gardens (Dupon 2014; Guggenheim 2016; Phaidon 2019). Several artists are also exploring the hybrid zones between scientific horticulture, ornament, delight, and contemporary art. One is the French botanist, Patrick Blanc (b. 1953), whose new approach to vertical gardens has transformed many civic spaces (Blanc 2008, 2016). Another is Mark Dion (b. 1961), whose Neukom Vivarium in Seattle is a hybrid between a gallery, ecological research chamber, and woodland garden.

The great flower spectacles of Keukenhof, the Netherlands, and Brussels continue to delight and have perhaps influenced an increasing hybridization between garden design and contemporary art as exemplified by the festivals at Chaumont sur Loire in France and Metis in Canada (Hunt 2015: 63–76). These are the fermentation chambers testing the new ideas of ornament and delight. Garden shows such as the Royal Horticultural Society's Chelsea Flower Show in London, the Philadelphia Flower Show, and the Singapore Garden Festival have expanded to become global attractions.

FIGURE 7.7 *Still Life of Flowers in a Stoneware Vase* by Jan Brueghel the Elder, 1568–1625 in front of an installation of five thousand dried flowers by British artist Rebecca Louise Law (b. 1980) at Sotheby's auction 2016. Ben A. Pruchnie/Stringer via Getty Images.

FIGURE 7.8 Auricula display in the Great Pavilion at the 2014 Chelsea Flower Show. Photo by Dan Kitwood/Getty Images.

PLANTS AS ORNAMENTS IN CONTEMPORARY GARDENS AND DESIGNED LANDSCAPES: TYPOLOGIES AND EXEMPLARS

Patterns and form—Ornament reflects systems—biophysical, social, cultural—as form and pattern. Working with form and pattern, humans can engage with and negotiate multiple systems and scales

Australia's National Arboretum in the national capital Canberra was designed by landscape architects TCL (Taylor Cullity Lethlean) and opened in 2013. The Arboretum incorporates patterns and forms across many scales to tell the stories of forest origins and engage the public with themes of environmental sustainability. The Arboretum has a hundred stands of the world's most endangered tree species, providing both an innovative seed source for the future and an important educational and recreational landscape. Pattern and form play a key role in the arboretum's design, which is organized by an overall grid with patterned planting unique to each stand. The inspiration for the design came from two existing forests on the 260-hectare site, one of cork oaks (*Quercus suber*) and the other Himalayan cedar (*Cedrus deodara*), where one can experience the quality of being enveloped within a singular stand of trees. The Arboretum's hundred "forests," each of 5–7 acres and varying from three hundred to two thousand trees per stand, are arranged within a grid across the undulating topography, linking the site to the geometry of the Walter Burley Griffin masterplan for Canberra. The layout of each forest is created specifically to reinforce a particular experiential quality, physical feature, or cultural story that is unique to each species. For example the stand of *Koelruteria paniculata* (golden rain tree) planting within a hexagonal grid referencing the beehive is a nod to German botanist Joseph Kölreuter (1733–1806), the first scientist to publish on the role of insects, particularly bees, in the pollination of trees.

The intent in these patterns is to convey cultural as well as environmental stories. While the patterns of the forest reflect the culture of the plants, as monocultural plantations, they also link to the lineage of forestry and scientific experiment, to expand the idea of the arboretum and its role in a twenty-first-century context. The arboretum will take a hundred years to mature and, during this process, it will function as an active testing ground, focused on conservation and resilience in the face of environmental change. All trees selected for the Arboretum meet the "threatened" criteria from the International Union for the Conservation of Nature (IUCN) classification (IUCN 2020). The stands also include endangered trees from global sources with national cultural significance or a unique character and presence. The design and implementation of the Arboretum is therefore a strategy, a program, and an ongoing event, rather than a design based chiefly on aesthetics. It grows out of contemporary issues of sustainability, biodiversity, and public environmental concern. The project has established connections across the world through the botanical value of its collection to institutions interested in preserving the species from extinction (Bourke 2013; Richardson et al. 2014).

At a smaller scale, plant installations have the ability to transform, at least temporarily, the surface and space of a site, as the pattern and form of plants and their arrangement interacts with their context. The Garden that Climbs the Stairs in Bilbao, Spain, designed by landscape architect Diana Balmori (1932–2016) (Balmori Associates 2009), and Grasslands by artist Linda Tegg (b. 1979) at the Victorian State Library, Melbourne, Australia (Lelouche 2014; Tegg 2014) are both garden installations that involve plants tumbling down a staircase. Part of an international competition of urban gardens, The Garden that Climbs the Stairs transformed a forbidding civic staircase into a public garden. While the competition brief requested a 100-square-meter (1,076 sq. ft.) garden, this area is distributed over a serpentine form, affecting the scale and pattern of the garden's context. The form of the garden and the pattern of its plantings—a slice of colorful perennials—in the designer's words, "performed a narrative of landscape which transformed the way this public corridor was perceived by users." In the case of Grasslands, Tegg's patterned plantings of pre-settlement Australian indigenous grasses reflect conditions that would have once characterized the library site, draped across the steps of one of the city's most iconic and symbolic civic buildings and linking it to the lost grasslands, managed by Aboriginal people for tens of thousands of years. This insertion of plants reconfigures the stairs from a place of transience to an inviting environment to linger within. The grass species and the pattern of their planting, informed by the archival content of the library and early colonial descriptions of the place, is also a political gesture—the garden re-colonizes the space, inviting memories of what came before.

Nature and aesthetics—Ornament links natural systems, both micro and macro, with cultural systems. Ornament (designs' patterns and details) can shape meaningful aesthetic relationships between people and place

The work of Chinese landscape architects Turenscape creates tensions between cultural and environmental systems to reveal relationships, historical and contemporary, between people and place. Yanweizhou Park, a riverfront project by the practice in Jinhua City, China, links the city's ecological systems—the riparian landscapes of the Wuyi and Yiwu River—with its cultural systems, the garden motifs and agricultural patterns of traditional Chinese landscape. The project's water-resistant terrain and plantings are designed to withstand monsoon floods; however, rather than high concrete abatement walls, the riparian edge is engineered as a series of terraced garden beds. These terraced gardens have native water plants that filter and cleanse urban storm water. The park's floodable paths meander between the terraces, allowing city residents the experience of being immersed in lush plantings of patterned tall grasses. The curvilinear forms of the park's paths and bridges reflect the local dragon dancing tradition of the city (Turenscape 2014).

At the Chinese rice farm at Chengtoushan Archaeological Park, another project by Turenscape, the relationship between ornamental plants and cultural-environmental systems is tactile and programmatic. The "outdoor museum" links visitors to the 6,500-year-old agricultural practices of the area, which were transformed into a display of historical gardens earlier in the twentieth century. The plants themselves, reflecting the traditional rice cultivars and subtle agricultural field patterns within the landscape, are showcased, as are the acts of cultivation and harvesting that shape the landscape, linking the site and visitors to its heritage (Padua 2010).

FIGURE 7.9 Yanweizhou Park designed by Turenscape. The project's water-resistant terrain and plantings are designed to withstand monsoon floods with the riparian edge engineered as a series of terraced garden beds only inaccessible during flooding. Photo courtesy of Kongjian Yu, Turenscape.

Abstraction and symbolism—Ornament abstracts natural systems in ways that make them legible and meaningful. Ornament allows humans to reveal, register, and form new connections with these systems

The Australian Garden at the Royal Botanic Gardens Cranbourne, Victoria, designed by Australian landscape architects TCL with horticulturist and planting designer Paul Thompson (b. 1945) opened fully in 2012. The design makes use of repetition, patterning, and the abstraction in the Australian landscape to explore and express the evolving relationship between Australians and the country's landscape and flora. The designers reimagined the Australian botanical garden and its cultural role, particularly as earlier botanical gardens in Australia had been mainly English in their origins and global in their exotic flora. As the Australian academic and polymath, George Seddon (1927–2007) wrote, "We [Australians] are still learning to see our own land and to forgive it for not being England" (Seddon 1970: 3).

As this Southern Hemisphere continent is molded by drought and floods, the overriding structure of the Australian Garden at Cranbourne explores the narrative of water (including the lack of water) and how water fundamentally sculpts and constructs the Australian landscape. This story is told through the abstraction of landscape elements and phenomena. For example, the patterning of the Sand Garden, with its staccato circles of grey-green plantings of the salt bush, *Rhagodia spinescens*, seeks to distil and intensify the sublime experience of the Australian desert, an environment that experiences minimal and intermittent rain. A sense of reference to the fragility of the desert is expressed by not allowing the visitor to enter the garden. As in a Japanese sand garden, the viewer keeps a respectful distance, and in that way does not interfere with the garden's powerful elemental experience. The garden's ornamental elements and abstraction of the continent's landscape challenge the cultural understandings of its beauty and redefine the role of the botanical garden in Australian society (Forbes 2012; Hunt 2015: 180–2).

Authenticity and adaptation—Ornament reflects tradition and innovation. It builds continuities of knowledge and practice while allowing for iteration. Working with a changeful, place-specific ecological and social medium, ornament encourages continuity and change, invention and adaptation, macro and micro

The landscape of Uluru-Kata Tjuta Aboriginal Cultural Centre in Australia's Northern Territory is another example of an iterative design process, involving adaptation and invention in response to place-specific ecological and cultural conditions. In 1990, landscape architects TCL were engaged to work on the cultural center with architect Gregory Burgess (b. 1945). The brief called for an expression of the rich living history of the site's traditional Aboriginal owners—to understand their deep and continuous commitment and care for, and connection to, their land, to one another and to their law, *Tjukurpa*, passed down as spirit stories (Underwood 1996). After much consultation, collaboration, and analysis, a preferred site was identified about 1 km (0.6 m) from the monolithic rock, Uluru. A sinuous snake-like building would enfold a large significant, though dead, desert oak (*Allocasuarina decaisneana*), as it was viewed by Anangu, the site's Aboriginal owners, as a strong pivotal point, one which expressed rejuvenation through the proliferation of young desert oaks sprouting under the parent tree's stark dead branches.

In the design of the landscape, a minimalist approach to intervention was adopted—an approach that arose out of consultation with the Anangu besides meeting the ambitions

FIGURE 7.10 Sand garden in the Australian Garden, Cranbourne, designed by TCL landscape architects with Paul Thompson included in Stage 1, 2006. Photo courtesy of TCL/Peter Hyatt.

of the brief. It was collectively decided that no trees would be removed, minimal removal of vegetation would occur, and no further vegetation would be planted. During the early consultative process, a number of park rangers wanted to include a range of the more spectacular plants from elsewhere in the park. This idea was eschewed by both the Anangu elders and the design team in favor of fully appreciating what was growing on this particular site, to respond to the subtle beauty of the immediate surrounding landscape.

There is a delicate microclimate created by the Rock, with the landscape being marked by the intermittent presence of water as it is shed from Uluru and disburses. As even micro-changes in grade affect the surrounding fragile landscape of shifting sand dunes it was decided there were to be no changes in levels and no construction of swales or curbs. Long, winding paths were constructed from compacted orange sand, and edges were marked with branches and brush collected from the park. The aim was to emphasize the experience of walking in the desert, of noticing the plants, the space between them and the animal tracks, so that people would slow down in preparation for what they would experience in the Aboriginal Cultural Centre.

Craft and connection—Ornament is practiced. It reflects engagement with and connection between the environment, the body, and a social and cultural context. The construct of craft encourages us to tone and master our skills, as well as adapt to new environmental and social conditions

Piet Oudolf (b. 1944) is a masterful Dutch nurseryman and garden designer who is world renowned for developing a naturalistic form of planting design known as the "New Perennial Movement." His finely honed skills, developed over more than thirty years, move beyond the use of flower or color to focus on longevity of form, seasonal change and structural characteristics, and, in his more recent projects, his gardens are also informed by ecology and place. This shift is particularly evident in his planting design for the High Line in the heart of New York City, a community-led project on a former derelict and elevated carriageway that has been rejuvenated into a 1.45-mile linear park, providing a natural retreat in one of the world's most populated cities. Here the planting palette was inspired by the plant species that reclaimed the site after the trains ceased operating in 1980, with the planting selection being predominately native and drought resistant. There are nuanced selections of these depending on the various microclimates created by elements such as strong winds off the Hudson River, or shade from adjacent buildings. In all there are twelve different and distinctive gardens. Oudolf has also influenced the ongoing horticultural management, encouraging the retention of dead plant stalks and seed heads to be retained through winter to allow appreciation of their acquired beauty, as well as providing habitat and food for wildlife (Lindner and Rosa 2017; Oudolf and Darke 2017).[5]

A kindred form of plant-based craft is practiced by British environmental artist Andy Goldsworthy (b. 1956), who attempts to make art every day, as an act of being connected to place, change, and time. His crafted works often employ the elements of repetition, multiples, structure, patterning, color, gradation, and decay. His materials are from the elemental world which invite the viewer to look anew and contemplate the beauty that lies within both the individual components and the overall composition. Of note are his ephemeral works with plants in all their configurations, components and seasonal display. These works which employ artistic skills, honed over many years, are constructed in isolation and photographed by the artist for exhibition. His more permanent works have a stronger social quality where he believes that the participation of people brings the work alive (Goldsworthy 2017: 6).

CONCLUSION

This chapter has re-examined traditional meanings of ornament and how this relates to, and can be reconfigured, in relation to the contemporary use of plants, as well as the design and cultivation of landscapes. Tracing a number of ornament's traditional key relationships, the chapter has attempted to highlight how these synergies have currency today, particularly in relation to important and continuing issues of social, cultural, and environmental sustainability, stewardship, and connection to place. Perhaps the future of plants as natural ornaments lies in the recognition of the fundamental importance of our relationship with plants, and the value of this relationship to our well-being and to social cohesion. The value of plants, and of gardens, might then be viewed within a broader reconciliation of our stewardship of the Earth as "our garden-home" (Francis 2020).

CHAPTER EIGHT

The Representation of Plants

GEOFF BIL

Modernity is a nebulous term. Without question, many of the features commonly associated with the modern period—secularization, industrialization, bureaucratization, and improvements in transportation and communication, to name a few—were well established before 1900. But there is no denying the watershed significance of the Great War (1914–18) in forging a world that, for many, seemed virtually unrecognizable in comparison with the pre-war era. It ushered in an era of American global hegemony, an increasingly global discourse of civil rights, and incredible technological innovation. At the same time, the sheer scope of violence and upheaval weakened earlier certainties in the West surrounding the position of "civilization" versus "nature," helping to entrench psychoanalysis as an authoritative means of coming to grips with suppressed inner conflicts, and spawning further wars and political movements of unprecedented ferocity. Most ominously, the war inaugurated a still-ongoing period of accelerated global modernization and ecological ruination that, in recent decades, has called into question the future viability of human life itself.

Representations of plants have reflected these wider shifts. They drew additional sustenance from advances in plant biology, themselves surely influenced by some of the factors described above. Chief among these was a series of works by Charles Darwin (1809–82) on climbing plants (1875), insectivorous plants (1875), and plant movement (1880), which underscored similarities between plants and animals. The Bengali scientist Jagadish Chandra Bose (1858–1937) gave these inquiries renewed importance with his research on plant sensory abilities and motor impulses. This culminated in his *The Nervous Mechanism of Plants* (1926), which found, most strikingly, that "the physiological mechanism of the plant is identical with that of the animal" (1926: ix).

This chapter examines plant representations in sculpture and architecture, fiction and film, and botanical illustration and art. In so doing, it has been necessary to abridge or exclude case studies (children's literature, poetry, magazines, nature photography, fashion, music, and videogames, say) that would doubtless corroborate or qualify this survey—as would, surely, a less Western-centric selection of material for consideration. These caveats notwithstanding, the patterns outlined here align broadly with what is often called the "crisis of representation" engendered by far-reaching changes in twentieth-century society and culture, which called into question the form and substance of earlier productions (Reynolds [2013] 2014: 158). In sculpture and architecture, biophilic principles came into prominence as part of a general move away from rigidly geometrical motifs reminiscent of anthropocentric domination over nature, and toward espousing nature–human kinship and entanglement. Although technological advances in photography during this period augured profound changes for scientific plant illustrators, the same era also witnessed the embrace of a more active role for the artist in scientific representation, while figurative depictions of plants have dealt more pointedly with emergent cultural concerns. Finally, while botanical imagery in literature and cinema since the Great War has of course reflected social tensions (as it did for analogous creations before this time), modern fiction is uniquely marked by an acute fascination with human–plant biological and ontological boundaries. Such a fascination—and at times, as we shall see, horror—speaks to a less certain conception of human nature, and also to a recognized need for a more equitable relationship with plants and non-human nature more broadly.

SCULPTURE AND ARCHITECTURE

The influence of botanical forms in sculpture can be traced to the Franco-German artist Jean Arp (1886–1966). Born in Strasbourg to a French mother and German father in the years immediately following Germany's annexation of Alsace-Lorraine, Arp identified fully with neither country when war broke out in 1914, and fled from Paris to Zurich the following year to take advantage of Swiss neutrality. It was here—in Ascona specifically—that Arp hit upon the idea of using organic motifs as a means of counteracting a stream of modernist works devoted to celebrating technological modernity. "I drew broken-off branches, roots, grasses and stones which the lake had washed up on its shore," he later described, "I simplified these forms and unified their essence in fluid ovals, symbols of eternal mutability and of the becoming of bodies" (Robertson 2006: 4–5, 23, 52). Results from these experimentations can be seen in a woodcut for an early issue of *Dada*, in which two images, adopting a generally plant-like form, suggest an organic process of evolution or growth from one to the other (Arp 1919). In the 1930s, Arp expressed this idea in sculptures he referred to—borrowing from geology—as "concretions" of organic and other environmental materials:

> Concretion signifies the natural processes of condensation, hardening, coagulation, thickening, growing together. Concretion designates the solidification of a mass. Concretion designates curdling, the curdling of the earth and the heavenly bodies. Concretion designates solidification, the mass of the stone, the plant, the animal, the man. Concretion is something that has grown. I want my work to find its humble place in the woods, the mountains, in nature.
>
> (Robertson 2006: 109)

Although pebble and stone motifs predominate in Arp's oeuvre, botanical "concretions" make an appearance as well. The most representative of these include a multi-part walnut wall relief entitled *Impish Fruit* (1943), which evinces both a human and floral process of generation; *Growth* (1938) and *Torso with Buds* (1961), free-standing structures that allude both to sprouting vegetation and a feminine torso; and his *Sculpture to be Lost in the Forest* (1932, cast 1953–8), fashioned from bronze, with its suggestion of a winding pathway between what could be trees, rocks, or shrubs. Arp intended these forms, as he described in 1948, "to cure human beings of the furious folly of genius and return them more modestly to their equitable place in nature" (Andreotti 1989: 2, 5, 193, 206; Read 1968: 143).

Such works have come to be known as *biomorphic*. As defined by the English poet Geoffrey Grigson (1905–85) in 1935, and elaborated by Alfred Barr (1902–81), director of the Museum of Modern Art, the following year, biomorphic representations presented a stark contrast with recent modernist productions. They were curvilinear, not rectilinear, emphasized emotion over intellect, and gave themselves over to forms that were dream-like, spontaneous, and irrational rather than ordered and mechanical (Mundy 2011: 62–3, 66–8). An exemplary point of reference for Grigson and Barr was the work of the English sculptor Henry Moore (1898–1986). Moore drew particular inspiration from continental Surrealism, with its emphasis on universal, yet unconscious, aspects of human experience (Nash 2001: 44–5). Spurning the alienating and frequently misogynistic temperament of modernism, Moore's oeuvre seeks a more general appreciation for vitality and growth in nature, synthesizing "ur-shapes" inspired by plants, bones, earth forms, and human beings alike (Cohen 2001: 264). His most representative work centers on the theme of the "reclining figure," in which the recumbent form of a woman's body coalesces with that of an undulating landscape.

Sculptures inspired by plants, leaves, and organic materials have their counterpart in architecture that draws on a generalized organic, fluid aesthetic—borrowing from plants, animals, and geographical phenomena—in self-conscious opposition to the rationalized, geometric configurations emphasized by architectural modernism. The Austrian-American architect Frederick Kiesler (1890–1965) expressed this philosophy using the concept of "correalism," or "the dynamics of continual interaction between man and his natural and technological environments." Achieving such an effect meant eschewing rectilinear and orthogonal forms in favor of emulating the "biotechnics" of nature (Kiesler 1939). Kiesler's own Endless House, realized only as a prototype miniature constructed for the Museum of Modern Art in 1960, applies these ideas in a seamless, flowing concrete structure designed to optimize the use of light and blur the distinction between interior and exterior spaces (Kiesler 1997). Biomorphic influences can also be seen in the Solomon R. Guggenheim Museum in Manhattan (1959) by Frank Lloyd Wright (1867–1949), Jørn Utzon's (1918–2008) Sydney Opera House (1973), and Grin-Grin Park (2005) in Fukuoka, Japan by Toyo Ito (b. 1941), which, fittingly, houses interior gardens beneath four large glass domes.

Indeed, biomorphic architecture can be seen as a subcategory of *biophilic* design more broadly, which seeks to cultivate physiological and psychological well-being through the use of naturalistic materials, lighting, colors, and motifs, as well as through proximity to plants and water—objectives also generally associated with *organic* structures, of which Wright's Fallingwater (1939) is perhaps most notable (Agkathidis 2017: 8; Kellert 2018). A biophilic aesthetic can also be seen in the ornamental façades of the Chicago-based architect Louis H. Sullivan (1856–1924), who modeled his patterned ornamental reliefs after flowing, flowering, intertwining vegetation emerging from seeds. Sullivan's earlier, monumental works would give way by the 1920s to a more generalized architectural style known as "Sullivanesque." Buildings in this fashion, including ordinary and smaller-scale structures, incorporate plant-inspired terracotta reliefs derived from standardized design plates. One such example is the M. Shooman Building (1922), designed by Rissman and Hirschfield architects in Chicago, the principal city for Sullivanesque architecture (Schmitt 2007: 4–7, 52–6, 251). Sullivan's reliefs bear a striking resemblance to the detailed and highly aestheticized plant and animal forms depicted in Ernst Haeckel's (1834–1919) *Kunstformen der Natur* (1904) (Agkathidis 2017: 12), the influence of which is likewise palpable in the work of the Spanish architect Antoni Gaudí (1852–1926), most notably his still-unfinished Sagrada Familia (1882–) (Ingersoll 2012: 576). Reliefs of flowers, palms, and apple-tree branches, interwoven with animals, stars, and ecclesiastical symbolism, adorn the building's Nativity façade. The Sagrada Familia's central nave, similarly, features central columns resembling trees and branches, with sunflower-like shapes etched into the ceiling. Forest canopy-like supporting structures remain common in contemporary works: a galvanized steel lattice shelter outside the entrance to a Marriot Hotel in Frankfurt, designed by Just/Burgeff Architekten and Asterios Agkathidis (b. 1974), and completed in 2001, is one of countless examples (Agkathidis 2017: 12, 80–1). The 82–164 foot (25–50 m) tall, steel, glass, and concrete "trees" sprouting in Singapore's Supertree Grove (2012), on the other hand, provide scaffolding for an array of photovoltaic cells and cooling exhausts for an adjoining conservatory complex, while also playing host to hundreds of thousands of climbing plants, orchids, ferns, and epiphytes. At first glance, the complex is eerily reminiscent of earlier, unabashedly colonial, representations of an exorbitantly fertile tropical nature. These postmodern, tree-like edifices, however, also enunciate a vision of nature harnessed by a fusion of eco-sustainability and neoliberal corporate efficiency, thereby representing a notable recent chapter in what has been a decades-long effort at national self-definition in the wake of postcolonial independence from Britain (Barnard 2016: 257; Lim 2014). More-than-aesthetic considerations are similarly at work in the recently completed Sandy Hook Elementary School building (2016) in Newtown, Connecticut. The building's wave-like wood-paneled exterior, with tree-like supports bolstering windowed atrium structures that overlook a rain garden and neighboring forest, has a designedly recuperative function, the previous school building being the site of one of the deadliest massacres in recent American history (Kellert 2018: 153). The rationale is akin to that of Arp's chastening, yet reassuringly organic, sculptures in the wake of the Great War.

Yet other structures embody a more taxonomically specific inspiration. The dramatic Lotus Temple (1986) in Delhi, for instance, designed by Fariborz Sahba (b. 1948), is modeled directly after the sacred lotus flower (*Nelumbo nucifera*). In conceiving a structure that was to be devoted to the syncretic Bahá'í faith, Sahba chose the lotus as a basis for the design owing to its symbolic prevalence in Hindu, Buddhist, Christian, Zoroastrian, and Islamic iconography alike. The composition is functional as well. Twenty-seven marble-clad "petals," arranged in concentric circles, work to diffuse light throughout the building's central auditorium, while the air inside is cooled by pools which surround the building's exterior and simultaneously mimic the water on which the lotus floats (Kazimi 2015; Sahba 2010). Taipei 101 (2003), a supertall office skyscraper designed by Chu-Yuan Lee (b. 1938) and built in Taipei, similarly borrows from the bamboo (Gramineae/Poaceae, subfamily Bambusoideae). The building's eight segments, which emerge successively from each other, correspond both to the traditional concept of *fa* (prosperity), as well as to the bamboo shoot's habit of growth—a symbol of hope for a new millennium (Cheng and Shih 2008: 194). The aluminum Rose Pavilion, constructed outside Eidgenössische Technische Hochschule (ETH) Zurich in 2014, likewise emulates the pentagonal structure of the *Rosa canina* or dog rose, with its double-curved petals (Agkathidis 2017: 98–9). Such examples, together with the others noted above, speak to the ubiquity of plants as symbols, as well as to their wider resonance in sculpture and architecture in helping to instill awareness of human-plant affinity and interdependency.

FIGURE 8.1 The Supertree Grove in Singapore's Gardens by the Bay. Photo by Ray in Manila. Photo courtesy of Creative Commons.

ILLUSTRATION AND ART

If twentieth- and twenty-first-century plastic arts espoused organismal and vegetational forms as an antidote to a mechanizing, anthropocentric modernity, plant illustrations can be seen to have performed a similar function. At the same time, visual depictions of plants benefited considerably from technological advancements, not least in the realm of photography. Early attempts at rendering color frequently involved hand-tinting black-and-white photographs. Autochrome technology, introduced by Lumière and Co. in France in 1907, provided a key innovation by using red-, green-, and blue starch-coated glass plates to achieve results that were truer to life and less reliant on operator interference (Elliot 1996). This is the approach adopted for the specimens portrayed in *Wild flowers of Kashmir* (1923–30) by B.O. Coventry, which display occasional browning, minor insect defoliation, and other naturalistic imperfections. A few years later, Torsten Lagerberg (1882–1964) and Jens Holmboe's (1880–1943) study of Scandinavian flora, *Våre ville planter* (1937–40), features detached specimens alongside field portraits, which foreground the plants in question against a blurred background (Elliot 1996). A superior example is furnished in Roger Phillips' (b. 1932) *Wild Flowers of Britain* (1977), which provides crisper images with greater depth of field. Tailored to the needs of the fieldworker, the guide categorizes plants according to their season of flowering, includes photographs of specimens isolated and *in situ*, cross-references similar-looking species, and includes only so much detail as is necessary for field identification, capturing specimens in their "natural," compact form. Photography, Phillips argues, thus provides a "better instant 'feel' of a specimen" in comparison with traditional illustrative techniques (Phillips [1977] 1978: 6–7; Saunders [1995] 2009: 136). Field guides for especially biodiverse regions must needs make further concessions to space and time constraints, as well as to a comparative dearth of reference literature. A recent field guide to Costa Rican plants, for instance, prioritizes common and conspicuous species, and, at times, "opportunistically" includes photographs of flowers or fruits that had already fallen to the ground. Only about 8 percent of the Costa Rican flora is thereby represented (Gargiullo 2008: xxxi–xxxii, xxxix).

Photography has proven especially useful when conjoined with other technologies. Scanners, for instance, facilitate the rendering of otherwise hidden parts of flowers, as with the striking images produced by Barbara (b. 1956) and Zafer Baran (b. 1955) in their *Flower Cabinet* (2004) (Saunders [1995] 2009: 149). Undoubtedly the most pivotal contribution of photography in recent years has been the digitization of herbarium sheets and associated materials. The New York Botanical Garden, which has led this effort, undertook to digitize its 7.8 million specimens beginning in 1995, including high-resolution close-ups of vital characters, field notes, and botanical illustrations or drawings. This project has since gone global in the form of the JSTOR Global Plants Initiative and the open-access World Flora Online (WFO 2012– present), which involves thirty institutional partners worldwide (Thiers, Tulig, and Watson 2016; WFO 2020). Digital photography has been similarly key for the Global Biodiversity Information Facility, which presently boasts over 1.3 trillion—and counting—"occurrence records" of species crowd-sourced from amateur and salaried participants alike (Global Biodiversity Information Facility n.d.).

Despite photography's efficacy as a scientific instrument, traditional forms of botanical representation have proven resilient. One reason for this is the attenuation of a nineteenth-century ideal of objectivity, which, via new technologies like photography, strove to eliminate all vestiges of mediation in the service of rendering vegetative forms according to "nature." Twentieth-century representational norms, on the other hand, commonly emphasized synthetic, expert intuition in rendering images that qualitatively balanced scientific utility, and often beauty, with fidelity to unaestheticized truth (Daston and Galison 2007: 309–61). This recuperation of scientific subjectivity vis-à-vis machines, akin to the depreciation of rationalized forms in the plastic arts, derived partly from the elevated position of scientists—in numbers, resources, and prestige—following the Great War (Galison 2015: 73), and no doubt from the obsolescence of Victorian ideals more broadly. Yet systematic botanists frequently prefer traditional illustrative approaches for more prosaic reasons. Photography, after all, remains ill-suited to capturing the fine detail of plant—especially flower—morphology, and to displaying individuals that are functionally representative of species, independent of variations owing to climate, season, habitat and growth histories. Modern cladistics, in particular, which classifies plants according to the proportion of shared phenotypic attributes, draws on a virtually infinite range of identifying characters, and thus calls for illustrations that are as comprehensive as possible (Sanders [1995] 2009: 12, 100). In capturing these elements, photography is useful primarily in enabling artists to preserve knowledge of a plant's habit, color, and three-dimensional appearance before the material wilts or fades, and later delineate these features in the comfort of their studios (De Bray 1989: 185; Rix 1981: 218; Simpson and Barnes 2008: 259).

The most pervasive approach to scientific plant illustration has been black-and-white line drawings, owing to the comparative ease and inexpensive means by which complex structures can thereby be reproduced. From a systematist's perspective, moreover, color is often a subsidiary concern: it seldom survives in dried herbarium specimens, pales in importance beside morphology in matters of plant identification, and is challenging to illustrate and describe with the necessary accuracy (Blunt [1950] 1994: 254; Hagerup and Petersson 1956–9: v; McNeill 2018; Sanders [1995] 2009: 15). One notable example of line drawing is the *Botanical Atlas* (1956–9) written and illustrated by the Danish naturalists Olaf Hagerup (1889–1961) and Vagn Petersson (1885–?). In an effort to "throw light upon the methods [plants] use in their struggle for life," the *Atlas* illustrates plants in all stages of growth, using microtome sections for flower and seed structures (Hagerup and Petersson 1956–9: v). Undoubtedly the most celebrated artist in this vein is Stella Ross-Craig (1906–2006). Ross-Craig produced color plates for *Curtis's Botanical Magazine* and numerous other publications (De Bray 1989: 173; Ward and Rix 2006), but is especially renowned for the 1,286 ink illustrations in her *Drawings of British Plants* (1948–73), the first British flora to provide life-size black-and-white portraits alongside descriptions that undertook to illustrate plants in their entirety—including roots, flowers, and leaves, with seeds and flower cross-sections magnified using a microscope (Buchan 2003; Saunders [1995] 2009: 132, 137; Stearn 1990: 36). Ross-Craig, as we shall see, has been one of numerous women botanical illustrators, building on a trend begun in the nineteenth century, when such a career served as an entrée to the otherwise male-dominated domain of science (Gates 1998: 74–83; Kohlstedt 1978: 86–9; Saunders [1995] 2009: 108; Shteir 1996: 178–82). This pattern continues today in the work of Alice Tangerini (b. 1949), a prolific line-drawing artist based at the Smithsonian Institution (McNeill 2018).

Color plant illustrations have also remained pertinent—for their scientific value, as well as their broader aesthetic appeal. Hand-coloring of lithographic plates was the dominant means of rendering plant colors in early-twentieth-century publications. This method's increased expense favored simpler coloring, enabling colorers to "work in hundreds of the same plate, one colour at a time and one plate after another," as Lovell Reeve & Co. described in the 1920s (Saunders 2009 [1995]: 102). Lilian Snelling (1879–1972), the lead artist at *Curtis's Botanical Magazine* from 1922 to 1952, was the foremost exponent of this method until it was replaced by four-color halftone images in 1948, shortly thereafter by photogravure, and later still by photolithography (Desmond 1987: 188–91). The Australian-born Margaret Stones (1920–2018), another notable recent artist, established herself as a freelance illustrator at Royal Botanic Gardens, Kew in the 1950s. Her six-volume *Endemic Flora of Tasmania* (1967–78), with descriptions by Winifred Curtis (1905–2005), portrays Tasmanian plants grown near Dublin, as well as at Kew, with 254 life-size watercolors, including dissections and enlargements with the aid of a microscope. Her *Flora of Louisiana* (1991), written by Lowell Urbatsch (b. 1942), makes more extensive use of microscope dissections, cross-sections, and amplifications over the course of 224 illustrations, reproduced both as black-and-white thumbnails and full-size color. Although Stones initially worked from herbarium specimens that had been rehydrated and dissected, her later watercolors derived exclusively from live plants in order to faithfully depict their color, form, and lifelike detail, spurning even the use of photographs as working aids (Smyth 2010). Stones' protégé, an Englishwoman named Pandora Sellars (1936–2017) who followed her mentor in eschewing photography, is particularly famous for her paintings of orchids in *Curtis's Botanical Magazine* and elsewhere. Although acclaimed for her taxonomical accuracy, some of Sellars' best-known works depart from abstracting a single species on a white background, in favor of portraying multiple species simultaneously—her *Paphiopedilum spicenarum, Calathea* [now *Goeppertia*] *roseopicta, Philodendron panduraforme* (1985) being one example (De Bray 1989: 181–5). Margaret Mee (1909–88), another English artist, took the ideal of true-to-life realism a step further: most of her paintings began with sketches of living plants in the field, with notes on color and other characteristics, followed by finishing work based on cuttings. Mee is renowned for her paintings of rare Amazonian species, produced during her travels throughout the region from the 1950s to the 1980s, which drew wide attention to its ecological vulnerability. She has several plants named after her, having been the first to make these species known to the West. Although Mee's earlier paintings center on isolated specimens, her later works frequently portray plants in their characteristic forest habitat, accompanied by other associated species (De Bray 1989: 173, 175; Mee 1988: 26, 206). Mee trained numerous Brazilian botanical painters before she succumbed to a car accident in 1988, following which Brazilian students have continued to train under Christabel King (b. 1930) and Anne Farrer (b. 1950). Both are renowned illustrators and travelers in their own right—to Uganda, where King painted flora and landscapes for Guy Yeoman's *Africa's Mountains of the Moon* (1989), and to Kashmir and Ladakh, where Farrer illustrated *Flowers of the Himalaya* (1984), with descriptions by Oleg Polunin (1914–85) and Adam Stainton (1921–91) (Sherwood 1996: 66, 112, 138–43).

The boundaries between scientific botanical painting and other kinds of plant illustration can be porous, however. Such fluidity of genre is personified in the work of the Scottish musician and artist Rory McEwen (1932–82). Descended directly from the botanist John Lindley (1799–1865), McEwen trained at Eton College, England, under the botanical artist and historian Wilfrid Blunt (1901–87) and drew early inspiration from classical luminaries of flower painting, even going so far as to paint on vellum. McEwen's paintings of the 1960s, however, departed from the idealized representations favored by systematists, toward an approach that emphasized plants as individuals. His *Anemone* series (1962–4), for instance, paid greater attention to stem curvature and to the incongruous, finely wrought veinery of leaves and petals, than it did to the comparatively fixed morphology of reproductive organs. Much of McEwen's later oeuvre, moreover, consists of withered fruits, dead plants, and plant fragments, frequently autumn-colored, diseased, and insect-eaten, suggestively human-like in their singularly dilapidated and imperfect states of decay, and in their idiosyncratic placement on the canvas. Macrophotography and cine-photography aided extensively in capturing these ephemeral aspects of plant decomposition (De Bray 1989: 181; Rix 2013: 78, 84–9, 92, 116–25, 166–85).

On the less naturalistic end of the spectrum, plant imagery has also figured in explorations of interior cognitive states. *Blumenmythos* (*Flower Myth*) (1918) by Paul Klee (1879–1940), for instance, features a blue flower at the center of a pink landscape palpably reminiscent of a dismembered female body. With a bulb corresponding to a womb, situated above a leaf-like figuration of a vagina, and below a suggestively phallic stem, flower, and bird, the painting can be seen as a contemplation—a disconcertingly misogynistic one—on the ways in which artistic creation mirrors the generative forces of nature and human sexuality (Aichele 2006: 73–4). Similarly, the stylized paintings of Georgia O'Keeffe (1887–1986) are less about providing lifelike depictions of plants than about contemplating emotive shapes, colors, and motifs from the natural world that, like music, resist expression in words. Among the most emblematic of her flower compositions is her six-painting *Jack-in-the-Pulpit* series (1930), which progresses from

FIGURE 8.2 Green-Fly Orchid (1991). Native Flora of Louisiana watercolor drawings by Margaret Stones. E.A. McIlhenny Natural History Collection, LSU Libraries, Baton Rouge, Louisiana.

a frame-filling close-up of the inflorescence's outward form to zero in on the spadix of tiny flowers, emphasizing throughout the dynamic contrast between the black and white of the inflorescence, green foliage, and azure blue and white, which could equally be water or sky. Interpretations of O'Keeffe's flowers often fixate on the visual resemblance they bear to women's sexual organs—a symbolism traceable to Freud (see below) and encouraged by her husband Alfred Stieglitz (1864–1946), but which O'Keeffe herself actively repudiated (Lynes 2009: 168; 2007: 18, 34, 143). A transformation in scale likewise characterizes *Le Tombeau des Lutteurs* (1960) by René Magritte (1898–1967), in which a gigantic red rose fills a room from floor to ceiling, with a window included for scale. Pat Steir's (b. 1940) *Nederland Landschaft* (1982) makes a similar point using a series of tulips, the leftmost iteration comprehensible as a distant object, toward a rightmost blossom that exceeds the frame, rendering it inscrutable and forcing viewers to confront it as more than a mere instrument of visual or aromatic pleasure (Sydney 1986: 54–5, 66–7). Floral symbolism can also convey a rootedness to place, as with *The Tree* (2003) by Ibrahim El-Salahi (b. 1930), inspired by the Hazan tree that grows along the Nile in his native Sudan, which combines stark vertical lines that connote both tree trunks and Islamic motifs, with floating, mist-like, horizontal bands of color (Tate Modern 2015). Such representations contrast markedly with imagery that caricatures human fascination with flowers as an extension of consumerist superficiality—whether the plastic-like tulips at the center of *Black Flowers* (1961) by Roy Lichtenstein (1923–97), the garishly hued blossoms in Andy Warhol's (1928–87) *Flowers* (1964), or the inflorescences of *Flower-piece I* (1971–4) and *Flower-piece II* (1973) by Richard Hamilton (1922–2011), adapted from picture postcards and juxtaposed with a roll of toilet paper and heap of turds respectively (Sydney 1986: 12–15).

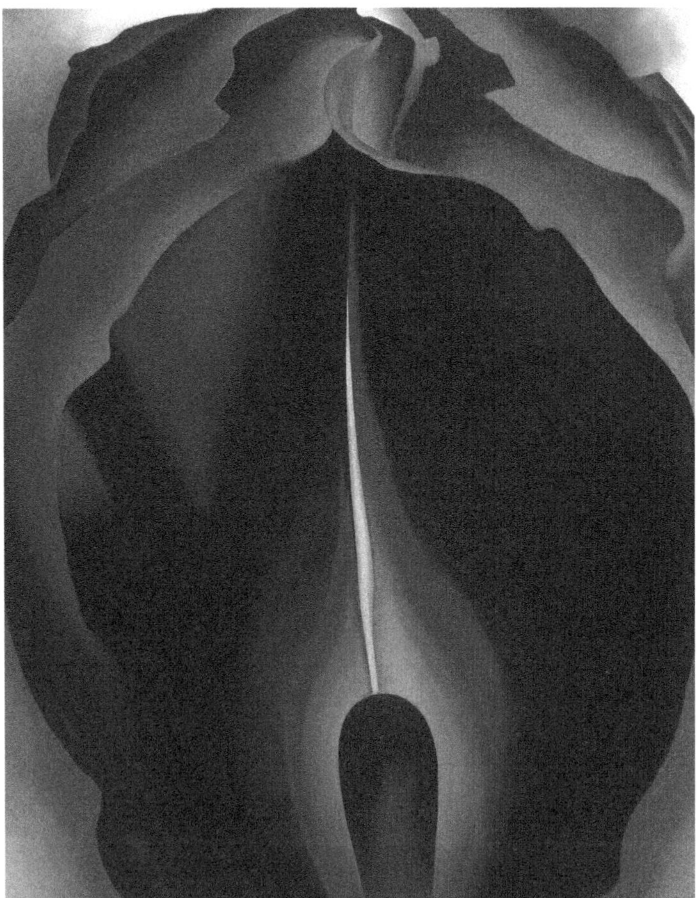

FIGURE 8.3 *Jack-in-the-Pulpit No. IV* (1930). Oil on canvas. Painted by Georgia O'Keeffe. Alfred Stieglitz Collection, Bequest of Georgia O'Keeffe, National Gallery of Art, Washington.

It is movement, however, that, perhaps more than anything else, defies the circumscription of plants as objects of scientific or consumerist desire. Such a quality is difficult to convey in a necessarily static format such as painting. Claude Monet's (1840–1926) water lily series (1896–1926) accomplishes this feat by blurring the distinction between plants, water, light, shadow, and other phenomena, thereby suggesting the ephemerality of natural forms in shifting and shimmering sunlight. *Bank of Crimson Flowers* (1984) by Jean Marie Toulgouat (1927–2006), a distant relative of Monet's, elicits a similar sense of movement with his rendering of a field of poppies and cornflowers, its variegated reds and pinks reminiscent of poppy petals caught in a breeze. Morris Louis' (1912–60) *Iris* (1954) and *Floral* series (1959–60) go so far as to dissolve the semblance of flowers almost entirely into changeable and indeterminate contemplations of color, light, and shadow. In the ethereal *A Fisherman* (1950) by Ota Janeček (1919–96) a foreground of well-defined grass husks, flowers, and stems, bending as if in a gentle gust, gives way to ever more shadowy forms, culminating in the shrouded silhouette of the painting's eponymous figure, scarcely visible on the left horizon (Sydney 1986: 12, 28–9, 34–5). Tu Hongtao (b. 1976) applies a kindred aesthetic to his phantasmal arrays of trees, ferns, and grasses—which, in the case of *Herbs* (2010), *Back to Youth* (2011), and *Walking through the Woods* (2014), also incorporate likenesses of human forms—to suggest the ephemeral and pictorial simultaneity of dreams or memories (Tu 2018).

In addition to problematizing the conception of plants as objects, painting has also worked to depict plants as agents, even subjects, with an ecological stature and staying-power equal to or above that of human beings. In the dynamic renderings of British Columbian landscapes by Emily Carr (1871–1945), mountains, clouds, tree-trunks, and foliage frequently swirl together, connoting mist, movement, growth, and shifting gradations of light. Shadowy forest canopies that enclose the view, paired with massive conifer trunks that taper upward, exceeding and by implication towering over the frame, are a recurring theme in Carr's work, which gestures toward a nature that is all-encompassing and anything but passive. The agency of plants reverberates similarly in the paintings of Frida Kahlo (1907–54), which emphasize the entanglement of humans with non-human nature. Her *Roots* (1943), for instance, portrays the artist herself reclining on the ground with leafy tendrils weaving inside and around her body, sprouting red rootlets in a clear analogy to the human circulatory system. *Flower of Life* (1944), on the other hand, features a phallic stem emerging from a cluster of leaves, penetrating a flower with human-like arms resembling ovaries and fallopian tubes, erupting into a display of stamens and pistils reminiscent of human reproductive fertilization (Udall 2000: 163–4, 261–2). *The Flower Arranger/Early Summer* (1982) by Anthony Green (b. 1939) broaches similar themes with its depiction of a man admiring a selection of cut flowers, encompassing the scene with an octagonal frame of grasses and daisies that suggests that he, too, is being contemplated by plants, and that his duration is similarly transitory (Sydney 1986: 48–9).

Fixed, two-dimensional construals of floral perspective cannot, of course, equal the achievements of literature and film in this respect. They come nearer the mark when they acquire a dynamic, participatory dimension, as a series of interactive exhibits by Laurent Mignonneau (b. 1967) and Christa Sommerer (b. 1964) have shown. In *Interactive Plant Growing* (1992), for instance, electrical signals generated from participant interactions with living plants were relayed via wired roots to computers and translated, using algorithms, to virtual plant growths on a screen, modulated in color, size, direction, and form of growth. *Data Tree* (2009) went even further in capturing information pertaining to humidity, soil pH, carbon dioxide, wind, human contact, and other variables, which

FIGURE 8.4 *The Flower Arranger/Early Summer* (1982). Oil on board. Painted by Anthony Green. Courtesy of Museum of Fine Arts, Boston.

was then projected back onto the tree itself, thereby allowing viewers to witness the plant's ecological interactions in real time (Sommerer et al. 2015: 237–42). An exhibition of paintings by Paul Wackers (b. 1978) entitled *Wait and Watch Awhile Go By* (2012) provides yet another vision of what a floral perspective might look like, by ushering visitors through a gallery hung with Wackers' paintings—themselves highly abstracted images of plants—and populated by houseplants situated so as to appear as spectators in their own right (Alice Gallery 2012).

FICTION AND FILM

In contrast to the dimensionally fixed quality of sculpture, architecture, and illustration, the temporal latitude made possible by literary and film media has ensured an especially efficacious role for plants as avatars of human concerns. Even while bloody stalemate gripped the Western front, flowers emerged as potent symbolic counterpoints to war and modernity. The corn poppy, in particular, came to encapsulate the ironic human cost of a war waged according to the purported virtue of human rationality, owing to its capacity to disseminate its seeds via the soil churning caused by artillery shelling, as well as its appetite for the nitrogen from explosives, lime from shattered buildings, and decaying bodies. Its brief period of bright red flowering also drew comparison with the blood of the countless soldiers cut down in the prime of their youth, an analogy immortalized in the opening verses of a 1915 poem by the Canadian Lieutenant-Colonel John McCrae (1872–1918): "In Flanders fields the poppies blow / Between the crosses, row on row" (Lewis-Stempel 2016: 233–7). *The Waste Land* (1922) by T.S. Eliot (1888–1965) resorts to similar imagery in considering April to be "the cruellest month" for engendering "Lilacs out of the dead land" (Keetley 2016: 1). The snapdragons, sunflowers, sweet-peas, roses, and pansies that flourished in an English POW camp behind German lines, as described in the poem "English Flowers in a Foreign Garden" (1917) by Lieutenant Frederick William Harvey (1888–1957), likewise suggest the possibility of regeneration amid the dystopian violence of war, while also serving as sentimental, and at times painful, reminders of home (Lewis-Stempel 2016: 226–9). The dichotomy between nature, plants, and gardening on the one hand, and warfare on the other, also appears in a 1917 poem for *Dada* by the abovementioned Jean Arp, which contrasts "man" who "kills animals, plants, his brothers," with the "artist," who "knows how to work in a way which becomes organic" and "looks after the garden of intentions, orders" (Andreotti 1989: 50). The obverse symbolism of plants and gardening versus war and other forms of oppression would persist: in the profuse biodiversity of the Middle Earth of J. R. R. Tolkien (1892–1973) opposite Sauron's industrial war machine dystopia (Hazell 2006: 4); in the "flower power" of the anti-Vietnam generation (McKay 2011: 106–33); and in the vibrant green and red sorghum fields of the novel (later film) *Red Sorghum* (1986) by Mo Yan (b. 1955) that simultaneously belie and portend the violence of the Second Sino-Japanese war (1937) (Meng-Lin 2019).

Plants, flowers in particular, have also long encoded stories and emotions—quite literally so, in the case of the Victorian "language of flowers," which Vanessa Diffenbaugh's (b. 1978) recent novel of the same title (2011), with accompanying dictionary (Kirkby 2011), recuperates. The modern—or rather, Modernist—reading of this symbolism can be traced to the fourth edition of Sigmund Freud's (1856–1939) widely read *Interpretation of Dreams* (1914), which analyzes dreams as expressions of repressed libidinal forces. He recounts one case study, for instance, in which a woman, "engaged to be married, but [with] hindrances in the way of the marriage which threatened to postpone it," dreams of arranging flowers on a table "for a birthday." Her use of lilies of the valley (*Convallaria majalis*), violets, and carnations are respectively symbolic, Freud argues, of chastity, "defloration," and—owing to the carnations' expense and their association with the word *carnal*—the woman's gift of her virginity in exchange for a "rich love-life." The birthday itself, meanwhile, signals the woman's wish for a baby. So entrenched is this symbolism, Freud argues, that to exchange flowers between lovers is nothing less than to exchange symbols of sex organs (Freud [1914] 1931: 255–8; Marder 2019: 107–8).

In many works of fiction, plants adopt a similarly—albeit less fixedly sexual—metaphorical function. The aspidistra (*Aspidistra eliator*), a houseplant widespread in Victorian British households, serves recurrently as an early-twentieth-century symbol of outmoded bourgeois conformity. *Keep the Aspidistra Flying* (1936) by George Orwell (1903–50), for instance, chronicles the experiences of a writer who self-consciously spurns the pursuit of respectability and financial well-being in his ill-fated attempt to carve out a living as a poet. When he finally relinquishes this goal and marries his girlfriend, one of their first purchases—against her objections—is an aspidistra for their front window (Orwell [1936] 1962). This symbolism is turned to comedic effect in "The Biggest Aspidistra in the World" (1938), a song performed by the English singer and actress Gracie Fields (1898–1979), about an aspidistra crossed with an oak: "The borough council told us that we've got to chop it down / It interferes with aeroplanes that fly above the town / So we sold it to a wood-yard for a lousy half a crown / It's the biggest aspidistra in the world" (Fields 2009). Decades later, the plant becomes an embodiment of innocence—like the lost Victorian world it represents—in "An Aspidistra in Babylon" (1960) by H. E. Bates (1905–74). "I was as dull as one of the many aspidistras that cluttered up ... our little boarding-house," the story's eighteen-year-old protagonist declares, "a female aspidistra and nothing more," when she falls for a man more than twice her age, who convinces her to steal for him before cheating on her with the boarding-house chambermaid (Bates [1960] 1974: 155).

More commonly, and in contrast to the hackneyed aspidistra, plants have served as emblems of authenticity and individuality. "Prelude" (1918) by New Zealand-born short-story author Katherine Mansfield (1888–1923) employs an *Aloe vera* for this purpose, its "thorny leaves" set apart from a conventional garden, with which the protagonist comes to identify as a consequence of her unhappy marriage (Mansfield 1921). The Māori-named karaka (*Corynocarpus laevigatus*) in "The Garden Party" (1921), meanwhile, align with the protagonist Laura's childlike empathy for a working-class family nearby, while their concealment behind a party tent foreshadows her fraught acclimatization to bourgeois class prejudice (Mansfield [1922] 1997). Similarly, in the novel *The Rainbow* (1915) by D. H. Lawrence (1885–1930), a woman's emancipation from her conservative upbringing gains momentum over the course of her botanical studies, during which she "[enters] into the life of the plants" and becomes "fascinated by the strange laws of the vegetable world." The analogy between plant growth and self-realization continues in *Women in Love* (1920), with Gudrun's false conjecture that socially advantageous marriage will prevent herself from "wither[ing] in the bud," while Ursula finds her "active living ... suspended, but underneath, in the darkness, something was coming to pass. If only she could break through the last integuments" (Lawrence 1920: 9–10; Mahood 2008: 189–90). Plants feature likewise in a pivotal scene in *Lady Chatterley's Lover* (1928), in which sexual passions supersede class differences between an aristocratic woman and the gamekeeper on her estate, who adorn their naked bodies with forget-me-nots, woodruff, creeping-jenny, hyacinth, and campion (Lawrence [1928] 1995: 226–36). For the protagonists of *Kangaroo* (1923) and *The Plumed Serpent* (1926), finally, it is the ability to correctly identify plants that augurs the possibility of psychological fulfillment and renewal in territories further afield—a form of intellectual colonization—in Australia and Mexico respectively (Bricout 2009).

As symbols of authenticity and self-realization, however, plants have proven equally suited to discourses of anticolonial criticism and resistance. Vegetation figures crucially in *Wide Sargasso Sea* (1966) by Jean Rhys (1890–1979), a reimagining of Charlotte Brontë's (1816–55) *Jane Eyre* (1847) from the standpoint of Antoinette, the creole Dominican wife of Jane's intended husband, whom he forcefully sequesters in the attic of his English manor, and whom Brontë's version of the story reduces to silence. The beautiful but disorderly array of ferns, mosses, lime trees, orchids, jasmine, honeysuckle, frangipani, and other plants that grow in and around Antoinette's Jamaican home contrast starkly with the violently patriarchal and imperial order represented by Edward Rochester—a relatively sympathetic figure in *Jane Eyre*. Floral imagery also connects Antoinette—an ostensible "madwoman"—at the moment of her emancipating fiery suicide, with the wistfully fecund landscape of her youth (Fincham 2010). Plants are likewise critical to *Ceremony* (1977) by Pueblo writer Leslie Marmon Silko (b. 1948). The novel follows the experience of Tayo, a Second World War veteran of mixed white-Laguna descent, who returns home traumatized, his Laguna identity marginalized in urban white America. Through the agency of a Navajo healer, Tayo reconnects with the landscape around him, including the touch, scent, and image of its plants—beeweed (*Cleome serrulata*), yucca, rabbit weed, cottonwood, apple and apricot blossoms, juniper, piñon, tamarisk, and others—which contrasts him both with the white ranchers fixated on logging and fencing the land, and with his erstwhile Laguna Pueblo drinking companions, who lack this grounding and are thus unable to cope with the violence and alienation of Euro-American modernity (Katawal 2013; Rice 2005). In *White Lies* (2013), a cinematic adaptation of a novella by the Te Aitanga-a-Māhaki author Witi Ihimaera (b. 1944), knowledge of plants is similarly vital to resisting the New Zealand colonial state. The film opens by showing the protagonist as a young Māori girl studying plants, who narrowly survives her family's massacre by a party of invading settlers. She nevertheless persists as an adult in gathering plants and using them to care for her people, becoming a skilled tohunga (traditional healer), despite the colonial government's prohibition of these practices—practices which, by the film's end, enable her to raise a child of her own in the same knowledge tradition.

Beyond their resonance as symbols of human identities and conflicts, plants, or plant-like humans, have also served as tools for rethinking human exceptionality, one incarnation of which has been termed "the botanification of the human" (Wampole 2016: 76). This shift is palpable in the now-entrenched comparison of the human brain to a tree, after the fashion of a Spanish neuroscientist named Santiago Ramón y Cajal (1852–1934). Using an advanced cell-staining method, Cajal's findings resulted in what would subsequently be known as the Neuron Doctrine, whereby "dendritic trees" relay signals between neurons, and the cerebral cortex resembles a dynamic garden of trees, roots, branches, flowers, and fruits (Newman et al. 2017: 8–9, 34–41). "Dialogue de l'arbre" (1943) by Paul Valéry (1871–1945) takes up this analogy with a scene in which Lucretius, modeled after the Roman philosopher, imagines himself as a tree, "improvising a dream of branches" (Wampole 2016: 92). The American poet Ronald Johnson (1935–98) appears to have thought in a related vein when he coined the term "nervetree" in his *Ark* (1970–96) to describe how nature—archetypified as plant life—might see itself in human thought (L'Abbé 2013: 184, 315). Cognate themes punctuate the film *The Fountain* (2006), directed by Darren Aronofsky (b. 1969). In one scene the protagonist, a doctor trying to cure his wife Izzi of brain cancer, imagines himself as a character in her story about a sixteenth-century conquistador charged by Queen Isabella with searching out the Tree of Life in Central America. Isabella herself appears, personified as Izzi, the patterns on her dress merging with the roots of the tree. The entire world of Pandora in James Cameron's (b. 1954) film *Avatar* (2009), meanwhile, comprises a neural network anchored by the Tree of Souls, to which the Na'vi connect via long, synaptic braids (Pisters 2012: 181, 206).

Without denying the significance of botanified humans in these and other works, a more conspicuous role in modern fiction has been acted by *plants* that refuse the alterity, silence, and subordination traditionally ascribed to them. An early envisioning of this plant-centered nature is provided in the short story "Mark on the Wall" (1917) by

Virginia Woolf (1882–1941), in which the narrator reflects on the tree from which the titular object of her gaze derives. "For years and years they grow," she muses,

> without paying any attention to us .... I like to think of the tree itself: first the close dry sensation of being wood; then the grinding of the storm; then the slow, delicious ooze of sap .... The song of birds must sound very loud and strange in June; and how cold the feet of insects must feel upon it, as they make laborious progresses up the creases of the bark, or sun themselves upon the thin green awning of the leaves, and look straight in front of them with diamond-cut red eyes.
>
> (1921: 89–90)

Woolf's "Kew Gardens" (1919) similarly adopts a snail's perspective as it navigates minute obstacles between flower stalks, while the garden's human visitors dissolve into indeterminate colors and forms (1921: 68–78). It is no accident, surely, that Woolf selected Kew—the epicenter of Enlightenment mastery over botanical nature—to stage this inversion of anthropocentrism (Swanson 2011: 30–1). Jean-Paul Sartre's (1905–80) *Nausea* (1938) spells out the existential consequences of this move when the novel's protagonist encounters a chestnut tree, the material qualities of which—the color, scent, and texture of its bark, its motionless-to-the-naked-eye movement and growth, its purposeless reproduction—stubbornly resist human attempts to categorize and comprehend them, and mirror, from the tree's perspective, a human existence equally inscrutable and meaningless (Laist 2016: 164–70).

Early cinema afforded new opportunities for visualizing a plant-centered universe. F. W. Murnau's (1888–1931) *Nosferatu* (1922) shows a Venus fly trap (*Dionaea muscipula*) closing on an insect—"Like a vampire, no?" the caption reads—and in so doing elicits what is surely one of the first depictions of a quintessentially zoomorphic plant in action (Miller 2014: 472). The more technologically cutting-edge *Plants of the Pantry* (1927), directed by the British naturalist and filmmaker Frank Percy Smith (1880–1945), uses stop-motion- and micro-photography, as well as animation, to depict animal-like moulds that search out nutrients, greet each other with hand-like branches, and "consummate" their "marriage" by producing "special growths" when they are offspring of different parents. Smith's *Floral Co-operative Societies* (1927) employs similar techniques to show how plants pool resources to attract pollinators, produce seeds, and differentiate into specialized "advertising" and reproductive "worker" portions. In selecting for colorful petals at the expense of reproductive parts, however, humans have produced "a gigantic bluff—a society of 'Advertisers' without a single 'Worker.'" From a purely anthropocentric standpoint, such nuances are easily missed. Smith's French counterpart, the microbiologist Jean Comandon (1877–1970), likewise portrays the growing motions of dandelions, daisies, cyclamen, lilies, and other plants in his *La croissance de végétaux* (Plant Growth) (1929), interspersing these with shots of a clock operating at hyperspeed to emphasize the accelerated temporality enabled by time-lapse microcinematography (Meeker and Szabari 2020: 121–3). *Das Blumenwunder* (The Miracle of Flowers) (1926) by Max Reichmann (1884–1958) puts the case most unambiguously of all. After opening with a personification of the goddess Flora instructing children to see plants as living beings rather than objects of spectatorship and possession, the film segues between time-lapse footage of moving plants, racing blood cells, pulsing veins, and dancers emulating plant movements (Janzen 2016: 95–127). More recent usages of time-lapse cinematography include the botanist Gordon Rowley's (1921–2019) *Cactus Polonaise* (1960), in which cacti dance to the music of Tchaikovsky, or, in a more pedagogical vein, the BBC's six-part documentary series *The Private Life of Plants* (1995) presented by David Attenborough (b. 1926), in which plants and fungi grow, sporulate, disseminate seeds, fertilize, parasitize, and obtain nourishment. In a different mood entirely, the surrealist thirty-second horror short *Flora* (1989) by Jan Švankmajer (b. 1934) centers on a human-like figuration of the goddess, tied to a bed, comprised of decaying and efflorescing plant growths.

Anthropomorphized plants are legion in fantasy literature. They feature throughout the *Chronicles of Narnia* series (1950–6) by C. S. Lewis (1898–1963), for instance. Most notably, in *Prince Caspian* (1951), the "pale birch-girls," "willow-women," "queenly beeches," "shaggy oak-men, lean and melancholy elms, shock-headed hollies (dark themselves, but with their wives all bright with berries) and gay rowans" help to rout the occupying Telmarines (Lewis [1951] 2008: 116–17). Bilbo from Tolkien's *The Hobbit* (1937) similarly discerns "a sort of watching and waiting feeling" inside Mirkwood Forest, while "all the trees leaned over them and listened" as he and the rest of the company traverse the narrow, winding path (127, 129). The trees do more than watch and listen when Frodo and his companions enter the Old Forest in *The Fellowship of the Ring* (1954). One of the party members surmises that "all the trees were whispering to each other, passing news and plots along in an unintelligible language: and the branches swayed and groped without any wind." The trees also possessed the ability to move, had attacked in the past, and harbored resentment toward hobbits for their ruthless logging of the forest (Tolkien 1954a: 153–4). These premonitions come true when the party encounters Old Man Willow, who sedates the hobbits, attempts to drown and stifle them to death, and then laughs at their struggles while the surrounding forest muffles their cries for help (161–7; Ryan 2015). The Fanghorn Forest in Tolkien's *The Two Towers* (1954b), however, is home to a very different kind of arborescent figure: Treebeard the Ent, an anthropomorphic figure with eyes and fluent speech. He eventually allies with the hobbits and their cause against Saruman in revenge for forest clear-cutting, and rallies his fellow ents to subdue the wizard's stronghold with floodwaters (66–90, 176).

If the idea of anthropoid vegetation invites the comforting prospect of human–plant communion, it also evokes the specter of a botanical agency indifferent or actively opposed to human existence. Indeed, plants' ambivalent affinities with human nature, and their seeming capacity for limitless growth and reasonless action, make them powerful villains when they refuse to serve as passive instruments for human concerns (Keetley 2016). One character in *The Vortex* (1924) by Colombian author José Eustasio Rivera (1888–1928), for instance, envisions the trees of the Amazon rainforest as "paralyzed giants" gesticulating and conversing at night. "They complained of the hand that scored them," Rivera writes, alluding to the region's pernicious rubber trade, "the ax that felled them. They were condemned to flourish, flower, grow, perpetuate their formidable species unfructified, unfecundated, uncomprehended by man." The novel's protagonists ultimately become collateral victims of botanical vengeance, "devoured" by the jungle through which they fled (Jaramillo 2016: 99–101; Rivera [1924] 2018: 219; Vieira 2019: 228–30). More conspicuously, the giant carnivorous plants at the center of John Wyndham's (1903–69) novel *Day of the Triffids* (1951) can walk, communicate, think, and learn, leaving humans with the sole advantage of sight (Keetley 2016: 11; Matthews 2016: 113–19)—which dissipates when a meteor shower blinds virtually everyone on earth. Lest the moral be missed, the triffids' first victim in the novel's 1962 film adaptation by Steve Sekely (1899–1979) is a security guard working the night shift inside the Palm House at Kew. Botanical monstrosity takes a more patently human form in the film *The Thing from Another World* (1951), based on John W. Campbell's (1910–71) novel *Who Goes There?*

(1938), which features a bloodthirsty alien, played by James Arness, comprised entirely of plant tissue, and thus immune to bullets, but which walks, vocalizes, uses objects, and, with its superior intellect, even engineers its own interstellar transportation. The 1982 and 2011 film adaptations abandon the monster's anthropoid outer form in favor of an amorphous, tendril-like entity that thrives by invading and taking on the appearance of people (Knee 2016: 147–9; Scherer 2016: 44–9). Invasive plants are likewise prominent in Sam Raimi's (b. 1959) horror film *Evil Dead* (1981), where a young woman is raped by demonically possessed vines; *Children of the Corn* (1984), directed by Fritz Kiersch (b. 1951) and based on a 1977 short story by Stephen King (b. 1947), in which a protagonist is ensnared by corn stalks, and in the later version (2009), by vines (Tenga 2016: 66–9); and especially Carter Smith's (b. 1971) *The Ruins* (2008), an adaptation of a 2006 novel by Scott Smith (b. 1965), where the victims encounter vines at a Mayan temple that imitate human sounds and consume the outside and inside of their bodies (Keetley 2016; Scherer 2016: 41, 44–9). The plants in both the Don Siegel (1912–91) and Philip Kaufman (b. 1936) versions of *Invasion of the Body Snatchers* (1956, 1978), based on a 1954 novel by Jack Finney (1911–95), infiltrate human society by taking peoples' outer forms while they sleep and killing the originals (Knee 2016: 149–51). In the novel *Greener than You Think* (1947) by Ward Moore (1903–78), and Thomas Disch's (1940–2008) *The Genocides* (1965), plant colonization goes so far as to precipitate the extinction of humanity—via, respectively, an omnivorous cultivar of Bermuda grass (*Cynodon dactylon*), and a monoculture of giant trees maintained by extraterrestrial farmers who exterminate humans as agricultural pests (Anderson 2016). *The Ruins* hints similarly by closing with a lone survivor escaping the temple while vines writhe beneath her face. Botanical revenge in M. Night Shyamalan's (b. 1970) film *The Happening* (2008) likewise has a global scope, albeit conditionally: the neurotoxins released by plants as a defence mechanism are by implication reversible, provided that humanity adopt a more equitable posture toward non-human nature (Williams 2016: 233–40).

Plants have also, more subtly, worked to buttress narratives that problematize dichotomous, anthropocentric, and at least implicitly Judeo-Christian worldviews. The strait-laced Christian policeman in Robin Hardy's (1929–2016) *The Wicker Man* (1973), for instance, encounters an island culture that has reverted to the paganistic worship of their ancestors. His diminutive appearance throughout the film—in relation both to the island's bountiful vegetation, and to the islanders themselves, most of whom are named after plants or plant-derived entities—foreshadows his own sacrifice in the name of ensuring a copious agricultural harvest (Tenga 2016: 62–6). The journey upriver into the jungle in *Apocalypse Now* (1979), directed by Francis Ford Coppola (b. 1939), on the other hand, serves as an allegory for the obverse of the purportedly virtuous intentions underlying America's war in Vietnam. Colonel Kurtz, an officer-turned-madman and demagogue who inflicts countless horrors from his isolated position, "took his orders from ... the jungle," the protagonist remarks, before killing him (Melbye 2010: 118–21). The Ghostwood forest surrounding the town of Twin Peaks—rendered in eerie, slow-motion close-ups of mist-enshrouded Douglas firs (*Pseudostuga menziesii*) blowing in the wind—likewise serves as a fitting backdrop for the initial run of Mark Frost (b. 1953) and David Lynch's (b. 1946) uncanny portrayal of murder, incest, and demonic possession in the eponymous TV series (1990–1). "There's a sort of evil out there," a local sheriff warns, "something very, very strange in these old woods. Call it what you want. A darkness, a presence." Confronting this evil requires enlisting the help of characters who repudiate Western binarisms: a Native American policeman, and more pointedly, a woman who communicates directly with a log that bears witness to the events unfolding in the forest (Bil 2016). In *Life of Pi* (2001) by Yann Martel (b. 1963), finally, the narrator encounters a carnivorous algae island that survives by turning acidic at night and digesting human beings and other life forms. The novel is widely seen as an exploration of the boundary between reason and imagination, and the relationship between human consciousness and the non-human world (Ilić 2017).

The sheer prepotency of plant horror narratives has spawned its own spoof subgenre. The eponymous villains of John De Bello's film (b. 1952) *Attack of the Killer Tomatoes* (1978) swim, hurl themselves, crush their victims, and converse in a kind of indecipherable burble, with no semblance of realism. Yet the fact that the malefactors derive from rationalized agriculture, emblematic of human dominion over nature—a point reinforced when a sentient carrot rises from the ground once the tomatoes have gone—aligns with the soberer content above (Parker 2016: 219). The giant paper-mâché-looking Venus flytrap named Audrey Jr. (gendered masculine but named after the film's love interest) in *The Little Shop of Horrors* (1960), directed by Roger Corman (b. 1926), meanwhile, goes so far as to devour people wholesale. Shouting "Feed me!," the plant hypnotizes a blundering young man into providing him with human food, until the man revolts and loses his life in the process. In Frank Oz's (b. 1944) cartoonish musical remake (1986), Audrey II is larger and more menacing than before, possesses exaggerated red lips in the vein of racist black caricature, and sings R&B while he wields his corrupting influence. The film thus interweaves concerns regarding botanical vengeance with equally entrenched white anxieties surrounding American race relations in the 1980s (Farnell 2016: 188–92; Jensen 2008). Most unsparingly of all, *Otesánek* (Little Otik) (2000), a Czech film based on a folk-tale by Karel Jaromír Erben (1811–70) and directed by Jan Švankmajer, segues from newborn babies appearing inside newly cut melons, and netted and weighed on scales like fish at a market stall, to the story of Otik himself: a garden stump with roots for hands, wished into existence as a newborn for a childless couple, dressed and speaking like a human infant, with a single eye that occasionally appears in place of his circular, tooth-filled mouth, who winds up eating his own parents. Otik's garden origin, to be sure, bespeaks a familiar parable about anthropocentric hubris (Parker 2016).

In *The Secret Life of Plants* (1973), a book by American journalists Peter Tompkins (1919–2007) and Christopher Bird (1928–96), the interrogation of human–plant boundaries reaches the point of self-parody. A motley combination of theosophy, parapsychology, hypnotism, psychokinesis, radionics, and other esoteric fields, alongside the work of recognized authorities such as Darwin and Bose, the book claims that plants feel emotions, count, can empathize with humans, and, most exorbitantly, are endowed with the capacity for interstellar communication (Laird 2018: 61–3; Tompkins and Bird 1973). A documentary film adapted from the book (1978) uses time-lapse photography to dramatize plant movement and agency, while electronic screen displays and whirring sounds underline the purportedly scientific nature of the research. More successfully perhaps, *Parallel Botany* (1977) by Leo Lionni (1910–99) adumbrates a similarly dubious floral universe with its tongue planted securely in cheek. The "para-material" or "matterless" species described therein—including "trees which produce animals," and the "*Anaclea taludensis*," which is "impossible to pick" because it "vaporize[s] instantly on contact with a hand"—"[elude] chemical analysis," contradict "all known laws of physics," and survive "unconstrained by any known laws of nature" (Lionni 1977: 6, 12, 52).

Late-twentieth- and early-twenty-first-century plant science, however, puts speculative forays into the extraordinary, and at times uncannily human-like, abilities of plants on

FIGURE 8.5 Audrey II, from *Little Shop of Horrors* (1986). Photo by Murray Close. Photo by Murray Close/Sygma/Sygma/Getty Images.

sounder footing. Plants, researchers have shown, "see" by distinguishing qualities and quantities of light and darkness; they communicate through "smell" using airborne organic compounds; they respond to "touch" by closing or withdrawing in response to a perceived threat or prey, or by winding around objects; they "hear" by sensing vibrations through their roots; they "taste" via roots that pursue nutrients in the soil, or through the development of above-ground carnivorous or protocarnivorous modalities (Mancuso and Viola [2013] 2015: 45–80); they even "learn from association" in modifying their anticipative behavior in response to conditioning stimuli (Gagliano 2018: 79–85). Assimilating some of these insights, a paleobotanist in Michael Crichton's (1942–2008) novel *Jurassic Park* (1990) hints at a connection between the hubris of resurrecting antediluvian reptiles for a theme park, and a zoocentric conception of plants—in this case a highly poisonous fern—as scenery embellishments. "People who imagined that life on earth consisted of animals moving against a green background seriously misunderstood what they were seeing," she reflects.

> That green background was busily alive. Plants grew, moved, twisted, and turned, fighting for the sun; and they interacted continuously with animals—discouraging some with bark and thorns; poisoning others; and feeding still others to advance their own reproduction, to spread their pollen and seeds. It was a complex, dynamic process which she never ceased to find fascinating. And which she knew most people simply didn't understand.
>
> (Chrichton 1990: 85–6)

A dendrologist in *Overstory* (2018) by Richard Powers (b. 1957) applies similar findings in her realization that trees pool resources, support each other, communicate warnings via aerosol signals, and share an extensive, densely interconnected rhizospheric network underground (Powers 2018: 112–44). Sue Burke's (b. 1957) science fiction novel *Semiosis* (2018) gives this idea fantastical expression in her exploration of how a party of interplanetary colonists learn to communicate with a plant, which—and shortly, whom—they refer to as a rainbow bamboo. Their very existence, in fact, hinges on reciprocating the bamboo's efforts to converse using pigment, odor, acidity, and electricity—a hypothesis corroborated in sections of narrative delivered from the first-person perspective of the plant itself. The humans and bamboo—who eventually acquires a name, Stevland—thereby forge a mutualistic relationship, eventually learning to correspond via a pictographic language, and later spoken English, once Stevland develops the means of vocalizing and processing human speech.

CONCLUSION

The foregoing surveys only a fraction of the ways that plants have figured in fiction, visual illustrations, and plastic arts from the early twentieth century to the present. To judge from even this limited selection of case studies, however, the agency of the modern period in forging new ways of representing plants is indelible: stimulating insights in plant biology; revolutionizing film, photographic and communications technologies; shaping new standards of scientific representation; viewing plants by analogy with emergent social and psychological conflicts; and spurring material, visual, cinematic, and literary forms that put human–plant relationships on a less anthropocentric footing. These developments have spawned an array of novel philosophical approaches to understanding human–plant and human–nature relationships, from posthumanism (Haraway [1985] 2004; Latour 2004) to object-oriented ontology (Bennett 2009; Morton 2010), multi-species ethnography (Kohn 2013; Tsing 2015), critical plant studies (Gagliano et al. 2019; Marder 2013) and others. Such insights are consonant with what numerous Indigenous writers have observed regarding the kinship and interdependence of human beings with plants, animals, and other elements of the non-human environment (Kimmerer 2013; Salmon 2012). These and further tools will doubtless shape future ways of seeing, depicting, and describing plants, as twenty-first-century climatic developments repudiate human-centered fantasies with a vengeance.

ACKNOWLEDGMENTS

The author is indebted to the helpful staff at the University of Delaware Interlibrary Loan office. This chapter likewise benefited from discussions with Robert Brain, Nadia Chiu, Makalé Faber Cullen, Lowell Duckert, Sonnet L'Abbé, Brittany Luby, and Jaipreet Virdi. I am additionally grateful for the patience and encouragement of Stephen Forbes, David Mabberley, Tristan Palmer, and especially Annette Giesecke. All errors and shortcomings are of course my own.

NOTES

Chapter 1

1 A bit less than half of livestock feed is provided by grazing, much of it on natural pastures that are unsuited for cultivation (Mottet et al. 2017).
2 What is agrobiodiversity? FAO Factsheet from Training Manual "Building on Gender, Agrobiodiversity and Local Knowledge." FAO, 2004. Available online: http://www.fao.org/3/a-y5609e.pdf (accessed October 16, 2020).
3 Maize and sugarcane use the C4 photosynthetic pathway with inherently higher yield potential than other predominantly C3 grasses.
4 Throughout, this chapter uses food and agricultural statistics from the UN Food and Agricultural Organization, FAOSTAT at http://www.fao.org/faostat, and population statistics are from the UN Population Division. Unless otherwise stated, most nutritional statistics are taken from the 2016 Global Nutrition Report published by the International Food Policy Research Institute.
5 The term "mono-cropping" is widely used to refer to specialization in one or two crops. However, technically mono-cropping refers to the growing of one crop in a field as opposed to intercropping of mixtures of crops.
6 However, a large number of processed foods depend on ingredients obtained from GM maize and soybean.
7 Important exceptions are intensive livestock rearing in much of the rich world which is now carried out in large corporate operations, as is some horticultural production.
8 Based on data from Government of South Australia Statistical Register, 1920–1 and http://www.agriculture.gov.au/abares/research-topics/surveys/farm-survey-data#summary-data-tables (accessed April 4, 2020). Farm size includes land in fallow or temporary pastures.
9 Estimation methods have varied over time, so data from before 1990 is not strictly comparable.
10 High fructose corn syrup commonly includes 55 percent fructose and the rest as glucose.
11 For example, such as JBC from Brazil in meat, Bimbo from Mexico in bread, and Olam and Wilmar from Singapore in edible oils and rice are now major players in developed country markets.
12 The term "agri-culture wars" was coined by Kerry Trueman, https://www.theatlantic.com/health/archive/2010/12/culture-wars-how-junk-food-and-obesity-became-politicized/67841/ (accessed April 4, 2020).
13 "Sales Fall Again in Mexico's Second Year of Taxing Soda," *The New York Times*, February 22, 2017, https://www.nytimes.com/2017/02/22/upshot/soda-sales-fall-further-in-mexicos-second-year-of-taxing-them.html (accessed April 4, 2020).

Chapter 2

1 This paragraph draws heavily on the business management and marketing article by Berthon et al. (2009).
2 Various kinds of plantains are starchy bananas that serve as a staple in many tropical regions. Easy to grow and highly productive, they became a popular slave food

(Soluri 2005: 6, 16). Export trade focused on sweet bananas which were initiatively novel outside the tropics prior to commercialization of "Gros Michel" and later "Cavendish," the current dominant commercial cultivar in global trade.

3 For example, during the Irish potato famine, in some regions of Ireland absentee landowners were exporting wheat to England while Irish farmers and their family members died of starvation and disease.

4 For more discussion of some of the environmental threats involved, see Lobell et al. (2006); Marelli et al. (2019); Money (2007); Parker and Abatzoglou (2018); and Sternberg (2015).

5 Note that the word luxury is derived from the Latin *luxus* which means "'excess, extravagance,' and even 'vicious indulgence' ... and more recently ... [an] escape from, or cure of, the ordinary and the struggle for betterment" (Berthon et al. 2009: 46).

6 In the US now, even purveyors of discount foods such as Costco, Grocery Outlet, and Walmart sell superfoods cheaply. Until fairly recently, many of these foods marketed as medicinal and/or nutrient-dense, like chia (*Salvia columbariae* and other *S.* spp.), hemp hearts (*Cannabis sativa*), maca (*Lepidium meyenii*), and quinoa (*Chenopodium quinoa*), could be obtained only from specialty grocers.

7 Sunkist remains a non-profit marketing cooperative made up of citrus grower members.

8 Both of these citrus plants are classified as different cultivars of the same hybrid *Citrus X aurantium.*

9 Oranges often served as symbolic proof of California as an American agricultural paradise during the late nineteenth and early twentieth centuries.

10 Starbucks was founded in 1971 in Seattle and owes its early success to Peet's generosity in sharing his roasting practices (Marshall 2007). In 2016, Peet's Coffee gave $250,000 to launch the Coffee Center at the University of California, Davis, a research hub dedicated to the multidisciplinary study of coffee (Pereira 2016).

11 Due to similar morphology and ecology, *Sparassis radicata* was previously confused with *S. crispa* which occurs in Europe and Britain, and rarely in eastern North America. *S. radicata* is limited to western North America. For a discussion on current nomenclature, see "Plants of the World Online" hosted by KewScience (http://powo.science.kew.org/taxon/urn:lsid:indexfungorum.org:names:228848 [accessed April 29, 2021]).

12 In 2007, the EU adopted "precautionary chemical legislation, called REACH ... which [placed] the burden of proof of safety on the chemical industry" (MacKendrick 2018: 29). Relative to the US, many more chemicals used in food production and body products are restricted in the EU. See MacKendrick (2018) for more discussion on this topic.

13 "Approximately 44% of hospitalized elderly worldwide are nutritionally deficient" in ways that contribute to negative outcomes including longer recovery or premature decline (Birch and Bonwick 2019: 1477).

14 *Manoomin* is the word for wild rice in the Ojibwe language.

15 This is true of many crops that humans depend on, including staples like wheat.

16 According to Marelli et al. (2019: 1340), "disease spread and management are further exacerbated by the ... poverty faced by most smallholder cacao farmers" who often do not have the resources to implement phytosanitary measures.

17 The confectionery enterprise, Mars Incorporated, helped to fund a gathering of cacao scientists that then led to the writing of the Marelli et al. (2019) article. Mars also contributed funding to genomic research on cacao. Together, these raise concerns about possible conflict of interest and undue corporate influence on scientific research and food and agricultural policy recommendations based on that research.

Chapter 5

1 The term "active principle" refers to the intrinsic chemical substance which induces pharmacological activity—see Capasso et al. (2003: 31).

2 Madaus' journal *Biologische Heilkunde* even included some articles on Ayurveda and its importance for herbal medicine.

3 This section is based on an investigation conducted with the anthropologist Laurent Pordié.

4 One of the most eloquent cases is that of the quinquina (*Cinchona* spp.) bark, massively employed against fevers and malaria from the seventeenth century onward, for which, in contrast to the plants included in treatises like the Hortus, its native origins and local uses are barely mentioned in the work of Europeans and never associated with names or specific circumstances.

Chapter 7

1 1. A thing used or serving to make something look more attractive but usually having no practical purpose, especially a small object such as a figurine. Synonyms: knick-knack, trinket, bauble, piece of bric-a-brac, bibelot, gewgaw, gimcrack, furbelow, object, accessory. 2. The accessories of worship. Of interest is an earlier definition from the *Oxford English Dictionary* 3rd edition that retains the primacy of "ornament" to practice—"1. a. An accessory or adjunct, primarily functional, but often also fancy or decorative" (*OED* n.d.-b).

2 Loos' lecture is often quoted as 1908. However, Christopher Long argues that the lecture was given in 1910—see Long (1997: 440–5).

3 Container-grown ornamental plants are variously referred to as nursery stock or P4Ps—an acronym for Plants for Planting.

4 Jim Nau et al. (1999) outlines the history of key American ornamental plant breeders in the modern era and Johan Van Huylenbroeck (2018) has edited a volume of recent case studies in breeding and production for a range of plants in key ornamental genera.

5 Reichl (2016) saw the High Line in the context of a democratic public space but the increasing value of surrounding real estate has now created real challenges to that notion.

BIBLIOGRAPHY

Ackerknecht, E. H. 1976. *Medicine at the Paris Hospital, 1794–1848*. Baltimore: Johns Hopkins University Press.

Adams, W. M. 2004. *Against Extinction: The Story of Conservation*. London: Earthscan.

Agence France-Presse in Chibuto. 2017. "'Miracle' Baby Born in Tree during Mozambique Floods to Turn 17." *The Guardian*, February 28. Available online: https://www.theguardian.com/world/2017/feb/27/miracle-baby-born-tree-mozambique-floods-turns-17 (accessed July 20, 2020).

Agkathidis, Asterios. 2017. *Biomorphic Structures*. London: Laurence King.

Agrawal, A., and C. C. Gibson. 1999. "Enchantment and Disenchantment: The Role of Community in Natural Resource Conservation." *World Development*, 27(4): 629–49.

Ahloowalia B. S., M. Maluszynski, and K. Nichterlein. 2004. "Global Impact of Mutation-Derived Varieties." *Euphytica*, 135: 187–204.

Aichele, K. Porter. 2006. *Paul Klee, Poet/Painter*. Rochester, NY: Camden House.

Ainsworth, P. 1930. "The Meredith Publications." *The Palimpsest*, 11(6): 256–65.

Alice Gallery. 2012. *Wait and Watch Awhile Go By*. Paintings by Paul Wackers. Available online: https://alicebxl.com/index.php/exhibitions/wait-and-watch-awhile-go-by (accessed May 16, 2021).

Alon-Mozes, Tal. 2013. "Landscape as a National Text: The Biblical Landscape Reserve of Neot Kedumim, Israel." *Studies in the History of Gardens & Designed Landscapes*, 33(4): 305–20.

Anderson, Jill E. 2016. "The Revenge of the Lawn: The Awful Agency of Uncontained Plant Life in Ward Moore's *Greener than you Think* and Thomas Disch's *The Genocides*." In Dawn Keetley and Angela Tenga (eds.), *Plant Horror: Approaches to the Monstrous Vegetal in Fiction and Film*, 129–43. London: Palgrave Macmillan.

Andreotti, Margherita. 1989. *The Early Sculpture of Jean Arp*. Ann Arbor, MI: UMI Research.

Ankeny, Rachel A. 2012. "Food and Ethical Consumption." In Jeffrey M. Pilcher (ed.), *The Oxford Handbook of Food History*, ch. 26, 461–80. New York: Oxford University Press.

Antonelli, A., R. J. Smith, and M. S. J. Simmonds. 2019. "Unlocking the Properties of Plants and Fungi for Sustainable Development." *Nature Plants*, 5(11): 1100–2.

Apocalypse Now. 1979. [Film] Dir. and Prod. Francis Ford Coppola. Beverly Hills: United Artists.

Appleby, M. 2017. "The Effect of Rising Costs on Single-Plant Societies." *Horticulture Week* (May 26), 6.

Araus J. L., and J. E. Cairns. 2014. "Field High-Throughput Phenotyping: The New Crop Breeding Frontier." *Trends in Plant Science*, 19: 52–61.

Arets, E. J. M. M., P. J. van der Meer, C. C. Verwer, G. -J. Nabuurs, G. M. Hengeveld, G. W. Tolkamp, and M. Van Oorschot. 2010. *Global Wood Production: Assessment of Industrial Round Wood Supply from Different Management Systems in Different Global Regions*. Wageningen, Alterra: Alterra-rapport 1808. Available online: https://edepot.wur.nl/196265 (accessed May 16, 2021).

Aronson, J. 2014. "The Ecological Restoration Alliance of Botanic Gardens: A New Initiative Takes Root." *Restoration Ecolology*, 22: 713–15.

Arora, David. 1986. *Mushrooms Demystified: A Comprehensive Guide to the Fleshy Fungi*. Berkeley, CA: Ten Speed Press.

Around the World in 80 Gardens. 2008. [TV program] BBC Two. Presenter Monty Don, Series Producer Kerry Richardson. Available online: https://www.bbc.co.uk/programmes/b008x9bh (accessed May 16, 2021).

Arp, Jean. 1919. Woodcut. *Dada*, 4/5: 19.

Ashton, P. S. 1981. Forest Conditions in the Tropics of Asia and the Far East. In V. H. Sutlive, N. Altshuler, and M. D. Zamora (eds.). 1981. *Where Have All the Flowers Gone? Deforestation in the Third World*. Studies in Third World Societies 13. College of William and Mary, Virginia.

Asner, G. P., T. K. Rudel, T. M. Aide, R. DeFries, and R. Emerson. 2009. "A Contemporary Assessment of Change in Humid Tropical Forests." *Conservation Biology*, 23(6): 1386–95.

Attack of the Killer Tomatoes (1978). [Film] Dir. John DeBellow. Prod.John DeBello and J. Stephen Pearce. NAI Entertainment.

Atwood, M. 2012. "Margaret Atwood: Rachel Carson's Silent Spring, 50 Years On." *The Guardian*, December 8. Available online: https://www.theguardian.com/books/2012/dec/07/why-rachel-carson-is-a-saint (accessed May 16, 2021).

Avatar. 2009. Dir. James Cameron. Prod. Cameron and John Landau. Los Angeles: Twentieth-Century Fox.

Bachman, S., E. M. Nic Lughadha, and M. C. Rivers. 2018. "Quantifying Progress Toward a Conservation Assessment for All Plants." *Conservation Biology*, 32(3): 516–24.

Baker, Keiligh. 2016. "'They are Trivialising the Sacrifice of Millions': BBC Slammed for 'Disgraceful' Decision to Pin a Poppy on the Cookie Monster during One Show Appearance." *Daily Mail*, November 8. Available online: https://www.dailymail.co.uk/news/article-3916416/BBC-slammed-disgraceful-decision-pin-poppy-Cookie-Monster.html (accessed May 16, 2021).

Balding, M., and K. J. H. Williams. 2016. "Plant Blindness and the Implications for Plant Conservation." *Conservation Biology*, 30(6): 1192–9.

Bale, Rachael. 2016. "The Disturbing Secret Behind the World's Most Expensive Coffee." *National Geographic*, April 29. Available online: https://www.nationalgeographic.com/news/2016/04/160429-kopi-luwak-captive-civet-coffee-Indonesia/ (accessed May 16, 2021).

Balmford, A., L. Clegg, T. Coulson, and J. Taylor. 2002. "Why Conservationists Should Heed Pokémon." *Science*, 295: 2367.

Balmori Associates. 2009. "The Gardens that Climbs the Stairs 2009—Bilbao, Spain." Available online: http://www.balmori.com/portfolio/the-garden-that-climbs-the-stairs (accessed March 3, 2020).

Balmori, D., and M. Morton. 1993. *Transitory Gardens, Uprooted Lives*. New Haven, CT: Yale University Press.

Banerjee, M. 2009. *Power, Knowledge, Medicine: Ayurvedic Pharmaceuticals at Home and in the World*. Hyderabad: Orient Blackswan.

Barnard, Timothy P. 2016. *Nature's Colony: Empire, Nation and Environment in the Singapore Botanic Gardens*. Singapore: NUS Press.

Barrett, W. (ed.). 2018. "The 100 Largest US Charities 2018 #20 The Nature Conservancy: America's Top Charities 2018." *Forbes Magazine*, December 11. Available online: https://www.forbes.com/companies/nature-conservancy/?list=top-charities#46a8930341d2 (accessed May 16, 2021).

Barron, Cheryll Aimee. 1988. "Madam Cocolat." *The New York Times*, September 25, Section 6, 31. Available online: https://www.nytimes.com/1988/09/25/magazine/madam-cocolat.html (accessed September 22, 2019).

Barthlott, W., A. Hostert, G. Kier, W. Küper, H. Kreft, J. Mutke, M. D. Rafiqpoor, and H. Sommer. 2007. "Geographic Patterns of Vascular Plant Diversity at Continental to Global Scales." *Erdkunde*, 61: 305–15.

Bates, H. E. [1960] 1974. "An Aspidistra in Babylon." In *The Grapes of Paradise*, 153–90. Hardmondsworth: Penguin.

Baxter, D. E., and L. G. Pelletier. 2019. "Is Nature Relatedness a Basic Human Psychological Need? A Critical Examination of the Extant Literature." *Canadian Psychology*, 60(1): 21–34.

Bebber, D. P., M. A. Carine, J. R. I. Wood, A. H. Wortley, D. J. Harris, G. T. Prance, G. Davidse, J. Paige, T. D. Pennington, N. K. B. Robson, and R. B. Scotland. 2010. "Herbaria are a Major Frontier for Species Discovery." *PNAS*, 107(51): 22169–71.

Beck, T. E. 2002. "Gardens as a 'Third Nature': The Ancient Roots of a Renaissance Idea." *Studies in the History of Gardens and Designed Landscapes*, 22(4): 327–34.

Beech, E., M. Rivers, S. Oldfield, and P. Smith. 2017. "GlobalTreeSearch: The First Complete Global Database of Tree Species and Country Distributions." *Journal of Sustainable Forestry*, 36(5): 454–89.

Belasco, Warren. 2008. *Food: The Key Concepts*. Oxford: Berg.

Belhaj, K., A. Chaparro-Garcia, S. Kamoun, N. J. Patron, and V. Nekrasov. 2015. "Editing Plant Genomes with CRISPR/Cas9." *Current Opinion Biotech*, 32: 76–84.

Bennett, Jane. (2009). *Vibrant Matter: A Political Ecology of Things*. Durham: Duke University Press.

Bentley, Amy. 2013. "Historical Overview: World War II." In Andrew F. Smith (ed.), *The Oxford Encyclopedia of Food and Drink in America*, 2nd edn., vol. 2, 227–33. New York: Oxford University Press.

Bentley, Amy. 2016. "Introduction." In Amy Bentley (ed.), *A Cultural History of Food: In the Modern* Age, vol. 6, 1–25. London: Bloomsbury.

Berger, A. 2006. *Drosscape: Wasting Land in Urban America*. New York: Princeton Architectural Press.

Berger, P. L. (1999). "The Desecularization of the World: A Global Overview." In P. L. Berger (ed.), *The Desecularization of the World: Resurgent Religion and World Politics*, 1–18. Grand Rapids, Michigan: William B. Eerdmans Publishing Company.

Bergren, K. 2015. "Localism Unrooted: Gardening in the Prose of Jamaica Kincaid and William Wordsworth." *ISLE Interdisciplinary Studies in Literature and Environment*, 22(2): 303–25.

Berry, Wendell. 2002. "Think Little." In N. Wirzba (ed.), *The Art of the Commonplace: The Agrarian Essays of Wendell Berry*, 81–90. Berkeley, CA: Counterpoint.

Berry, Wendell. 2018. "For Love of Place: Reflections of an Agrarian Sage." *Great Transition Initiative*. Available online: https://greattransition.org/publication/for-love-of-place (accessed July 19, 2020).

Berthon, Pierre, Leyland Pitt, Michael Parent, and Jean-Paul Berthon. 2009. "Aesthetics and Ephemerality: Observing and Preserving the Luxury Brand." *California Management Review*, 52(1): 45–66.

Betsky, Aaron. 2002. *Landscrapers: Building with the Land*. New York: Thames & Hudson.

BGCI. 2020. PlantSearch. Available online: https://www.bgci.org/plant_search.php (accessed May 16, 2021).

Biberg, Isaac. 1749. "*Linnaeus, Carl, The Oeconomy of Nature, 1749*." In Benjamin Stillingfleet (trans.) and J. Dodsley (ed.), *Miscellaneous Tracts Relating to Natural History Husbandry and Physick*. London: Arno Press, 1977.

Bil, Geoff. 2016. "Tensions in the World of Moon: *Twin Peaks*, Indigeneity and Territoriality." *Senses of Cinema*, 79. Available online: http://sensesofcinema.com/2016/twin-peaks/twin-peaks-indigeneity-territoriality/ (accessed May 16, 2021).

Billock, J. 2017. "These Fake Trees Were Used as Spy Posts on the Front Lines of World War I." Smithsonianmag.com. Available online: http://www.smithsonianmag.com/travel/fake-spy-trees-WWI-180962179/ (accessed July 19, 2020).

Bilz, M., S. P. Kell, N. Maxted, and R. V. Lansdown. 2011. *European Red List of Vascular Plants*. Luxembourg: Publications Office of the European Union.

Birch, C. S., and G. A. Bonwick. 2019. "Ensuring the Future of Functional Foods." *International Journal of Food Science and Technology*, 54: 1467–85.

Birdlife International. 2019. Our History BirdLife—the World's Oldest International Conservation Organisation. Available online: https://www.birdlife.org/worldwide/partnership/about-birdlifeand https://www.birdlife.org/worldwide/partnership/our-history (accessed May 16, 2021).

Blackmore, S. 2009. *Gardening the Earth: Gateways to a Sustainable Future*. Edinburgh: Royal Botanic Garden.

Blackmore, S., and S. F. Oldfield (eds.). 2017. *Plant Conservation Science and Practice. The Role of Botanic Gardens*. Cambridge: Cambridge University Press.

Blackmore, S., and S. F. Oldfield. 2017. "Mounting a Fundamental Defence of the Plant Kingdom." In S. Blackmore and S. F. Oldfield (eds.), *Plant Conservation Science and Practice. The Role of Botanic Gardens*, ch. 1. Cambridge: Cambridge University Press.

Blanc, Patrick. 2008. *The Vertical Garden: From Nature to the City*. New York: W. W. Norton & Co.

Blanc, Patrick. 2016. "Patrick Blanc—Vertical Gardens." Available online: http://www.inexhibit.com/case-studies/patrick-blanc-vertical-gardens/ (accessed May 16, 2021).

Blunt, Wilfrid. [1950] 1994. Assisted by William T. Stearn. *The Art of Botanical Illustration: An Illustrated History*, 4th edn. New York: Dover.

Bode, M. 2008. *Taking Traditional Medicine to the Market: The Modern Image of the Ayurvedic and Unani Industry 1980–2000*. Hyderabad: Orient Longman.

Boettcher, M., and M. T. McManus. 2015. "Choosing the Right Tool for the Job: RNAi, TALEN, or CRISPR." *Molecular Cell*, 58: 575–85.

Booker, A., D. Johnston, and M. Hiendrich. 2012. "Value Chains of Herbal Medicines—Research Needs and Key Challenges in the Context of Ethnopharmacology." *Journal of Ethnopharmacology*, 140(3): 624–33.

Bose, Jagadish, Chandra. 1926. *The Nervous Mechanism of Plants*. London: Longmans, Gren and Co.

Bosker, Bianca. 2017. "Why Should a Melon Cost as Much as a Car?" *Roads and Kingdoms*, March 27. Available online: https://roadsandkingdoms.com/2017/why-should-a-melon-cost-as-much-as-a-car/ (accessed May 16, 2021).

Bosma, U., and J. Curry-Machado. 2013. "Turning Javanese: The Domination of Cuba's Sugar Industry by Java Cane Varieties." *Itinerario*, 37(2): 101–20.

Boumediene, S. 2016. *La colonisation du savoir. Une histoire des plantes médicinales du "Nouveau Monde" (1492–1750)*. Vaulx en Velin: Les Editions des Mondes à Faire.

Bourke, M. 2013. "The National Arboretum Canberra: An Historic Concept with an Australian Twist." *Australian Garden History*, 24(4): 23–7.

Bowcutt, Frederica. 2015. *The Tanoak Tree: An Environmental History of a Pacific Coast Hardwood*. Seattle, WA: University of Washington Press.

Boyes, R. 2000. "Nazi Symbol Falls to the Chainsaw." *The Times*, December 5, 21.

Bozzano, M., R. Jalonen, E. Thomas, D. Boshier, L. Gallo, S. Cavers, S. Bordács, P. Smith, and J. Loo (eds.). 2014. *Genetic Considerations in Ecosystem Restoration Using Native Tree Species. State of the World's Forest Genetic Resources—Thematic Study*. Rome: FAO and Biodiversity International.

Braidwood, Ellen. 2015. "If You Don't Wear a Poppy, You're Sent to the Social Guillotine. Is This What British Soldiers Died For?" *The Independent*, November 4. Available online: http://www.independent.co.uk/voices/if-you-dont-wear-a-poppy-youre-sent-to-the-social-guillotine-is-this-what-british-soldiers-died-for-a6720786.html (accessed May 16, 2021).

Bramwell, D. 2011. "Introduction: Islands and Plants." In D. Bramwell and J. Caujupé Castells (eds.), *The Biology of Island Floras*, ch. 1, 1–10. Cambridge: Cambridge University Press.

Brenchley R., M. Spannagl, M. Pfeifer, G. L. A. Barker, R. D'Amore, A. M. Allen, N. McKenzie, M. Kramer, A. Kerhornou, D. Bolser, S. Kay, D. Waite, M. Trick, I. Bancroft, Y. Gu, N. Huo, M. -C. Luo, S. Sehgal, B. Gill, S. Kianian, O. Anderson, P. Kersey, J. Dvorak, W. R. McCombie, A. Hall et al. 2012. "Analysis of the Bread Wheat Genome Using Whole-Genome Shotgun Sequencing." *Nature*, 491: 705–10.

Bricout, Shirley. 2009. "From Babel to Eden: D. H. Lawrence's Quest Abroad." *Études britanniques contemporaines*, 37: 27–39.

Brockway, Lucile H. 2002. *Science and Colonial Expansion: The Role of the British Royal Botanic Gardens*. New Haven, CT: Yale University Press.

Brown, L. R. 1966. *Cómo Aumentar La Producción Mundial De Alimentos: Problemas Y Perspectivas*. Mexico City: Uteha.

Bruland, S. 2013. "Transgressing Religious Boundaries: The Power of Aesthetics in Tamil Catholic and Hindu Worship." *Material Religion*, 9(4): 418–40.

Brummitt, N., and S. Bachman. 2010. *Plants Under Pressure—a Global Assessment: The First Report of the IUCN Sampled Red List Index for Plants*. Royal Botanic Gardens, Kew, UK.

Brummitt, N. A., S. P. Bachman, J. Griffiths-Lee, M. Lutz, J. F. Moat, A. Farjon, J. S. Donaldson, C. Hilton-Taylor, T. R. Meagher, S. Albuquerque, E. Aletrari et al. 2015. "Green Plants in the Red: A Baseline Global Assessment for the IUCN Sampled Red List Index for Plants." *PloS one*, 10(8): e0135152. Available online: https://journals.plos.org/plosone/article?id=10.1371/journal.pone.0135152 (accessed May 16, 2021).

Buchan, Ursula. 2003. "Stella Ross-Craig, Botanical Artist." *New Plantsman*, n.s. 2(3): 152–9.

Buchanan, M., R. A. Burton, K. S. Dhugga, J. A. Rafalski, S. V. Tingey, N. J. Shirley, and G. B. Fincher. 2012. "Endo-(1,4)-β-Glucanase Gene Families in the Grasses: Temporal and Spatial Co-transcription of Orthologous Genes." *BMC Plant Biology*, 12: 235. DOI: 10.1186/1471-2229-12-235.

Burgess, P. F. 1989. "Asia." In D. Poore (ed.), *No Timber Without Trees. Sustainability in the Tropical Forest. A Study for ITTO*, ch. 5. London: Earthscan Publications Ltd.

Bürgi, M. 2011. *Pharmaforschung im 20. Jahrhundert—Arbeit an der Grenze zwischen Hochschule und Industrie*. Zurich: Chronos Verlag.

Burke, Sue. 2018. *Semiosis*. New York: Tor Books.

Busch, Lawrence, William B. Lacy, Jeffrey Burkhardt, Douglas Hemken, Jubel Moraga-Rogel, Timothy Koponen, and José De Souza Silva. 1995. *Making Nature, Shaping Culture: Plant Biodiversity in Global Context*. Lincoln: University of Nebraska Press.

Byerlee, D. 1987. "The Political Economy of Third World Food Imports: The Case of Wheat." *Economic Development and Cultural Change*, 35(2): 307–28.

Byerlee, D. 2021. "The Super State: The Political Economy of Phosphate Fertilizer Use in South Australia, 1880–1940." *Jahrb. f. Wirtschaftsg* [*German Economic History Yearbook*], 262 (1): 99–128.

Byerlee, D., and H. J. Dubin. 2009. "Crop Improvement in the CGIAR as a Global Success Story of Open Access and International Collaboration." *International Journal of the Commons*, 4(1): 452–80.

Byerlee, D., W. P. Falcon, and R. L. Naylor. 2017. *The Tropical Oil Crop Revolution: Food, Feed, Fuel, and Forests*. Oxford: Oxford University Press.

Cactus Polonaise. 1960. [Film] Dir. and Prod. Gordon Rowley. UK.

Cairns, Kate, Josée Johnston, and Shyon Baumann. 2015. "Caring about Food: Doing Gender in the Foodie Kitchen." In Josée Johnston and Shyon Baumann (eds.), *Foodies: Democracy and Distinction in the Gourmet Foodscape*, 2nd edn., 177–201. New York: Routledge.

Caldwell, D. G., N. McCallum, P. Shaw, G. J. Muehlbauer, D. F. Marshall, and R. Waugh. 2004. "A Structured Mutant Population for Forward and Reverse Genetics in Barley (Hordeum vulgare L.)." *Plant Journal*, 40: 143–50.

Callaway, E. (2018) "CRISPR Plants Now Subject to Tough GM Laws in European Union. Top Court's Ruling Threatens Research on Gene-Edited Crops in the Bloc." *Nature*, 560: 16. DOI: 10.1038/d41586-018-05814-6.

Callicott, J. B. 1993. "The Land Ethic Today." *Topoi*, 12(1): 41–51.

Calonje, M., D. W. Stevenson, and L. Stanberg. 2013–17. *The World List of Cycads*, online edn. Available online: http://www.cycadlist.org (accessed July 29, 2017).

Cameron, B. 2017. *Gardening in Eastern Africa*. Nairobi: Kenya Horticultural Society.

Capasso, F., T. S. Gaginella, G. Grandolini, and A. A. Izzo. 2003. "Active Principles." In *Phytotherapy*, ch. 8, 31–4. Berlin, Heidelberg: Springer.

Carder, G., H. Proctor, J. Schmidt-Burbach, and N. D'Cruze. 2016. "The Animal Welfare Implications of Civet Coffee Tourism in Bali." *Animal Welfare*, 25: 199–205.

Carrithers, M. 2001. *The Buddha: A Very Short Introduction*. Oxford: Oxford University Press.

Carson, Rachel. 1962. "Of Man and the Stream of Time." Speech on *Silent Spring* at Scripp's College, June 12. Available online: https://loa-shared.s3.amazonaws.com/static/pdf/Carson_Stream_Time.pdf (accessed May 16, 2021).

Carson, Rachel, Lois Darling, and Louis Darling. 1962. *Silent Spring*. Cambridge, MA: Houghton Mifflin Co.

Carvalho, W. D., K. Mustin, R. R. Hilário, I. M. Vasconcelos, V. Eilers, and P. M. Fearnside. 2019. "Deforestation Control in the Brazilian Amazon: A Conservation Struggle Being Lost as Agreements and Regulations are Subverted and Bypassed." *Perspectives in Ecology and Conservation*, 17(3): 122–30. DOI: 10.1016/j.pecon.2019.06.002.

Carver, S. 2013. "(Re)Creating Wilderness: Rewilding and Habitat Restoration." In P. Howard, I. Thompson, E. Waterton, and M. Atha (eds.), *The Routledge Companion to Landscape Studies*, 383–94. Abingdon: Routledge.

Casanova-Hernadez. 2011. "Housing Beekbergen Apeldoorn, Netherlands 2007–11."Available online: http://www.casanova-hernandez.com/CH_PROJECTS/A_ARCHITECTURE/CHRONOLOGY/A055 (accessed May 16, 2021).

Castaneda, C. 1968. *The Teachings of Don Juan: A Yaqui Way of Knowledge*. Berkeley: University of California Press.

Castañeda-Álvarez, N. P., C. K. Khoury, H. A. Achicanoy, V. Bernau, H. Dempewolf, R. J. Eastwood, L. Guarino, R. H. Harker, A. Jarvis, N. Maxted, J. V. Müller, Julian Ramirez-Villegas, C. C. Sosa, P. C. Struik, H. Vincent, and J. Toll 2016. "Global Conservation Priorities for Crop Wild Relatives." *Nature Plants*. DOI: 10.1038/NPLANTS.2016.22.

Chakrabarti, P. 2010. *Materials and Medicine. Trade, Conquest and Therapeutics in the XVIIIth Century*. Manchester: Manchester University Press.

Chakraborty, S., and A. C. Newton. 2011. “Climate Change, Plant Diseases and Food Security: An Overview.” *Plant Pathology*, 60: 2–14.

Chamovitz, D. 2012. *What a Plant Knows*. New York: Scientific American/ Farrar, Straus and Giroux.

Chandler, P. M., J. A. Zwar, J. V. Jacobsen, T. J. Higgins, and A. S. Inglis. 1984. “The Effects of Gibberellic Acid and Abscisic Acid on α-amylase mRNA Levels in Barley Aleurone Layers Studies Using an α-amylase cDNA Clone.” *Plant Molecular Biology*, 3: 407–18.

Chauveau, S. 2009. *L’invention pharmaceutique. La pharmacie française entre l’État et la société au XXe siècle*. Paris: Les empêcheurs de penser en rond.

Chen, F., and B. Fan. 2012. “Historical Changes of Ginkgo biloba L. Culture.” *Asian Agricultural Research*, 4(11): 102–5.

Chen, J., C. H. Cannon, and H. B. Hu. 2009. “Tropical Botanical Gardens: At the In Situ Ecosystem Management Frontier.” *Trends in Plant Science*, 14: 584–9.

Chen, L., L. Hao, M. A. J. Parry et al. 2014. “Progress in TILLING as a tool for functional genomics and improvement of crops.” *J. Integr. Plant Biol.* 56: 425–443.

Cheng, Hsing-Chen, and Shih Chih-Ming. 2008. “Reinterpreting Traditional Chinese Architecture in the Office Building and Skyscraper: The Architecture of C. Y. Lee.” *Journal of Architectural and Planning Research*, 25(3): 181–202.

Children of the Corn. 1984. [Film] Dir. Fritz Kiersch. Prod. Donald P. Borchers and Terrence Kirby. Atlanta: New World Pictures.

Children of the Corn. 2009. [Film] Dir. and Prod. Donald P. Borchers. Beverly Hills: Anchor Bay Entertainment.

Chimeli, A. B., and R. R. Soares. 2017. “The Use of Violence in Illegal Markets: Evidence from Mahogany Trade in the Brazilian Amazon.” *American Economic Journal: Applied Economics*, 9(4): 30–57.

China News Service. 2014. “Make a Wish Under the Ginkgo Tree,” December 29. Available online: http://www.ecns.cn/visual/hd/2014/12-29/55797.shtml (accessed July 19, 2020).

Christenhusz, M. J. M., J. L. Reveal, A. Farjon et al. 2011. “A New Classification and Linear Sequence of Extant Gymnosperms.” *Phytotaxa*, 19: 55–70.

Cicero. 1933. *De Natura Deorum*. Trans. H. Rackham. London: Loeb Classical Library, W. Heinemann.

Cohen, C. M. 2009. “The Unique Representation of Trees in *The Lord of the Rings*.” *Tolkien Studies*, 6(1): 91–125.

Cohen, David. 2001. “Who’s Afraid of Henry Moore?” In Dorothy Kosinski (ed.), *Henry Moore: Sculpting the 20th Century*, 263–75. New Haven, CT: Yale University Press.

Colchester, M. 2010. *Land Acquisition, Human Rights Violations and Indigenous Peoples on the Palm Oil Frontier*. Moreton-in-Marsh, UK: Forest Peoples Programme and International Land Coalition.

Collingham, Lizzie. 2017. *The Hungry Empire: How Britain’s Quest for Food Shaped the Modern World*. London: The Bodley Head.

Comme des Garçons Parfums. 2017. “Comme 3.” Available online: http://comme-des-garcons-parfum.com/perfumes/comme-3/ (accessed July 19, 2020).

Conan, M. 1999. “From Vernacular Gardens to a Social Anthropology of Gardening.” In M. Conan (ed.), *Perspectives on Garden Histories*. Dumbarton Oaks Colloquium on the History of Landscape Architecture, vol. 21, 181–204. Washington, DC: Dumbarton Oaks Research Library and Collection.

Conan, M. 2007. *Contemporary Garden Aesthetics, Creations and Interpretations*. Dumbarton Oaks Colloquium on the History of Landscape Architecture vol. 29. Washington, DC: Dumbarton Oaks Research Library and Collection.

Conneway, Renee L. 2013. "Evaluating Biodegradable Containers as Alternatives to Plastic Pots." PhD diss., West Virginia University, Morgantown.

Connor, D. J., 2013. "Organically Grown Crops Do Not a Cropping System Make and Nor Can Organic Agriculture Feed the World." *Field Crops Research*, 1444: 145–7.

Cook, H. J. 2007. *Matters of Exchange. Commerce, Medicine and Science in the Dutch Golden Age*. New Haven, CT: Yale University Press.

Corlett, R. T. 2016. "Plant Diversity in a Changing World: Status, Trends and Conservation Needs." *Plant Diversity*, 1: 11–18.

Coventry, B. O. 1923–30. *Wild Flowers of Kashmir*. 3 vols. London: Raithby, Lawrence & Co.

Cragg, G. M., and D. J. Newman. 2005. "Biodiversity: A Continuing Source of Novel Drug Leads." *Pure and Applied Chemistry*, 77(1): 7–24.

Creasy, G. L., and L. L. Creasy. 2018. *Grapes*. 2nd edn. Oxfordshire: CABI.

Creative Review. 2018. "Top 20 Slogans 11: FTD (1917)—Say It With Flowers: Patrick O'Keefe." Available online: https://www.creativereview.co.uk/say-it-with-flowers/ (accessed July 17, 2020).

Crocker, R. 2016. *Somebody Else's Problem: Consumerism, Sustainability and Design*. Sheffield: Greenleaf Publishing Ltd.

Crutzen, P. J. 2002. "Geology of Mankind." *Nature*, 415: 23.

Dafni, A. 2002. "Why are Rags Tied to the Sacred Trees of the Holy Land?" *Economic Botany*, 56: 315–27.

Dale, James, Jean-Yves Paul, Benjamin Dugdale, and Robert Harding. 2017. "Modifying Bananas: From Transgenics to Organics?" *Sustainability*, 9(3): 333. DOI: 10.3390/su9030333.

Darwin, Charles. 1875. *Insectivorous Plants*. London: John Murray.

Darwin, Charles. 1875. *The Movements and Habits of Climbing Plants*, 2nd edn. London: John Murray.

Darwin, Charles. 1880. *The Power of Movement in Plants*. London: John Murray.

Das Blumenwunder. 1926. [Film] Dir. Max Reichmann. Ludwigshafen and Berlin, BASF AG, Unterrichtsfilm GmbH, and Verlag wissenschaftlicher Filme.

Das, L., E. Bhaumik, U. Raychaudhuri, and R. Chakraborty. 2012. "Role of Nutraceuticals in Human Health." *Journal of Food Science and Technology*, 49(2): 173–83. DOI: 10.1007/s13197-011-0269-4.

Daston, Lorraine, and Peter Galison. 2007. *Objectivity*. New York: Zone Books.

Daum. 2019. "Amber Ginkgo Perfume Bottle." Available online: https://www.daum.fr/en/shop/flacon-de-parfum-ginkgo/ (accessed May 16, 2021).

Davis, Brenda, and Melina Vesanto. 2014. *Becoming Vegan: The Complete Reference to Plant-based Nutrition,* Comprehensive edn. Summertown, Tennessee: Book Publishing Company.

Davis, W. 1998. *Plants of the Gods. In Shadows in the Sun: Travels to Landscapes of Spirit and Desire*. New York: Broadway Books.

De Bray, Lys. 1989. *The Art of Botanical Illustration: The Classic Illustrators and their Achievements from 1550 to 1900*. Secaucus, New Jersey: Wellfleet.

De Carvalho, L. M. M. 2011. "The Symbolic Uses of Plants." In E. N. Anderson, D. Pearsall, E. Hunn, and N. Turner (eds.), *Ethnobiology*, 351–69. Hoboken, New Jersey: Wiley-Blackwell.

de Klemm, C. 1993. *Biological Diversity conservation and the Law. Legal Mechanisms for Conserving Species and Ecosystems*. IUCN Environmental Policy and Law Paper No. 29. IUCN, Switzerland.

de Menezes, Jack. 2016. "England Fined £35,000 by Fifa for Wearing Poppies Against Scotland but Will Appeal Sanction," December 19. Available online: https://www.independent.co.uk/sport/football/international/fifa-poppy-ban-england-fine-35000-scotland-a7484296.html (accessed May 16, 2021).

Deakin, R. 2007. *Wildwood: A Journey Through Trees*. London: Hamish Hamilton.

DeFelice, Stephen L. 1995. "The Nutraceutical Revolution: Its Impact on Food Industry R&D." *Trends in Food Science & Technology*, 6(2): 59–61.

Deininger, K., and D. Byerlee, 2012. "The Rise of Large Farms in Land Abundant Countries: Do They Have a Future?" *World Development*, 40: 701–14.

Delaporte, F. 1990. *Le savoir de la maladie: Essai sur le choléra de 1832 à Paris*. Paris: Presses Universitaires de France.

Delgado, C. L. 2003. "Rising Consumption of Meat and Milk in Developing Countries Has Created a New Food Revolution." *The Journal of Nutrition*, 133(11): 3907S–3910S.

Descola, P. 2014. "All Too Human (Still): A Comment on Eduardo Kohn's How Forests Think." *HAU: Journal of Ethnographic Theory*, 4(2): 267–73.

Desmond, Ray. 1987. *A Celebration of Flowers: Two Hundred Years of* Curtis's Botanical Magazine. London: Collingridge.

Desrochers, Pierre, and Hiroko Shimizu. 2012. *The Locavore's Dilemma: In Praise of the 10,000-Mile Diet*. New York: Public Affairs.

di Castri, F. 1989. "History of Biological Invasions with Emphasis on the Old World." In J. Drake, F. di Castri, R. Groves, F. Kruger, H. A. Mooney, M. Rejmanek, and M. Williamson (eds.), *Biological Invasions: A Global Perspective*. New York: Wiley.

Dicenta, Federico, Raquel Sánchez-Pérez, Ignasi Batlle, and Pedro Martinéz-Gómez. 2017. "Late-blooming Cultivar Development." In Rafel Socias i Company and Thomas M. Gradziel (eds.), *Almonds: Botany, Production and Uses*, 168–87. Oxfordshire: CABI.

Dickinson, E. 2013. "The Misdiagnosis: Rethinking 'Nature-deficit Disorder.'" *Environmental Communication*, 7(3): 315–35.

Dietrich, O., M. Heun, J. Notroff, K. Schmidt, and M. Zarnkov. 2012. "The Role of Cult and Feasting in the Emergence of Neolithic Communities. New Evidence from the Göbekli Tepe, South-eastern Turkey." *Antiquity*, 86: 674–95.

Diffenbaugh, Vanessa. [2011] 2012. *The Language of Flowers: A Novel*. New York: Ballantine.

Dilk, A., and D. A. Savaiano. 2017. "Sugar Price Supports and Taxation: A Public Health Policy Paradox." *Nutrition today*, 52(3): 143–50. DOI: 10.1097/NT.0000000000000217.

Dinges, M. (ed.). 1996. *Medizinkritische Bewegungen im Deutschen Reich* (ca. 1870–ca. 1933). Stuttgart: Franz Steiner Verlag.

Dirk, N. B. 2001. *Castes of Mind: Colonialism and the Making of Modern India*, Princeton, NJ: Princeton University Press.

Donaldson, J. (ed.) 2003. *Cycads: Status Survey and Conservation Action Plan*, IUCN-SSC Cycad Specialist Group. Available online: https://portals.iucn.org/library/node/8203

Donaldson, J. S. 2009. "Botanic Gardens Science for Conservation and Global Change." *Trends in Plant Science*, 14: 608–13.

Downey, P. O., and D. M. Richardson. 2016. "Alien Plant Invasions and Native Plant Extinctions: A Six-Threshold Framework." *AoB PLANTS*, 8, plw047. DOI: 10.1093/aobpla/plw047.

Downie, H. F., M. O. Adu, S. Schmidt, W. Otten, L. X. Dupuy, P. J. White, and T. A. Valentine. 2015. "Challenges and Opportunities for Quantifying Roots and Rhizosphere Interactions Through Imaging and Image Analysis." *Plant, Cell and Environment*, 38: 1213–32.

Drayton, R. 2000. *Nature's Government. Science, Imperial Britain and the "Improvement" of the World*. New Haven, CT, Yale University Press.

Dreisigacker, S., M. Kishii, J. Lage, and M. Warburton. 2008. "Use of Synthetic Hexaploid Wheat to Increase Diversity for CIMMYT Bread Wheat Improvement." *Australian Journal of Agricultural Research*, 59: 413–20.

Dubin, H. J., and S. Rajaram. 1982. "The CIMMYT International Approach to Breeding Disease-Resistant Wheat." *Plant Disease*, 66: 967–71.

Dupon, Olivier. 2014. *Floral Contemporary: The Renaissance of Flower Design*. London: Thames & Hudson Ltd.

Dye, Alan. 1994. "Avoiding Holdup: Asset Specificity and Technical Change in the Cuban Sugar Industry, 1899–1929." *Journal of Economic History*, 54(3): 628–53.

Dyson, C. 2016. "Rethinking Australian Natural Gardens and National Identity, 1950–1979." *Studies in the History of Gardens & Designed Landscapes*, 36(1): 53–64.

Egan, D., and W. H. Tishler. 1999. "Jens Jensen, Native Plants, and the Concept of Nordic Superiority." *Landscape Journal*, 18(1): 11–29.

Einarson, Benedict. 1976. "Introduction" to Theophrastus *De Causis Plantarum*, vol. I, Books 1–2. Translated by Benedict Einarson and George K. K. Link. Loeb Classical Library 471, Cambridge, MA: Harvard University Press.

Elliot, Brent. 1996. "A Question of Image: Botanical Illustration in the Lindley Library." *The Garden*, 121(8): 486–9.

Ens, E. J., P. Petina, P. Clarke, et al. 2015. "Indigenous Biocultural Knowledge in Ecosystem Science and Management: Review and Insight from Australia." Available online: https://www.nintione.com.au/resources/rao/indigenous-biocultural-knowledge-in-ecosystem-science-and-management-review-and-insight-from-australia/ (accessed May 16, 2021).

ENSCONET. 2015. The European Native Seed Conservation Network. Available online: http://ensconet.maich.gr/About.htm (accessed May 16, 2021).

Eschen, R., et al. 2015. "International Variation in Phytosanitary Legislation and Regulations Governing Importation of Plants for Planting." *Environmental Science and Policy*, 51: 228–37. DOI: 10.1016/j.envsci.2015.04.021.

Eschen, R., et al. 2017. "A Risk Categorisation and Analysis of the Geographic and Temporal Dynamics of the European Import of Plants for Planting." *Biological Invasions*, 19(11): 3243–57.

Evans, M. C., J. E. M. Watson, R. A. Fuller, O. Venter, S. C. Bennett, P. R. Marsack, and H. Possingham. 2011. "The Spatial Distribution of Threats to Species in Australia." *BioScience*, 61(4): 281–9.

Evenson, R. E., and D. Gollin. 2003. "Assessing the Impact of the Green Revolution, 1960 to 2000." *Science*, 300: 758–62.

Evil Dead. 1981. [Film] Dir. Sam Raimi, Prod. Robert Tapert. Burbank: New Line Cinema.

Fahlgren, N., M. A. Gehan, and I. Baxter. 2015. "Lights, Camera, Action: High-Throughput Plant Phenotyping is Reading for a Close-up." *Current Opinion in Plant Biology*, 24: 93–9.

Fairwild. 2021. "Mission." Available online: https://www.fairwild.org/our-mission (accessed August 15, 2021).

FAO. 2002. "Small-Scale Palm Oil Processing in Africa." *FAO Agricultural Services Bulletin*, 148: 1010–365. Rome, Italy: FAO.

FAO. 2010. "Global Forest Resources Assessment, 2010—Main Report." *FAO Forestry Paper 163*. Rome, Italy: FAO.

FAO. 2016. "Global Forest Products: Facts and Figures 2016." Available online: http://www.fao.org/3/I7034EN/i7034en.pdf (accessed May 16, 2021).

FAO. 2019. "The State of Food Security and Nutrition in the World." Available online: http://www.fao.org/state-of-food-security-nutrition/en/ (accessed April 8, 2020).

FAOSTAT. 2016. "Crops; Indonesia; Oil, Palm and Oil, Palm Fruit; Average Production of Commodity; 1990–2014." Available online: http://www.fao.org/faostat/en/#data/QC/visualize (accessed May 16, 2021).

Farjon, A., and C. N. Page (eds.). 1999. *Conifers: Status Survey and Conservation Action Plan.* IUCN-SSC Conifer Specialist Group.

Farmer, Jared. 2013. *Trees in Paradise: A California History*. New York: W. W. Norton & Company.

Farnell, Gary. 2016. "What do Plants Want?" In Dawn Keetley and Angela Tenga (eds.), *Plant Horror: Approaches to the Monstrous Vegetal in Fiction and Film*, 179–96. London: Palgrave Macmillan.

Farquhar, J., and K. S. Rajan. 2014. "Introduction." *East Asian Science Technology and Society*, 8(4): 383–9.

Fearnside, P. M. 2017. "Deforestation of the Brazilian Amazon." *Oxford Research Encyclopedia of EnvironmentalScience*, 58: 1–52. DOI: 10.1093/acrefore/9780199389414.013.102.

Feeny, D. 1984. *Agricultural Expansion and Forest Depletion in Thailand, 1900–1975.* Discussion Paper No. 458. Economic Growth Center, Yale University, New Haven, CT.

Ferriss, S. 2014. *An Analysis of Trade in five CITES-Listed Taxa*. London: Chatham House/TRAFFIC.

Fields, Gracie. 2009. *Gracie Fields: Sing as We Go*. Vanilla OMP, GBQRF0832772.

Fincham, Gail. 2010. "The Mind's Eye: Focalizing 'Nature' in *Jane Eyre* and *Wide Sargasso Sea*." *English Academy Review*, 27(1): 14–23.

Fincher, G. B. 1989. "Molecular and Cellular Biology Associated with Endosperm Mobilization in Germinating Cereal Grains." *Annual Review of Plant Biology*, 40: 305–46.

Fischer, Tony, Derek Byerlee, and Greg Edmeades. 2014. *Crop Yields and Global Food Security: Will Yield Increase Continue to Feed the World?* Canberra: ACIAR. Available online: http://aciar.gov.au/publication/mn158 (accessed May 16, 2021).

Fitzgerald, D. K. 2003. *Every Farm a Factory: The Industrial Ideal in American Agriculture.* New Haven, CT: Yale University Press.

Fitzherbert, E. B., M. J. Struebig, A. Morel, F. Danielsen, C. A. Brühl, P. F. Donald, and B. Phalan. 2008. "How will Oil Palm Expansion Affect Biodiversity?" *Trends in Ecology and Evolution*, 23(10): 539–45.

Fletcher, R. 2017. "Connection with Nature is an Oxymoron: A Political Ecology of 'nature-deficit disorder'." *Journal of Environmental Education*, 48(4): 226–33.

Flora. 1989. [Film] Dir. Jan Švankmajer, Prod. Švankmajer and Jaromír Kallista. New York: MTV.

Floral Co-operative Societies. 1927. [Film] Dir. Percy Smith, Prod. British Instructional Films. London: British Film Institute.

Food and Agriculture Organisation (FAO) of the United Nations. 2009. "Declaration of the World Summit on Food Security, Rome, 16–18 November 2009." Available online: http://www.fao.org/fileadmin/templates/wsfs/Summit/Docs/Declaration/WSFS09_Draft_Declaration.pdf (accessed May 16, 2021).

Food and Agriculture Organisation (FAO) of the United Nations. 2017. *Statistical Year Book. (FAOSTAT)*. Available online: http://www.fao.org/faostat/en/#data/QC (accessed May 16, 2021).

Forbes, S. J. 2012. "Growing Literacy in Australian Flora." *The Adelaide Review*. Available online: https://www.adelaidereview.com.au/latest/2012/11/12/growing-literacy-in-australias-flora/ (accessed May 16, 2021).

Forbes, S. J. 2016a. "The Oldest Historical Tree in the World." Available online: https://stephenjforbes.wordpress.com/2016/12/26/the-oldest-historical-tree-in-the-world/ (accessed May 16, 2021).

Forbes, S. J. 2016b. "White Bread and Densuke Melons—Plants as Luxury Foods."Available online: https://stephenjforbes.wordpress.com/2016/03/29/white-bread-and-densuke-melons-plants-as-luxury-foods/ (accessed August 22, 2019).

Forbes, S. J., D. Cooper, and T. Kendle. 1997. "The History of Development of Ecological Landscape Styles." In T. Kendle, and S. J. Forbes (eds.), *Urban Nature Conservation: Landscape Management in the Urban Countryside*. London: Spon.

Foresight. 2011. *Global Food and Farming Futures: Synthesis Reports*. London: Government Office for Science. Department for Business, Skills and Innovation.

Foucault, M. 1963. *Naissance de la clinique*. Paris: Presses Universitaires de France.

Fox, James. 2014. "Poppy Politics: Remembrance of Things Present." In Constantine Sandis (ed.), *Cultural Heritage Ethics: Between Theory and Practice*, 23–8. Open Book Publishers.Available online: http://books.openedition.org/obp/2129?lang=en (accessed July 19, 2020).

Francis. 2015. *Encyclical Letter Laudato si' of the Holy Father Francis*. 1st edn. [Vatican City]: [Vatican website]. Available online: https://laudatosi.com/watch (accessed July 19, 2020).

Francis. 2020. "Pope at Audience on Earth Day: Common Action Needed to Protect 'our garden-home'."Available online: https://www.vaticannews.va/en/pope/news/2020-04/pope-francis-general-audience-earth-day-common-home-vulnerable.html (accessed April 22, 2020).

Freedman, Paul (ed.). 2007. *Food: The History of Taste*. Berkeley: University of California Press.

Freud, Sigmund. [1914] 1931. *Interpretation of Dreams*, 4th edn. Translated by A. A. Brill. London: Allen and Unwin.

Friedmann, Harriet. 1982. "The Political Economy of Food: The Rise and Fall of the Postwar International Food Order." *American Journal of Sociology*, 88: S248–S286.

Friedmann, Harriet. 2004. "Feeding the Empire: The Pathologies of Global Agriculture." In Colin Leys and Leo Panitch (eds.), *The Empire Reloaded: Socialist Register*, 124–43. London: Merlin.

Fuller, D. Q. 2007. "Contrasting Patterns in Crop Domestication and Domestication Rates: Recent Archaeobotanical Insights from the Old World." *Annals of Botany* (London), 100: 903–24.

Gagliano, Monica. 2018. *Thus Spoke the Plant: A Remarkable Journey of Groundbreaking Scientific Discoveries and Personal Encounters with Plants*. Berkeley: North Atlantic Books.

Gagliano, Monica, John C. Ryan, and Patricia Vieira (eds.). 2019. *The Language of Plants: Science, Philosophy, Literature*. Minneapolis: University of Minnesota Press.

Galison, Peter. 2015. "The Journalist, the Scientist, and Objectivity." In Flavia Padovani, Alan Richardsona, and Jonathan Y. Tsou (eds.), *Objectivity in Science: New Perspectives from Science and Technology* Studies, 57–75. Cham, Switzerland: Springer.

Gargiullo, Margaret B. 2008. *A Field Guide to Plants of Costa Rica*. Photos by Barbara L. Magnuson and Larry D. Kimball. Oxford: Oxford University Press.

Gates, Barbara. 1998. *Kindred Nature: Victorian and Edwardian Women Embrace the Living World*. Chicago: University of Chicago Press.

Gaudillière, J. 2010. "Une marchandise scientifique? Savoirs, industrie et régulation du médicament dans l'Allemagne des années trente." *Annales. Histoire, Sciences Sociales*, 65(1): 89–120.

Gaudillière, J-P. 2005. "Better Prepared than Synthesized: Adolf Butenandt, Schering AG and the Transformation of Sex Steroids into Drugs." *Studies in History and Philosophy of Science*, 36(4): 612–44.

Gaudillière, J-P, and V. Hes (eds.), 2013. *Ways of Regulating Drugs in the 19th and 20th Centuries*. Basingstoke and New York: Palgrave Macmillan.

Gaulthier, F. 2017. "Religion is Not What it Used to Be. Consumerism, Neoliberalism, and the Global Reshaping of Religion." Available online: https://blogs.lse.ac.uk/religionglobalsociety/2017/10/religion-is-not-what-it-used-to-be-consumerism-neoliberalism-and-the-global-reshaping-of-religion/ (accessed March 8, 2020).

Gauthier, F. 2018. "From Nation-state to Market: The Transformations of Religion in the Global Era, as Illustrated by Islam." *Religion*, 48(3): 382–417.

Gauthier, F., L. Woodhead, and T. Martikainen. 2013. "Introduction: Consumerism as the Ethos of Consumer Society." In François Gauthier and Tuomas Martikainen (eds.), *Religion in Consumer Society: Brands, Consumers and Markets*, 1–24. Farnham, Surrey: Ashgate.

Gaynor, A. 2020. "Lawnscaping Perth: Water Supply, Gardens, and Scarcity, 1890–1925." *Journal of Urban History*, 46(1): 63–78.

Gearys. 2018. "Daum Ginkgo Perfume Bottle." Available online: https://www.gearys.com/ginkgo-perfume-bottle.html (accessed May 16, 2021).

Gessert, G. 1996. "Bastard Flowers." *Leonardo*, 29(4): 291–8.

Gessert, G. 2010. *Green Light: Toward an Art of Evolution*. Cambridge, MA: MIT Press.

Glass, R., and C. Foster. 1975. "'Les Cedres': The Private Botanical Garden of Julien Marnier Lapostolle." *Journal of the Cactus and Succulent Society* (Los Angeles), 47(2): 51–3.

Global Biodiversity Information Facility. n.d. "What is GBIF?" *Global Biodiversity Information Facility*. Available online: http://gbif.org/what-is-gbif (accessed October 27, 2019).

Global Burden of Diseases 2015 Obesity Collaborators. 2017. "Health Effects of Overweight and Obesity in 195 Countries over 25 Years." *New England Jornal of Medicine*, 377(1): 13–27.

Global Panel on Agriculture and Food Systems for Nutrition. 2016. *Food Systems and Diets: Facing the Challenges of the 21st Century*. London, UK: IFPRI. Available online: https://www.ifpri.org/publication/food-systems-and-diets-facing-challenges-21st-century (accessed May 16, 2021).

Goettsch, B., C. Hilton-Taylor, G. Cruz-Piñón, J. P. Duffy, A. Frances, H. M. Hernández, R. Inger, C. Pollock, J. Schipper, M. Superina, and N. P. Taylor. 2015. "High Proportion of Cactus Species Threatened with Extinction." *Nature Plants*, 1: 15142.

Goldstein, Joyce, with Dore Brown. 2013. *Inside the California Food Revolution: Thirty Years That Changed Our Culinary Consciousness*. Berkeley: University of California Press.

Goldsworthy, Andy. 2017. *Projects*. New York: Abrams.

Goody, J. 1993. *The Culture of Flowers*. Cambridge: Cambridge University Press.

Gottschalk, Molly. 2016. "Why Certain Photographs Quickly Come to Define a Movement. Artsy," July 12. Available online: https://www.artsy.net/article/artsy-editorial-why-certain-photographs-quickly-come-to-define-a-movement (accessed October 18, 2020).

Gould, S. J. 1998. "An Evolutionary Perspective on Strengths, Fallacies, and Confusions in the Concept of Native Plants." *Arnoldia* (Spring): 3–11.

Greene, O. 2005. "Environmental Issues." In J. Baylis and S. Smith (eds.), *The Globalization of World Politics. An Introduction to International Relations*, 3rd edn., 451–78. Oxford: Oxford University Press.

Greenhalgh, S. 1999. "Our Lady of Flowers: The Ambiguous Politics of Diana's Floral Revolution." In A. Kear and D. L. Steinberg (eds.), *Mourning Diana: Nature, Culture and the Performance of Grief*, 40–59. London: Routledge.

Grese, R. E. 2000. "Jens Jensen: The Landscape Architect as Conservationist." In W. H. Tishler (ed.), *Midwestern Landscape Architecture*. ch. 7, 117–41. Urbana and Chicago: University of Illinois Press.

Grey, Sam, and Leonore Newman. 2018. "Beyond Culinary Colonialism: Indigenous Food Sovereignty, Liberal Multiculturalism, and the Control of Gastronomic Capital." *Agriculture and Human Values*, 35: 717–30.

Grey, Sam, and Raj Patel. 2015. "Food Sovereignty as Decolonization: Some Contributions from Indigenous Movements to Food System and Development Politics." *Agriculture and Human Values*, 32: 431–44.

Groffman, P. M., P. Kareiva, S. Carter, N. B. Grimm, J. Lawler, M. Mack, V. Matzek, and H. Tallis. 2014. "Ecosystems, Biodiversity and Ecosystem Services." In J. M. Melillo, T. C. Richmond, and G. W. Yohe (eds.), *Climate Change Impacts in the United States: The Third National Climate Assessment*. Washington, DC: US Global Change Research Programme.

Grogan, J., P. Barreto, and A. Veríssimo. 2002. *Mahogany in the Brazilian Amazon: Ecology and Perspectives on Management*. Belém, Brazil: Imazon.

Gröning, G., and J. Wolschke-Bulmahn. 1992. "Some Notes on the Mania for Native Plants in Germany." *Landscape Journal*, 11(2): 116–26.

Gröning, G., and J. Wolschke-Bulmahn. 2003. "The Native Plant Enthusiasm: Ecological Panacea or Xenophobia?" *Landscape Research*, 28(1): 75–88.

Groombridge, B. (ed.). 1992. *Global Biodiversity. Status of the Earth's Living Resources*. London: Chapman and Hall.

Grosser, J. W., and F. G. Gmitter. 2011. "Protoplast Fusion for Production of Tetraploids and Triploids: Application for Scion and Rootstock Breeding in Citrus." *Plant Cell, Tissue Organ Culture*, 104: 343–57.

Groth, P. 1994. "Review of *The Lawn: A History of an American Obsession* by Virginia Scott Jenkins and *Front Yard America: The Evolution and Meanings of a Vernacular Domestic Landscape* by Fred E.H. Schroeder." *Winterthur Portfolio*, 29(4): 296–9.

Grove, R. H. 1992. "Origins of Western Environmentalism." *Scientific American*, 267(1): 42–7.

Grove, R. H. 1995. *Green Imperialism: Colonial Expansion, Tropical Island Edens and the Origins of Environmentalism, 1600–1860*. Cambridge: Cambridge University Press.

Guarino, Ben. 2020. "The Audacious Effort to Reforest the Planet." *The Washington Post*, January 22. Available online: https://www.washingtonpost.com/graphics/2020/climate-solutions/trillion-tree-reforestation-climate-change-philippines/ (accessed March 11, 2020).

Gubler, F., R. Kalla, J. K. Roberts, and J. V. Jacobsen. 1995. "Gibberellin-Regulated Expression of a Myb Gene in Barley Aleurone Cells: Evidence for Myb Transactivation of a High-pI Alpha-amylase Gene Promoter." *Plant Cell*, 7: 1879–91.

Guggenheim. 2016. "Jeff Koons—Puppy." Available online: https://www.guggenheim.org/artwork/48 (accessed March 29, 2020).

Gupta, S. M. 1971. *Plant Myths and Traditions in India*, 3rd edn. New Dehli, India: Munshiram Manoharla Publishers.

Hagerup, Olaf, and Vagn Petersson. 1956–9. *A Botanical Atlas*. Vol. 1: *Angiosperms*. Translated by Gilberg-Carter. Copenhagen: Ejnar Munksgaard.

Hamlin, C. 2009. *Cholera: The Biography*. Oxford: Oxford University Press.

Haraway, Donna. [1985] 2004. "A Manifesto for Cyborgs: Science, Technology, and Socialist Feminism in the 1980s." In *The Haraway* Reader, 7–45. New York: Routledge.

Harlan, J. R. 1992. *Crops and Man*, 2nd edn. Madison, Wisconsin: American Society of Agronomy.

Harlan, J. R., J. M. J. De Wet, and E. G. Price. 1973. "Comparative Evolution of Cereals." *Evolution*, 27: 311–25.

Harrison, K. D. 2007. "An Extinction of (Ideas about) Species." In *When Languages Die: The Extinction of the World's Languages and the Erosion of Human* Knowledge, 23–60. Oxford: Oxford University Press.

Harrison, K. D. 2019. "Languages, Plants, and People." *Language Magazine* (March): 28–32.

Harrison, M. 1994. *Public Health in British India. Anglo-Indian Preventive Medicine, 1859–1914*. Cambridge: Cambridge University Press.

Harrison, M. 2010. *Medicine in an Age of Commerce and Empire*. Oxford: Oxford University Press.

Hartley, S., F. Gillund, L. van Hove, and F. Wickson. 2016. "Essential Features of Responsible Governance of Agricultural Biotechnology." *PLoS Biology*, 14: e1002453. DOI: 10.1371/journal.pbio.1002453.

Havens, K., P. Vitt, M. Maunder, E. O. Guerrant, and K. Dixon. 2006. "Ex-situ Plant Conservation and Beyond." *BioScience*, 56: 525–31.

Hawkins, B. 2008. *Plants for Life: Medicinal Plant Conservation and Botanic Gardens*. Richmond, UK: Botanic Gardens Conservation International.

Hawkins, B., S. Sharrock, and K. Havens. 2008. *Plants and Climate Change: Which Future?* Richmond, UK: Botanic Gardens Conservation International.

Hazell, Dinah. 2006. *The Plants of Middle-Earth: Botany and Sub-Creation*. Kent, Ohio: Kent State University Press.

Hazell, P., and S. Wood. 2008. "Drivers of Change in Global Agriculture." *Philosophical Transactions of the Royal Society B*, 363: 495–515.

Hedblom, M., F. Lindberg, E. Vogel, J. Wissman, and K. Ahrné. 2017. "Estimating Urban Lawn Cover in Space and Time: Case Studies in Three Swedish Cities." *Urban Ecosystems*, 20(5): 1109–19.

Hefferon, Kathleen L., and Ronald J. Herring. 2017. "The End of the GMO? Genome Editing, Gene Drives and New Frontiers of Plant Technology." *Review of Agrarian Studies*, 70(1). Available online: http://ras.org.in/7d7edc7ecf0c4d5a04189e9cad6e829f (accessed April 25, 2020).

Heisey, P. W. 2015. "The Structure of US Agricultural and Food Research, with an Emphasis on Seed-Biotechnology Research." *NABC Report*, 27: 153.

Hendel, A., R. O. Bak, J. T. Clark, A. B. Kennedy, D. E. Ryan, S. Roy, I. Steinfeld, B. D. Lunstad, R. J. Kaiser, A. B. Wilkens, R. Bacchetta, A. Tsalenko, D. Dellinger, L. Bruhn, and M. H. Porteus. 2015. "Chemically Modified Guide RNAs Enhance CRISPR-Cas Genome Editing in Human Primary Cells." *Nature Biotechnology*, 33: 985–9.

Hercus, L. 2012. "Trees from the Dreaming." *The Artefact: The Journal of the Archaeological and Anthropological Society of Victoria*, 35: 40–9.

Herrero, M., P. K. Thornton, B. Power, J. R. Bogard, R. Remans, S. Fritz, J. S. Gerber, G. Nelson, L. See, K. Waha, and R. A. Watson. 2017. "Farming and the Geography of Nutrient Production for Human Use: A Transdisciplinary Analysis." *The Lancet Planetary Health*, 1(1): e33–e42.

Heywood, V. H. 2011. "Ethnopharmacology, Food Production, Nutrition and Biodiversity Conservation: Towards a Sustainable Future for Indigenous Peoples." *Journal of Ethnopharmacology*, 137: 1–15.

Heywood, V. H. 2017. "Plant Conservation in the Anthropocene—Challenges and Future Prospects." *Plant Diversity*, 39(6): 314–30.

Hickey, L. T., S. E. German, S. A. Pereyra, J. E. Diaz, L. A. Ziems, R. A. Fowler, G. J. Platz, J. D. Franckowiak, and M. J. Dieters. 2017. "Speed Breeding for Multiple Disease Resistance in Barley." *Euphytica*, 213: 64. DOI: 10.1007/s10681-016-1803-2.

Higginbotham, J. A. 1990. "Four Centuries of Planting and Progress: A History of the US Nursery Industry." *American Nurseryman*, 17: 36–59.

Hing, L. W., and Mohammed Nor Salleh. 1989. "Malaysia—a Responsible Supplier of Tropical Forest Products." In M. J. McDermott (ed.), *The Future of the Tropical Rain Forest*.

Proceedings of an International Conference 27 28 June 1988, 151–81. Oxford: Oxford Forestry Institute.

Hinsley, A., T. E. Lee, J. R. Harrison, and D. L. Roberts. 2016. “Estimating the Extent and Structure of Trade in Horticultural Orchids via Social Media.” *Conservation Biology*, 30(5): 1038–47.

Hird, Abby. 2014. “2013 BGANZ Collections Assessment.” *The Botanic Gardener*, 40 (Nov.): 45–7. Available online: https://issuu.com/bganz/docs/tbg_iss40_nov2014_final_141105/49 (accessed May 16, 2021).

Hiroi, Kenichiro, et al. 2013. “Time-lapse Imaging of Self- and Cross-pollinations in *Brassica rapa*.” *Annals of Botany*, 112(1): 115–22.

Hodgkin, T., N. Demers, and E. Frison. 2013. “The Evolving Global System of Conservation and Use of Plant Genetic Resources for Food and Agriculture: What Is It, and Where Does the Treaty Fit In.” In M. Halewood, N. I. López., and S. Louafi (eds.), *Crop Genetic Resources as a Global Commons: Challenges in International Law and Governance*, ch. 18, 344–67. London: Routledge.

Hoge, Warren. 1997. “Millions Say Farewell to their Princess.” *The New York Times*, September 7. Available online: http://www.nytimes.com/1997/09/07/world/millions-say-farewell-to-their-princess.html (accessed May 16, 2021).

Holmes, C. 2019. “The Amazon is Burning: 4 Essential Reads on Brazil’s Vanishing Rainforest.” Available online: https://theconversation.com/the-amazon-is-burning-4-essential-reads-on-brazils-vanishing-rainforest-122288(accessed June 9, 2020).

Holt-Giménez, Eric. 2017. *A Foodie’s Guide to Capitalism: Understanding the Political Economy of What We Eat*. New York: Monthly Review Press and Food First Press.

Howard, R. A. 1994. “The Role of Botanists during World War II in the Pacific Theatre.” *Botanical Review*, 60(2): 197–257.

Howard, R. A. 2000. “The Role of Botanists during World War II in the Pacific Theatre.” In R. M. MacLeod (ed.), *Science and the Pacific War: Science and Survival in the Pacific, 1939–1945*, 83–118. Dordrecht: Kluwer.

Hunt, J. D. 2000. *Greater Perfections: The Practice of Garden Theory*. Philadelphia: University of Pennsylvania Press.

Hunt, J. D. 2015. *The Making of Place: Modern and Contemporary Gardens*. London: Reaktion Books.

Huxley, A. 1954. *The Doors of Perception*. London: Chatto & Windus.

Iles, J. 2008. “In Remembrance: The Flanders Poppy.” *Mortality*, 13(3): 201–21.

Ilić, Biljana Vlašković. 2017. “Ecocriticism and Anthropocentrism in Yann Martel’s *Life of Pi*.” *Issues in Ethnology and Anthropology*, n.s., 12(3): 883–903.

Imort, M. 2000. *Forestopia: The Use of the Forest Landscape in Naturalizing National Socialist Ideologies of Volk, Race, and Lebensraum, 1918–1945*. Thesis/diss., Queen’s University, Canada.

Imperial War Museum. 2014. “Why We Wear Poppies on Remembrance Day.” Available online: https://www.iwm.org.uk/history/why-we-wear-poppies-on-remembrance-day (accessed May 16, 2021).

Imperial War Museum. 2015. “The Journey of the Camouflage Tree.” Available online: http://www.iwm.org.uk/history/the-journey-of-the-camouflage-tree (accessed July 19, 2020).

Imperial War Museum. 2018. “5 Facts about Camouflage in the First World War.” Available online: https://www.iwm.org.uk/history/5-facts-about-camouflage-in-the-first-world-war (accessed May 16, 2021).

Ingersoll, Richard. 2012. “The Ecology Question and Architecture.” In C. Greig Crysler, Stephen Cairns, and Hilde Heynen (eds.), *The SAGE Handbook of Architectural Theory*, 573–89. London: SAGE.

International Assessment of Agricultural Knowledge, Science and Technology for Development. 2009. *International Assessment of Agricultural Knowledge, Science and Technology for Development*. Washington, DC: Island Press.

International Rice Genome Sequencing Project (IRGC) and Takuji Sasaki. 2005. "The Map-based Sequence of the Rice Genome." *Nature*, 436: 793–800.

International Wheat Genome Sequencing Consortium (IWGSC). 2018. "Shifting the Limits in Wheat Research and Breeding Using a Fully Annotated Reference Genome." *Science*, 361. DOI: 10.1126/science.aar7191.

IOC. 1999. "Athens 2004: The Athens Olive Wreath. (2004 Olympic Games)." *Olympic Review*, 26(29) (Oct.—Nov.): 4. Available online: https://digital.la84.org/digital/collection/p17103coll1/id/36657/rec/1 (accessed May 16, 2021).

Iriarte-Chiapusso, M. J., A. Arrizabalaga, and G. Cuenca-Bescós. 2015. "The Vegetational and Climatic Contexts of the Lower Magdalenian Human Burial in El Mirón Cave (Cantabria, Spain): Implications Related to Human Behavior." *Journal of Archaeological Science*, 60: 66–74.

ISAAA. 2017. *Global Status of Commercialized Biotech/GM Crops in 2017: Biotech Crop Adoption Surges as Economic Benefits Accumulate in 22 Years*. ISAAA Brief No. 53. ISAAA: Ithaca, NY.

IUCN. 2014. "Historical IUCN Red Data Books and Red Lists." Available online: https://web.archive.org/web/20140629072452/http://www.iucnredlist.org/about/publication/historical-red-lists#red_lists 29 June 2014 Internet Archive Wayback Machine (accessed November 15, 2019).

IUCN. 2016. "The IUCN Red List of Threatened Species (2015)." Available online: www.iucnredlist.org (accessed May 16, 2021).

IUCN. 2019. "IUCN: A Brief History." Available online: https://www.iucn.org/about/iucn-a-brief-history (accessed May 16, 2021).

IUCN. 2020. "The IUCN Red List of Threatened Species." Available online: https://www.iucnredlist.org/ (accessed April 23, 2020).

IUCN/SSC MPSG. 2007. "International Standard for Sustainable Wild Collection of Medicinal and Aromatic Plants (ISSC-MAP)." *BfN Skripten* 195. Bundesamt fur Naturschutz (BfN), Bonn, Germany.

Jack, Ian. 1997. "Those who Felt Differently." *Granta 60: Unbelievable* (Winter): 9–35.

Jaffee, Daniel. 2007. *Brewing Justice: Fair Trade Coffee, Sustainability, and Survival*. Berkeley: University of California Press.

Jamieson, Dale. 2008. *Ethics and the Environment: An Introduction*. Cambridge: Cambridge University Press.

Janzen, Janet. 2016. *Media, Modernity and Dynamic Plants in Early 20th Century German Culture*. Boston: Brill.

Jaramillo, Camilo. 2016. "Green Hells: Monstrous Vegetations in Twentieth-Century Representations of Amazonia." In Dawn Keetley and Angela Tenga (eds.), *Plant Horror: Approaches to the Monstrous Vegetal in Fiction and* Film, 91–109. London: Palgrave Macmillan.

Jayakodi, M., S. Padmarasu, G. Haberer, et al. 2020. "The Barley Pan-genome Reveals the Hidden Legacy of Mutation Breeding." *Nature*, 588: 284–9. DOI: 10.1038/s41586-020-2947-8.

Jenkins, V. S. 1994. *The Lawn: A History of an American Obsession*. Washington, DC: Smithsonian Institution Press.

Jenkins, Virginia Scott (revised by Andrew F. Smith). 2013. "Bananas." In Andrew F. Smith (ed.), *The Oxford Encyclopedia of Food and Drink in America*, 2nd edn., vol. 1, 114–15. New York: Oxford University Press.

Jenson, Marc. 2008. "'Feed Me!': Power Struggles and the Portrayal of Race in *Little Shop of Horrors*." *Cinema Journal*, 48(1): 51–67.

Jex-Blake, A. J. 1934. *Gardening in East Africa: A Practical Handbook*. London: Longmans, Green & Co.

Jobey, L. 2017. "Silent Blooms: Artworks by Taryn Simon." *The Financial Times*, May 19. Available online: https://www.ft.com/content/192d59ec-39bf-11e7-ac89-b01cc67cfeec (accessed May 16, 2021).

Johnson, S. 2011. "I Was an Under-age Semiotician." *The New York Times*, October 14. Available online: https://www.nytimes.com/2011/10/16/books/review/i-was-an-under-age-semiotician.html (accessed May 16, 2021).

Johnston, Josée, and Shyon Baumann. 2015. *Foodies: Democracy and Distinction in the Gourmet Foodscape*, 2nd edn. New York: Routledge.

Jones, Owen. [1856] 2016. *The Grammar of Ornament: A Visual Reference of Form and Colour in Architecture and the Decorative Arts*. Princeton, NJ: Princeton University Press.

Joppa, L. N., D. L. Roberts, N. Myers, and S. L. Pimm. 2011. "Biodiversity Hotspots House Most Undiscovered Plant Species." *PNAS*, 108(32): 13171–6.

Joppa, L. N., P. Visconti, C. N. Jenkins, and S. L. Pimm. 2013. "Achieving the Convention on Biological Diversity's Goals for Plant Conservation." *Science*, 341: 1100. DOI: 10.1126/Science.1241706.

Judd, W. S., and G. A. Judd. 2017. *Flora of Middle-Earth Plants of J. R. R. Tolkien's Legendarium*. New York: Oxford University Press.

Julier, Alice P. 2005. "Hiding Gender and Race in the Discourse of Commercial Food Consumption." In Arlene Voski Avakian and Barbara Haber (eds.), *From Betty Crocker to Feminist Food Studies: Critical Perspectives on Women and* Food, 163–84. Amherst: University of Massachusetts Press.

Jurassic Park. 1990. [Film] Dir. Michael Crichton. New York: Ballantine.

Jütte, R. 1996. *Geschichte der alternativen Medizin: Von der Volksmedizin zu den unkonventionelleren Therapien von Heute*. Munich: Verlag C. H. Beck.

Kahn, P. H. 1997. "Developmental Psychology and the Biophilia Hypothesis: Children's Affiliation with Nature." *Developmental Review*, 17(1): 1–61.

Kandeler, R., and W. R. Ullrich. 2009. "Symbolism of Plants: Examples from European-Mediterranean Culture Presented with Biology and History of Art." *Journal of Experimental Botany*, 60(9): 2461–4.

Kapferer, Jean-Noël. 1997. "Managing Luxury Brands." *The Journal of Brand Management*, 4 (4): 251–60.

Karp, David. 2013. "Citrus." In Andrew F. Smith (ed.), *The Oxford Encyclopedia of Food and Drink in America*, 2nd edn., vol. 1, 418–19. New York: Oxford University Press.

Katawal, Ubaraj. 2013. "The Smell of Cottonwood Leaves: Plants and Taylo's Healing in Silko's *Ceremony*." In *Plants and Literature: Essays in Critical Plant Studies*, ed. Randy Laist, 147–63. Amsterdam: Rodopi.

Katz, S. H., and M. M. Voigt. 1986. "Bread and Beer: The Early Use of Cereals in the Human Diet." *Expeditions*, 28: 23–34.

Kazimi, Mithaq. 2015. "In Conversation with Fariborz Sahba, Renowned Architect." *Medium*, July 11. Available online: https://medium.com/@kazimi/fariborz-sahba-is-a-humble-yet-passionate-man-in-his-mid-sixties-he-has-grayish-white-hair-a-9782e72bbf4c (accessed November 3, 2019).

Keetley, Dawn. 2016. "Introduction: Six Theses on Plant Horror; or, Why are Plants Horrifying?" In Dawn Keetley and Angela Tenga, *Plant Horror: Approaches to the Monstrous Vegetal in Fiction and Film*, 1–30. London: Palgrave Macmillan.

Kell, S. P., N. Maxted, and M. Bilz. 2012. "European Crop Wild Relative Threat Assessment: Knowledge Gained and Lessons Learnt." In N. Maxted, M. E. Dulloo, B. V. Ford-Lloyd, L. Frese, J. M. Iriondo, and M. A. A. Pinheiro De Carvalho (eds.), *Agrobiodiversity Conservation: Securing the Diversity of Crop Wild Relatives and Landraces,* ch. 28, 218–42. Wallingford: CAB International.

Keller, D. R. 2008. "Deep Ecology." In J. B. Callicott and R. Frodeman (eds.), *Encyclopedia of Environmental Ethics and Philosophy*, 206–11. Detroit: Macmillan Reference.

Kellert, Stephen R. 2018. *Nature by Design: The Practice of Biophilic Design*. New Haven, CT: Yale University Press.

Kennedy, David O. 2014. *Plants and the Human Brain.* Oxford: Oxford University Press.

Kesebir, S., and P. Kesebir. 2017. "A Growing Disconnection from Nature is Evident in Cultural Products." *Perspectives on Psychological Science*, 12(2): 258–69.

Khoury, C. K., H. A. Achicanoy, A. D. Bjorkman,C. Navarro-Racines, L. Guarino, X. Flores-Palacios, J. M. Engels, J. H. Wiersema, H. Dempewolf, S. Sotelo, and J. Ramírez-Villegas. 2016. "Origins of Food Crops Connect Countries Worldwide." *Proceedings of the Royal Society B*, 283 (1832): 20160792.

Khoury, C. K., A. D. Bjorkman, H. Dempewolf, J. Ramirez-Villegas, L. Guarino, A. Jarvis, L. H. Rieseberg, and P. C. Struik. 2014. "Increasing Homogeneity in Global Food supplies and the Implications for Food Security." *Proceedings of the National Academy of Sciences*, 111(11): 4001–6. Available online: http://www.pnas.org/content/111/11/4001.full.pdf (accessed May 16, 2021).

Kiesler, Frederick. 1939. "On Correalism and Biotechnique: A Definition and Test of a New Approach to Building Design." *Architectural Record*, 86(3): 60–75.

Kiesler, Frederick. 1997. *Inside the Endless House*. Valencia: IVAM Centre Julio González.

Kimmerer, Robin Wall. 2013. *Braiding Sweetgrass: Indigenous Wisdom, Scientific Knowledge, and the Teachings of Plants*. Minneapolis: Milkweed.

Kingsbury, N. 2009. "Ornament: Furnishing our Gardens." In *Hybrid: The History and Science of Plant Breeding,* ch. 13, 329–63. Chicago: University of Chicago Press.

Kingsbury, N. 2013. Daffodil: *The Remarkable Story of the World's Most Popular Spring Flower*. Portland, Oregon: Timber Press.

Kirkby, Mandy. 2011. *A Victorian Flower Dictionary: The Language of Flowers Companion*. New York: Ballantine.

Klepeis, P., I. A. Orlowska, E. F. Kent, C. L. Cardelús, P. Scull, A. Wassie Eshete, and C. Woods. 2016. "Ethiopian Church Forests: A Hybrid Model of Protection." *Human Ecology*, 44(6): 715–30.

Knee, Adam. (2016). "Vegetable Discourses in the 1950s US Science Fiction Film." In Dawn Keetley and Angela Tenga (eds.), *Plant Horror: Approaches to the Monstrous Vegetal in Fiction and* Film, 145–62. London: Palgrave Macmillan.

Knees, S. G., and Gardner, M. F. 1983. "Mahoganies: Candidates for the Red Data Book." *Oryx*, 17: 88–92.

Knight, T., and M. Rose. 2017. *With Honourable Intent. A Natural History of Fauna & Flora International*. London: William Collins.

Koch, E. 1936. "Biologische Untersuchungen." *Jahrbuch Madaus*, 1936: 42–58.

Kohlstedt, Sally Gregory. 1978. "In from the Periphery: American Women in Science, 1830–1880." *Signs*, 4(1): 81–96.

Kohn, Eduardo. 2013. *How Forests Think: Toward an Ecology Beyond the Human*. Berkeley: University of California Press.

Kojo, Daisuke. 2006. "Importance of the Geographic Origin of Agricultural Products: A Comparison of Japanese and American Approaches." *Journal of Environmental and Sustainability Law, Missouri Environmental Law and Policy Review*, 14(2): 275–321.

Konishi, S., T. Izawa, S. Y. Lin, K. Ebana, Y. Fukuta, T. Sasaki, and M. Yano. 2006. "An SNP Caused Loss of Seed Shattering during Rice Domestication." *Science*, 312: 1392–6.

Kowarik, I. 2005. "Wild Urban Woodlands: Towards a Conceptual Framework." In Ingo Kowarik and Stefan Körner (eds.), *Wild Urban Woodlands. New Perspectives for Urban Forestry*, 1–32. Berlin: Springer.

Krementsov, N. 2010. "Darwinism, Marxism and Genetics in the Soviet Union." In D. Alexander and R. Numbers (eds.), *Biology and Ideology from Descartes to Dawkins*, 215–46. Chicago: University of Chicago Press.

Krishna, Nanditha, and M. Amirthalingam. 2014. *Sacred Plants of India*. Gurgaon: Penguin, India.

Kuck, Lorraine E., and R. C. B. Tongg. 1936. *The Tropical Garden: Its Design, Horticulture, and Plant Materials*. New York: The Macmillan Company.

Kuck, Lorraine E., and R. C. B. Tongg. 1960. *The Modern Tropical Garden: Its Design, Plant Materials and Horticulture*. Honolulu, Hawaii: Tongg Pub. Co.

Kuhn, A. 1936. "Die Aufbereitung von Arzneipflanzen." *Jahrbuch Madaus*, 1936: 29–36.

Kuligowska, K., H. Lutken, and R. Muller. 2016. "Towards Development of New Ornamental Plants: Status and Progress in Wide Hybridisation." *Planta*, 244: 1–17.

Kutner, L. S., and L. E. Morse. 1996. "Reintroduction in a Changing Climate." In D. A. Falk, C. I. Millar, and M. Olwell (eds.), *Restoring Diversity. Strategies for Reintroduction of Endangered Plants*, ch. 2, 23–38. Washington, DC: Island Press.

L'Abbé, Sonnet Lynn. 2013. "Green Men, Plant Brains and Nervetrees: Ronald Johnson's Object-Oriented Poetics of Embodied Mind." PhD thesis, University of British Columbia, Vancouver.

La Croissance des végétaux. 1929. [Film] Dir. and Prod. Jean Comandon. Paris: Pasteur Institute.

La Force, T. 2016. "Fleurs du mal: How Artist Taryn Simon Constructs Political Bouquets." The *Guardian*, February 26. Available online: https://www.theguardian.com/artanddesign/2016/feb/24/taryn-simon-flowers-floral-arrangements-photography-politics (accessed May 16, 2021).

LaDuke, Winona. 2005. *Recovering the Sacred: The Power of Naming and Claiming*. Cambridge, MA: Sound End Press.

Laird, S. A. 1999. "Forests, Culture and Conservation." In D. A. Posey (ed.), *Cultural and Spiritual Values of Biodiversity. A Complimentary Contribution to the Global Biodiversity Assessment*, 345–96. London and Nairobi: Intermediate Technology Publications & UNEP.

Laird, Tessa. 2018. "Spores from Space: Becoming the Alien." In *Covert Plants: Vegetal Consciousness and Agency in an Anthropocentric World*, ed. Prudence Gibson and Baylee Brits, 61–77. Santa Barbara: Brainstorm Books.

Laist, Randy. 2016. "Sartre and the Roots of Plant Horror." In Dawn Keetley and Angela Tenga (eds.), *Plant Horror: Approaches to the Monstrous Vegetal in Fiction and Film*, 163–78. London: Palgrave Macmillan.

Lam, Sophie. 2004. "In Search of an Old Romantic." *The Independent*, March 13. Available online: https://www.independent.co.uk/travel/uk/in-search-of-an-old-romantic-8847190.html (accessed May 16, 2021).

Lambert, J., J. Srivastava, and N. Vietmeyer. 1997. "Medicinal Plants: Rescuing a Global Heritage." World Bank Technical Paper.

Lamson-Scribner, F. 1897. “Lawns and Lawnmaking.” In *Yearbook of the United States Department of Agriculture*. Washington, DC: US Government Printing Office.

Lanahan, M. B., T. Ho, S. W. Rogers, and J. C. Rogers. 1992. “A Gibberellin Response Complex in Cereal Alpha-amylase Gene Promoters.” *Plant Cell*, 4, 203–11.

Larsen, Christian A. 2015. *Philodendron: From Pan-Latin Exotic to American Modern*. Miami Beach: The Wolfsonian Museum, Florida International University.

Latour, Bruno. 2004. *Politics of Nature: How to Bring the Sciences into Democracy*. Translated by Catherine Porter. Cambridge, MA: Harvard University Press.

Laudan, Rachel. 2015. *Cuisine and Empire: Cooking in World History*. Berkeley: University of California Press.

Laurance, W. F. 2019. “The Anthropocene.” *Current Biology*, 29(19): R953–R954.

Lawrence, D. H. 1920. *Women in Love*. New York: Thomas Seltzer.

Lawrence, D. H. [1928] 1995. *Lady Chatterley’s Lover*. New York: Barnes and Noble.

Le Roux, P., H. Flachowsky, M. Hanke, C. Gessler, and A. Patocchi. 2012. “Use of a Transgenic Early Flowering Approach in Apple (Malus x domestica Borkh.) to Introgress Fire Blight Resistance from Cultivar Evereste.” *Mol Breeding*, 30: 857–74.

Le Vasseur, T., and A. Peterson. 2016. “Introduction.” In T. Le Vasseur and A. Peterson (eds.), *Religion and Ecological Crisis: The Lynn White Thesis at Fifty*, 1–16. London and New York: Routledge.

Leary, T. 1964. *The Psychedelic Experience: A Manual Based on the Tibetan Book of the Dead*. New York: University Books.

Ledford, H. 2015. “CRISPR, the Disruptor.” *Nature*, 522: 20–4.

Leissle, Kristy. 2018. *Cocoa*. Cambridge, UK: Polity Press.

Lellouche, A. 2014. “Tegg’s Grasslands: The Beauty of Ideas.” Available online: http://www.lindategg.com/text_grasslands.html (accessed May 16, 2021).

Léonard, J. 1981. *La médecine entre les savoirs et les pouvoirs: Histoire intellectuelle et politique de la médecine française au XIXème siècle*. Paris: Aubier.

Leopold, A. 1949. *A Sand County Almanac, and Sketches Here and There*. New York: Oxford University Press.

Lesch, J. E. 2007. *The First Miracle Drugs. How the Sulfa Drugs Transformed Medicine*. Oxford: Oxford University Press,.

Lewis, C. S. [1951] 2008. *Prince Caspian: The Return to Narnia*. London: Harper Collins.

Lewis, J. 1985. “The Birth of the EPA.” *EPA Journal*. Available online: https://archive.epa.gov/epa/aboutepa/birth-epa.html (accessed November 24, 2019).

Lewis-Stempel, John. 2016. *Where Poppies Blow: The British Soldier, Nature, the Great War*. London: Weidenfeld & Nicholson.

Li, Y., G. Zhang, and B. A. Pfeifer. 2014. “Current and Emerging Options for Taxol Production.” In J. Schrader and J. Bohlmann (eds.), *Biotechnology of Isoprenoids*, vol. 148. Cham: Springer. DOI: 10.1007/10_2014_292.

Library of Congress. 2002. “The Evolution of the Conservation Movement, 1850–1920.” Available online: https://memory.loc.gov/ammem/amrvhtml/conshome.html (accessed May 16, 2021).

Lim, Eng-Beng. 2014. “Future Island.” *Third Text*, 28(4–5): 443–53.

Lindner, C., and B. Rosa (eds.). 2017. *Deconstructing the High Line: Postindustrial Urbanism and the Rise of the Elevated Park*. New Jersey: Rutgers University Press.

Lionni, Leo. 1977. *Parallel Botany*. Translated by Patrick Creagh. New York: Alfred A. Knopf.

Little Shop of Horrors. 1960. [Film] Dir. and Prod. Roger Corman. Los Angeles: American International Pictures.

Little Shop of Horrors. 1986. [Film] Dir. Frank Oz. Prod. Frank Geffen. Geffen Company.

Lobell, David B., Christopher B. Field, Nicholas Cahill Kimberly, and Celine Bonfils. 2006. "Impacts of Future Climate Change on California Perennial Crop Yields: Model Projections with Climate and Crop Uncertainties." *Agricultural and Forest Meteorology*, 141(2–4): 208–18. DOI: 10.1016/j.agrformet.2006.10.006.

Logan, A., and S. Prescott. 2017. "Astrofood, Priorities and Pandemics: Reflections of an Ultra-Processed Breakfast Program and Contemporary Dysbiotic Drift." *Challenges*, 8(2): 24.

Long, C. 1997. "Ornament, Crime, Myth, and Meaning." 85th ACSA Annual Meeting Proceedings, *Architecture: Material and Imagined.*

Long, J. C. 2009. "Rooting Diaspora, Reviving Nation: Zionist Landscapes of Palestine–Israel." *Transactions of the Institute of British Geographers*, 34(1): 61–77.

Loos, Adolf. [1910] 2002. "Ornament and Crime." Reprinted in Bernie Miller and Melony Ward (eds.), *Crime and Ornament: The Arts and Popular Culture in the Shadow of Adolf Loos*. Toronto: YYZ Books.

Louv, R. 2005. *Last Child in the Woods: Saving our Children from Nature-deficit Disorder.* Chapel Hill, NC: Algonquin Books of Chapel Hill.

Louv, R. 2012. *The Nature Principle: Reconnecting with Life in a Virtual Age*. Chapel Hill, NC: Algonquin Books of Chapel Hill.

Lovegren, Sylvia. 2013a. "Historical Overview: World War II to the Early 1960s." In Andrew F. Smith (ed.), *The Oxford Encyclopedia of Food and Drink in America*, 2nd edn., vol. 2, 233–40. New York: Oxford University Press.

Lovegren, Sylvia. 2013b. "From the 1960s to the Present." In Andrew F. Smith (ed.), *The Oxford Encyclopedia of Food and Drink in America*, 2nd edn., vol. 2, 241–50. New York: Oxford University Press.

Lovelock, J. 1979. *Gaia: A New Look at Life on Earth*. Oxford: Oxford University Press.

Low, T. 2005. "Preventing Alien Invasions: The Precautionary Principle in Practice in Weed Risk Assessment in Australia." In R. Cooney and B. Dickson (eds.), *Biodiversity and the Precautionary Principle*. London: Earthscan Publications Ltd.

Lowder, S. K., J. Skoet, and T. Raney. 2016. "The Number, Size, and Distribution of Farms, Smallholder Farms, and Family Farms Worldwide." *World Development*, 87: 16–29.

Lowe, P. 1989. "The Rural Idyll Defended: From Preservation to Conservation." In G. E. Mingay (ed.), *The Rural Idyll*, 113–31. London: Routledge.

Lucas, G., and S. Synge. 1978. *The IUCN Plant Red Data Book*. IUCN, Morges, Switzerland.

Lusser, M., C. Parisi, D. Plan, and E. Rodríguez-Cerezo. 2011. "New Plant Breeding Techniques: State-of-the-art and Prospects for Commercial Development." *Publications Office of the European Union* (ISBN 978-92-79-19715-4).

Lynes, Barbara Buhler. 2007. *Georgia O'Keeffe Museum Collections*. New York: Abrams.

Lynes, Barbara Buhler. 2009. "Georgia O'Keeffe and Abstraction: An Uneasy Peace." In Barbara Haskell (ed.), *Georgia O'Keeffe: Abstraction*, ch. 1, 1–17. New Haven, CT: Yale University Press.

Mabberley, D. J. 2009. "Exploring Terra Incognita." *Science*, 324(5926): 472.

Mabey, Richard. 2015. *The Cabaret of Plants: Botany and the Imagination*. London: Profile Books.

Macfarlane, Robert. 2018. "Word of the Day: 'witness-tree'." Twitter @RobGMacfarlane, January 21. Available online: https://twitter.com/robgmacfarlane/status/954979420883881984?lang=en (accessed May 16, 2021).

Mack, R. N., D. Simberloff, W. N. Londsdale, H. Evans, M. Clout, and F. A. Bazzaz. 2000. "Biotic Invasions: Causes, Epidemiology, Global Consequences and Control." *Ecological Applications*, 10(3): 689–710.

MacKendrick, Norah. 2018. *Better Safe than Sorry: How Consumers Navigate Exposure to Everyday Toxics*. Oakland: University of California Press.

Madaus, G. 1934. "Unsere Heilpflanzen heilen auch die Erde." *Jahrbuch Madaus*, 1934: 14–23.

Madaus, G. 1935. "Die Wirkung von Digitalis purpura und ihrer Abhängigkeit von der Düngung." *Jahrbuch Madaus*, 1935: 21.

Madaus, G. 1937–42. Sächsiche Staatsarchiv Dresden, 11610 Firma Dr. Madaus, Aktenordner 127 and 128, Pharmazutische Untersuchung Berichte.

Madaus, G. 1938. *Lehrbuch der biologischen Heilmittel*, vol. I, Leipzig. Georg Thieme.

Madaus, G. 1939. Sächsiche Staatsarchiv Dresden, 11610 Firma Dr. Madaus, Aktenordner Nummer 127, Pharmazeutische Untersuchung Berichte, 1937–42. "Bericht Dezember 1939."

Maha Viharaya—Bomaluwa Temple. 2018. "Sri Maha Bodhi—The Official Web Site of Maha Viharaya, Anuradhapura, Sri Lanka." Available online: http://srimahabodhi.lk/ (accessed May 16, 2021).

Mahood, M. M. 2008. *The Poet as Botanist*. Cambridge: Cambridge University Press.

Mancuso, Stefano, and Viola Alessandra. [2013] 2015. *Brilliant Green: The Surprising History and Science of Plant Intelligence*. Trans. Joan Benham. Washington, DC: Island Press.

Mangelsdorf, P. C. 1974. *Corn. Its Origin, Evolution and Improvement*, 2nd edn. Belknap Press of Harvard University Press.

Mansfield, Katherine. 1921. "Prelude." In *Bliss and Other Stories*, 1–70. New York: Alfred A. Knopf.

Mansfield, Katherine. [1922] 1997. "The Garden Party." In *The Garden Party and Other Stories*, 38–51. London: Penguin.

Marder, Michael. 2013. *Plant-Thinking A Philosophy of Vegetal Life*. New York: Columbia University Press.

Marder, Michael. 2017. "Phytosemiotics: A Fork in the Road." Available online: https://philosoplant.lareviewofbooks.org/?p=155(accessed November 22, 2019).

Marder, Michael. 2018. "Phytosemiotics." *Semiotic Review* [S. l.], (March 6). Available online: https://www.semioticreview.com/ojs/index.php/sr/article/view/39 (accessed November 22, 2019).

Marder, Michael. 2019. "To Hear Plants Speak." In Monica Gagliano, John C. Ryan, and Patricia Vieira (eds.), *The Language of Plants: Science, Philosophy, Literature*, 103–25. Minneapolis: University of Minnesota Press.

Marelli, Jean-Philippe, David I. Guest, Bryan A. Bailey, Harry C. Evans et al. 2019. "Chocolate Under Threat from Old and New Cacao Diseases." *Phytopathology*, 109: 1331–43.

Margulies, J. D., L. Bullough, A. Hinsley, D. J. Ingram et al. 2019. "Illegal Wildlife Trade and the Persistence of Plant Blindness.'" *Plants, People, Planet*, 1(3): 173–82. DOI: 10.1002/ppp3.10053.

Mark, J., A. Newton, S. Oldfield, and M. Rivers. 2014. *The International Timber Trade: A Working List of Commercial Timber Tree Species*. Richmond, UK: Botanic Gardens Conservation International.

Maroukis, T. C. 2010. *The Peyote Road: Religious Freedom and the Native American Church* (The Civilization of the American Indian Series, number 265). Norman: University of Oklahoma Press.

Marshall, Carolyn. 2007. "Alfred H. Peet, 87, Dies; Leader of a Coffee Revolution." *The New York Times*, September 3.

Martin, S. M. 2005. *The UP Saga*. Singapore: NIAS Press.

Martinelli, L., M. Batistella, R. Silva, and E. Moran. 2017. "Soy Expansion and Socioeconomic Development in Municipalities of Brazil." *Land*, 6: 62. DOI: 10.3390/land6030062.

Mascher, M., H. Gundlach, A. Himmelbach, and S. Beier et al. 2017. "A Chromosome Conformation Capture Ordered Sequence of the Barley Genome." *Nature*, 544: 427–33.

Mather, Victor. 2016. "Flowers a Thing of the Past at the Olympic Medal Ceremonies." *The New York Times*, August 13. Available online: https://www.nytimes.com/2016/08/14/sports/olympics/no-flowers-at-medal-ceremonies-rio-games.html (accessed May 16, 2021).

Matthews, Graham J. 2016. "What We Think About When We Think About Triffids: The Monstrous Vegetal in Post-War British Science Fiction." In Dawn Keetley and Angela Tenga (eds.), *Plant Horror: Approaches to the Monstrous Vegetal in Fiction and Film*, 111–27. London: Palgrave Macmillan.

Maunder, M. 2008. "Beyond the Greenhouse." *Nature*, 455: 596–7.

Maunder, M. 2015. "The Tropical Aroids: The Discovery, Introduction, and Cultivation of Exotic Icons." In Christian A. Larsen (ed.), *Philodendron: From Pan-Latin Exotic to American Modern*. Miami Beach: The Wolfsonian Museum, Florida International University.

Mayer, K. F. X., R. Waugh, P. Langridge et al. 2012. "A Physical, Genetical and Functional Sequence Assembly of the Barley Genome." *Nature*, 491: 711–16.

Mazza, Giuseppe. 2012. "Nong Nooch Tropical Botanical Garden." *Monaco Nature Encyclopedia*. Available online: https://www.monaconatureencyclopedia.com/nong-nooch-tropical-botanical-garden/?lang=en (accessed May 16, 2021).

McCauley, Martin. 1976. *Kruschev and the Development of Soviet Agriculture: The Virgin Land Program 1953–1964*. London: Macmillan.

McCouch, S., G. P. Baute, J. Bradeen, P. Bramel et al. 2013. "Agriculture: Feeding the Future." *Nature*, 499: 23–4.

McDonald, Bryan L. 2017. *Food Power: The Rise and Fall of the Postwar American Food System*. Oxford University Press.

McDonell, Emma. 2018. "The Quinoa Boom Goes Bust in the Andes." The North American Congress on Latin America website. Available online: https://nacla.org/news/2018/03/12/quinoa-boom-goes-bust-andes (accessed November 12, 2019).

McFeely, Mary Drake. 2000. *Can She Bake a Cherry Pie? American Women and the Kitchen in the Twentieth Century*. Amherst: University of Massachusetts Press.

McHarg, Ian L. 1969. *Design with Nature*. Garden City, NY: Natural History Press.

McKay, George. 2011. *Radical Gardening: Politics, Idealism and Rebellion in the Garden*. Toronto: Frances Lincoln.

McMichael, Philip. 2013. *Food Regimes and Agrarian Questions: Agrarian Change and Peasant Studies*. Nova Scotia, Canada: Fernwood Publishing.

McNeill, J. R., and P. Engelke. 2016. *The Great Acceleration: An Environmental History of the Anthropocene Since 1945*. Cambridge, MA: Harvard University Press.

McNeill, Leila. 2018. "The Botanical Artist Who Translates Plant Science into Beautiful Art." *Smithsonian.com*, June 10. Available online: https://www.smithsonianmag.com/science-nature/meet-woman-who-meticulously-illustrates-plants-name-science-180969544/(accessed October 26, 2019).

McQuaid, John. 2011. "Flower Power: Chances Are the Bouquet You're about to Buy Came from Colombia. What's Behind the Blooms?" *Smithsonian*, 41(10): 54.

McWilliams, James. 2013. *The Pecan: A History of America's Native Nut*. Austin: University of Texas Press.

Mee, Margaret. 1988. *Margaret Mee: In Search of the Flowers of the Amazon Forest*. Tony Morrison (ed.). Woodbridge, UK: Nonesuch Expeditions.

Meeker, Natania, and Antónia Szabari. 2020. *Radical Botany: Plants and Speculative Fiction*. New York: Fordham University Press.

Meilland, A., and G. Lambert. 1984. *Meilland: A Life in Roses*. Carbondale & Edwardsville: University of Southern Illinois Press.

Melbye, David. 2010. *Landscape Allegory in Cinema: From Wilderness to Wasteland*. New York: Palgrave Macmillan.

Melville, R. 1936. "A list of true and false mahoganies." *Kew Bull*, 3(20): 193–210.

Mendelson, Anne. 2013. "Historical Overview: World War I to World War II." In Andrew F. Smith (ed.), *The Oxford Encyclopedia of Food and Drink in America*, 2nd edn., vol. 2, 221–7. New York: Oxford University Press.

Meng-Lin, Chen. 2019. "A Corpus-based Study on Imagery and Symbolism in Goldblatt's Translation of *Red Sorghum*." *Babel*, 65(3): 399–423.

Midgley, Clare. 1996. "Slave Sugar Boycotts, Female Activism and the Domestic Base of British Anti-Slavery Culture." *Slavery and Abolition*, 17(3): 137–62.

Milesi, C., Steven Running, Christopher Elvidge et al. 2005. "Mapping and Modeling the Biogeochemical Cycling of Turf Grasses in the United States." *Environmental Management*, 36(3): 426–38. DOI: 10.1007/s00267-004-0316-2.

Millennium Ecosystem Assessment (MA). 2005. *Ecosystems and Human Well-being: Current State and Trends*. Washington, DC: Island Press.

Millennium Restaurant. 2019. "Eric Tucker, Executive Chef." Available online: https://www.millenniumrestaurant.com/eric-tucker (accessed October 20, 2019).

Miller, T. S. (2014). "Plants, Monstrous." In *The Ashgate Encyclopedia of Literary and Cinematic* Monsters, 470–5. Edited by Jeffrey Andrew Weinstock. Farnham: Ashgate.

Ministry of Health. 1948. "Report of the Committee on Indigenous Systems of Medicine." New Delhi: Government of India.

Mintz, Sidney W. 1985. *Sweetness and Power: The Place of Sugar in Modern History*. New York: Penguin Books.

Mitra, R., and L. D. Kapoor. 1976. "Kamala—the National flower of India—Its Ancient History and Uses in Indian Medicine." *Indian Journal of History of Science*, 11(2): 125–32.

Money, Nicholas P. 2007. *The Triumph of the Fungi: A Rotten History*. New York: Oxford University Press.

Monforte, A. J., A. Diaz, A. Cano-Delgado, E. Van Der Knaap et al. 2014. "The Genetic Basis of Fruit Morphology in Horticultural Crops: Lessons from Tomato and Melon." *Journal of Experimental Botany*, 65: 4625–37.

Monger, G. 1997. "Modern Wayside Shrines." *Folklore*, 108: 113–14.

Monger, G., and J. Chandler. 1998. "Pilgrimage to Kensington Palace." *Folklore*, 109: 104–9.

Monteiro, C. A., J. -C. Moubarac, G. Cannon et al. 2013. "Ultra-processed Products are Becoming Dominant in the Global Food System." *Obesity Reviews*, 14: 21–8.

Monteiro, C., G. Cannon, R. Levy et al. 2019. "Ultra-processed Foods: What They Are and How to Identify Them." *Public Health Nutrition*, 22(5): 936–41.

Monteiro, C., J. Moubarac, R. Levy, D. Canella, M. Louzada, and G. Cannon. 2017. "Household Availability of Ultra-processed Foods and Obesity in Nineteen European Countries." *Public Health Nutrition*, 21(1): 18–26.

Moodie, R., D. Stuckler, C. Monteiro et al. 2013. "Profits and Pandemics: Prevention of Harmful Effects of Tobacco, Alcohol, and Ultra-processed Food and Drink Industries." *The Lancet*, 381(9867): 670–9.

Moore, Jason W. 2010. "Cheap Food and Bad Money: Food, Frontiers, and Financialization in the Rise and Demise of Neoliberalism." *Review*, 33(2/3): 225–61.

Morton, Timothy. 2010. *The Ecological Thought*. Cambridge, MA: Harvard University Press.

Mottet, A., C. De Haan, A. Falcucci, G. Tempio, C. Opio, and P. Gerber. 2017. "Livestock: On our Plates or Eating at our Table? A New Analysis of the Feed/food Debate." *Global Food Security*, 14: 1–8.

Mugnozza, G. T. S. 2005. "The Contribution of Italian Wheat Geneticists: From Nazareno Strampelli to Francesco D'Amato." In R. Tuberosa, R. L. Phillips, and M. Gale (eds.), Proc International Congress "*In the Wake of the Double Helix: From the Green Revolution to the Gene Revolution*," May 27–31, 2003, 53–75. Bologna, Italy.

Mukharji, P. 2011. *Nationalizing the Body. The Medical Market, Print and Daktari Medicine*. London: Anthem Press.

Mulliken, T., and P. Crofton. 2008. "Review of the Status, Harvest, Trade and Management of Seven Asian CITES-listed Medicinal and Aromatic Plant Species." BfN—Skripten Report 227.

Mundy, Jennifer. 2011. "The Naming of Biomorphism." In Oliver A. I. Botar and Isabel Wünsche (eds.), *Biocentrism and Modernism*, 61–75. Farnham, England: Ashgate.

Murdock, A. 2019. "The Best Plant Stories of 2018." Medium, January 8. Available online: https://medium.com/@andy_murdock/the-best-plant-stories-of-2018-a2bb1d9735c9 (accessed May 16, 2021).

Murphy, M. K. 2019. "Scientific Argument Without a Scientific Consensus: Rachel Carson's Rhetorical Strategies in the Silent Spring Debates." *Argumentation and Advocacy*, 55(3): 194–210.

Nabhan, Gary Paul, Deja Walker, and Alberto Mellado Moreno. 2016. "Biocultural and Ecogastronomic Restoration: The Renewing America's Food Traditions Alliance." In Gary Paul Nabhan (ed.), *Ethnobiology for the Future: Linking Cultural and Ecological Diversity* ch. 11, 156–83. Tucson: The University of Arizona Press.

Nadel, D., A. Danin, R. C. Power, A. M. Rosen et al. 2013. "Earliest Floral Grave Lining from 13,700–11,700-y-old Natufian Burials at Raqefet Cave, Mt. Carmel, Israel." *PNAS*, 110(29) (July 16): 11774–8. Available online: https://www.pnas.org/content/110/29/11774 (accessed May 16, 2021).

Naess, Arne. 2012. "Ecophilosophy in a World of Crisis." In R. Bhaskar, P. Naess, and K. G. Høyer (eds.), *Critical Realism and the Nordic Contributions*, 84–98. Abingdon: Routledge.

Nash, Steven A. 2001. "Moore and Surrealism Reconsidered." In Dorothy Kosinski (ed.), *Henry Moore: Sculpting the 20th Century*, 43–51. New Haven: Yale University Press.

National Red List. 2016. Available online: www.nationalredlist.org (accessed May 16, 2021).

Nature Conservancy, The. 2019. "Who We Are—Our History." Available online: https://www.nature.org/en-us/about-us/who-we-are/ (accessed May 16, 2021).

Nau, J., J. Goldsmith, M. Shank et al. 1999. "History of US Floriculture: Breeding & Plant Selection." *Greenhouse Grower* (Fall): 28–37.

Nealon, J. T. 2015. *Plant Theory: Biopower and Vegetable Life*. Palo Alto, CA: Stanford University Press.

Nelson, Diane. 2019. "Saving Citrus: Researchers Race to Keep Citrus in our Diets and Communities." *Outlook Magazine* (Fall/Winter): 8–9.

Neot Kedumim Park. 2011. "History of Neot Kedumim Park." Available online: https://www.neot-kedumim.org.il/history/ (accessed July 17, 2020).

Nesbitt, M., and S. G. Prance. 2005. *The Cultural History of Plants*. New York: Routledge.

Nestle, Marion. 2018. *Unsavory Truth: How Food Companies Skew the Science of What We Eat*. New York: Basic Books.

New Yorkers for Parks. n.d. "The Daffodil Project." Available online: http://www.ny4p.org/the-daffodil-project#2189 (accessed July 17, 2020).

Newman, Eric A., Alfonso Araque, and Janet Dubinsky (eds.). 2017. *The Beautiful Brain: The Drawings of Santiago Ramón y Cajal*. New York: Abrams Books.

Newman, Kara. 2013. "Historical Overview: World War I." In Andrew F. Smith (ed.), *The Oxford Encyclopedia of Food and Drink in America*, 2nd edn., vol. 2, 219–21. New York: Oxford University Press.

Nong Nooch Tropical Garden. 2020. "History." Available online: http://www.nongnoochtropicalgarden.com/home/ (accessed May 16, 2021).

Nosferatu. 1922. [Film] Dir. F. W. Murnau, Prod. Enrico Dieckmann and Albin Grau. Berlin: Prana.

Nugteren, A. 2005. "Symbol and Sacredness: Trees in Ancient Religious Literature." In *Belief, Bounty, and Beauty: Rituals around Sacred Trees in India*, 1–71. Numen Books Series—Studies in the History of Religions, vol. 108. Leiden: Brill.

Nursery & Garden Industry of Australia. 2013. *National Plant Labelling Guidelines*, Version 2, Nursery & Garden Industry of Australia. Available online: http://yourlevyatwork.com.au/wp-content/uploads/2013/01/Plant_Labelling_Jan2013_V2_FINAL.pdf (accessed May 16, 2021).

Occupiedpalestine. 2012. "Israel Fraud the History by Building Talmudic Gardens around Al Aqsa." Blog, Feburary 26. Available online: https://occupiedpalestine.wordpress.com/2012/02/27/israel-fraud-the-history-by-building-talmudic-gardens-around-al-aqsa/ (accessed July 17, 2020).

Ohlrogge, John. 1999. "Plant Metabolic Engineering: Are We Ready for Phase Two? Editorial Overview." *Current Opinion in Plant Biology*, 2: 121–2.

Oldfield, S. F. 1988. *Rare Tropical Timbers*. Gland, Switzerland and Cambridge, UK: IUCN.

Oldfield, S. F. 2010. *Botanic Gardens: Modern Day Arks*. New Holland Publishers.

Oldfield, S. F. 2013. "The Evolving Role of CITES in Regulating the International Timber Trade." *RECIEL*, 22(3): 291–300.

Oldfield, S. F. and M. Jenkins. 2012. *Wild flora for improved rural livelihoods*. Case studies from Brazil, China, India and Mexico. Botanic Gardens Conservation International, Richmond, UK.

Oldfield, S. F., C. Lusty, and A. MacKinven. 1998. *The World List of Threatened Trees*. Cambridge: World Conservation Press.

Olmstead, Gracy. 2018. "Wendell Berry's Right Kind of Farming." *The New York Times*, October 1. Available online: https://www.nytimes.com/2018/10/01/opinion/wendell-berry-agriculture-farm-bill.html (accessed May 16, 2021).

Orczyk, W., J. Przetakiewicz, and A. Nadolska-Orczyk. 2003. "Somatic Hybrids of Solanum tuberosum—Application to Genetics and Breeding." *Plant Cell, Tissue Organ Culture*, 74: 1–13.

Ordonez, N., M. F. Seidl, C. Waalwijk, A. Drenth, A. Kilian, and B. P. H. J. Thomma. 2015. "Worse Comes to Worst: Bananas and Panama Disease—When Plant and Pathogen Clones Meet." *PLoS Pathogens*, 11(11): e1005197.

Orlean, S. 1998. *The Orchid Thief*. New York: Random House.

Orwell, George. [1936] 1962. *Keep the Aspidistra Flying*. Harmondsworth: Penguin.

Otesánek. 2000. [Film] Dir. Jan Šankmajer, Prod. Jaromir Kallista and Keith Griffiths. New York: Zeitgeist.

Oudolf, P., and R. Darke. 2017. *Gardens of the High Line: Elevating the Nature of Modern Landscapes*. Portland, Oregon: Timber Press.

OED. n. d.-a."Ornament, n1., n2." *Oxford English Dictionary Online*. Available online: https://www.lexico.com/definition/ornament (accessed March 2, 2020).

OED. n. d.-b. "Ornament, n1." *Oxford English Dictionary Online*. Available online: https://www.lexico.com/definition/ornament(accessed March 29, 2020).

Padua, M. 2010. "Touching the Good Earth—an Innovative Campus Design Reconnects Students to China's Agricultural Landscapes." Available online: https://www.turenscape.com/en/paper/detail/20.html (accessed May 16, 2021).

Pakenham, Thomas. 1997. *Meetings with Remarkable Trees*. London: Weidenfield & Nicolson.

Pakenham, Thomas. 2002. *Remarkable Trees of the World*. London: Weidenfield & Nicolson.

Papetti, Adele. 2019. "Nutraceuticals: Health Benefits and Government Regulations." *Current Research in Nutrition and Food Science*, 7(1): 1–4.

Parker, Elizabeth. 2016. "'Just a Piece of Wood': Jan Švankmajer's *Otesánek* and the EcoGothic." In Dawn Keetley and Angela Tenga (eds.), *Plant Horror: Approaches to the Monstrous Vegetal in Fiction and Film*, 215–25. London: Palgrave Macmillan.

Parker, Lauren E., and John T. Abatzoglou. 2018. "Shifts in the Thermal Niche of Almond Under Climate Change." *Climatic Change*, 147: 211–24.

Patrick, J., and J. Wade. 2012. *Australian Garden Design: Secrets of Leading Garden Designers Revealed*. Sydney, Australia: ABC Books.

Payyappallimana, U., Z. Fadeeva, and R. O'Donoghue. 2013. "Traditional Knowledge and Biodiversity within Regional Centres of Expertise on Education for Sustainable Development." In U. Payyappallimana, Z. Fadeeva, and A. Dirksen (eds.), *Innovation in Local and Global Learning Systems for Sustainability: Traditional Knowledge and Biodiversity—Learning Contributions of the Regional Centres of Expertise on Education for Sustainable Development*, editorial, 8–21. Yokohama, Japan: UnU-IAs.

Peace Pledge Union. 2020. "Remembrance and White Poppies." Available online: https://ppu.org.uk/remembrance-white-poppies (accessed July 19, 2020).

Pei, S. J., and H. Y. Huai. 2013 "Medicinal Plants." In D. Y. Hongand and S. Blackmore (eds.), *Plants of China. A Companion to the Flora of China*, ch. 18, 324–41. Beijing: Science Press and Cambridge and New York: Cambridge University Press.

Pereira, Alyssa. 2016. "With Grant from Peet's, UC Davis will Launch Research Center for Coffee." SFGate, September 8.

Pergams, O. R. W., and P. A. Zaradic. 2006. "Is Love of Nature in the US becoming Love of Electronic Media? 16-year Downtrend in National Park Visits Explained by Watching Movies, Playing Video Games, Internet Use, and Oil Prices." *Journal of Environmental Management*, 80: 387–93.

Perkins, J. H. 1997. *Geopolitics and the Green Revolution: Wheat, Genes, and the Cold War*. New York: Oxford University Press.

Petrenko, C., J. Paltseva, and S. Searle. 2016. "Ecological Impacts of Palm Oil Expansion in Indonesia." International Council on Clean Transportation. Washington, DC.

Petruzzello, Melissa. 2020. "David Fairchild." In *Encyclopædia Britannica*. Available online: https://www.britannica.com/biography/David-Grandison-Fairchild (accessed April 17, 2020).

Pettman, D. 2013. "The Noble Cabbage: Michael Marder's 'Plant Thinking'." *Los Angeles Review of Books*, July 28. Available online: https://lareviewofbooks.org/article/the-noble-cabbage-michael-marders-plant-thinking/ (accessed May 16, 2021).

Pevsner, Nikolaus. 2005. Pioneers of Modern Design: From William Morris to Walter Gropius (4th ed., rev. and expanded)., New Haven, Conn.; London: Yale University Press.

Pew Research Center. 2015. "The Future of World Religions: Population Growth Projections, 2010–2050." Available online: https://www.pewforum.org/2015/04/02/religious-projections-2010-2050/ (accessed April 8, 2020).

Phaidon editors. 2019. *Blooms: Contemporary Floral Design*. London: Phaidon Press.

Phillips, Roger. [1977] 1978. Assisted by Sheila Grant. Edited by Tom Wellsted. *Wild Flowers of Britain*. London: Ward Lock.

Phillips, T. 2015. "The Nature of Ornament: A Summary Treatise." *The New Bookbinder*, 35: 7–14

Pickstone, J. 2000. *Ways of Knowing*. Manchester: Manchester University Press.

Pilcher, Jeffrey M. 2016. "Food Production." In Amy Bentley (ed.), *A Cultural History of Food: In the Modern Age*, vol. 6, 27–45. London: Bloomsbury.

Pilgrim, S. E., L. C. Cullen, D. J. Smith, and J. Pretty. 2008. "Ecological Knowledge is Lost in Wealthier Communities and Countries." *Environmental Science and Technology*, 42(4): 1004–9.

Pimentel, D., C. Wilson, C. McCullum, R. Huang, P. Owen, J. Flack, Q. Tran, T. Saltman, and B. Cliff. 1997. "Economic and Environmental Benefits of Biodiversity." *BioScience*, 47: 747–57.

Pingali, P. 2007. "Westernization of Asian Diets and the Transformation of Food Systems: Implications for Research and Policy." *Food Policy*, 32(3): 281–98.

Pinto, R. S., and M. P. Reynolds. 2015. "Common Genetic Basis for Canopy Temperature Depression under Heat and Drought Stress Associated with Optimized Root Distribution in Bread Wheat." *Theoretical and Applied Genetics*, 128: 575–85.

Pisters, Patricia. 2012. *The Neuro-Image: A Deluzian Film-Philosophy of Digital Screen Culture*. Stanford: Stanford University Press.

Pittman, A. M., K. M. Carlson, L. M. Curran, and A. Ponette-Gonzalez. 2013. "NASA Satellite Data Used to Study the Impact of Oil Palm Expansion Across Indonesian Borneo." *The Earth Observer*, 25(5): 12–16.

Plantlife. 2019. "About Us." Available online: https://www.plantlife.org.uk/uk/about-us (accessed May 16, 2021).

Plants in Action. 1983. [Film] Dir. Roger Seccombe, Prod. David Morgan and Nick Alexander. Melbourne: Commonwealth Scientific and Industrial Research Organization.

Plants of the Pantry. 1927. [Film] Dir. Frank Percy Smith, Prod. British Instructional Films. London: British Film Institute.

Pollan, M. 1991. *Second Nature: A Gardener's Education*. New York: Atlantic Monthly Press.

Pollan, M. 2001. *The Botany of Desire: A Plant's Eye View of the World*. New York: Random House.

Pollan, M. 2008. *In Defence of Food: An Eater's Manifesto*. New York: Penguin Press.

Pollan, M. 2018. *How to Change Your Mind: What the New Science of Psychedelics Teaches Us About Consciousness, Dying, Addiction, Depression, and Transcendence*. New York: Penguin Press.

Pollan, M. 2020. "About Michael Pollan." Available online: https://michaelpollan.com/about/ (accessed July 23, 2020).

Popkin, B. M. 2001. "The Nutrition Transition and Obesity in the Developing World." *The Journal of nutrition*, 131(3): 871S–873S.

Pordié, L., and J. -P. Gaudillière. 2014. "The Reformulation Regime in Drug Discovery: Revisiting Polyherbals and Property Rights in the Ayurvedic Industry." *East Asian Science, Technology and Society*, 8: 57–79.

Pordié, Laurent. 2014. Pers. comm. Interview with the author and Laurent Pordié, February. Himalaya headquarter, Bangalore.

Pourkheirandish, M., G. Hensel, B. Kilian, G. Willcox et al. 2015. "Evolution of the Seed Dispersal System in Barley." *Cell*, 162: 527–39.

Powers, Richard. 2018. *The Overstory: A Novel*. New York: W. W. Norton & Company.

Prance, G. T. 2001. "Discovering the Plant World." *Taxon*, 50(2): 345–59.

Prance, G. T., H. Beentje, J. Dransfield, and R. Johns. 2000. "The Tropical Flora Remains Undercollected." *Annals of the Missouri Botanical Garden*, 87(1): 67–71.

Prouix, Annie. 2016. *Barkskins: A Novel*. New York, Scribner.

Purugganan, M. D., and D. Q. Fuller. 2009. "The Nature of Selection during Plant Domestication." *Nature*, 457: 843–8.

Raj, K. 2016. "Go-Betweens, Travelers and Cultural Translators." In B. Lightman (ed.), *The Blackwell Companion to the History of Science*, ch. 3, 39–57. Chichester: Wiley-Blackwell.

Rakestraw, L. 1955. "A History of Forest Conservation in the Pacific Northwest, 1891–1913." Thesis, University of Washington. Available online: https://foresthistory.org/wp-content/uploads/2017/01/A-HISTORY-OF-FOREST-CONSERVATION-IN-THE-PACIFIC-NORTHWEST.pdf (accessed May 16, 2021).

Rametsteiner, E., and M. Simula. 2003. "Forest Certification—an Instrument to Promote Sustainable Forest Management?" *Journal of Environmental Management*, 67: 87–98.

Ranasinghe, Lakshman. 2015. "How Sal Trees arrived in Sri Lanka." *The Nation*, March 1. Available online: https://web.archive.org/web/20150313140246/http:/www.nation.lk:80/edition/fine/item/38684-how-sal-trees-arrived-in-sri-lanka.html (accessed May 16, 2021).

Rapsomanikis, G. 2015. "The Economic Lives of Smallholder Farmers: An Analysis based on Household Data from Nine Countries." Available online: http://www.fao.org/3/a-i5251e.pdf (accessed April 30, 2017).

Ratsimbazafy, C., D. J. Newton, and R. Stéphane. 2016. *Timber Island: Rosewood and Ebony Trade of Madagascar*. Cambridge, UK: TRAFFIC.

RBG Kew. 2016. *The State of the World's Plants Report—2016*. Royal Botanic Gardens, Kew.

Read, Herbert. 1968. *The Art of Jean Arp*. New York: Harry N. Abrams.

Reardon T., and J. Berdegue 2006. "The Retail-led Transformation of AgriFood Systems and Its Implications for Development Policies." Background Paper, World Development Report 2008. Agriculture for Development, World Bank, Washington DC.

Reichl, A. J. 2016. "The High Line and the Ideal of Democratic Public Space." *Urban Geography*, 37: 6, 904–25.

Reynolds, David. [2013] 2014. *The Long Shadow: The Legacies of the Great War in the Twentieth Century*. New York: W.W. Norton & Company.

Rhizopoulou, S. 2004. "Symbolic Plant(s) of the Olympic Games." *Journal of Experimental Botany*, 55(403): 1601–6.

Rice, David A. 2005. "Witchery, Indigenous Resistance, and Urban Space in Leslie Marmon Silko's *Ceremony*." *Studies in American Indian Literatures*, 17(4): 114–43.

Richardson, Mark, Adam Burgess, and Lynne Sealie. 2014. "Global Strategy for Plant Conservation and the National Arboretum Canberra." *Australasian Plant Conservation: Journal of the Australian Network for Plant Conservation*, 22(4): 15–17.

Riley, Barry. 2017. *The Political History of American Food Aid: An Uneasy Benevolence*. New York: Oxford University Press.

Ripple, W. J., C. Wolf, and T. M. Newsome. 2019. "World Scientists' Warning of a Climate Emergency." *Bioscience Magazine*, 2000(X): 1–20.

Ritchie, H., and M. Roser. 2020. "Urbanization." *OurWorldInData.org*. Available online: https://ourworldindata.org/urbanization (accessed April 8, 2020).

Rivera, José Eustasio. [1924] 2018. *The Vortex: A Novel*. Durham: Duke University Press.

Rivers, M. 2017. "Conservation Assessments and Understanding the Impacts of Threats on Plant Biodiversity." In S. Blackmore and S. Oldfield (eds.), *Plant Conservation Science & Practice—The Role of Botanic Gardens*, ch. 3, 49–72. Cambridge: Cambridge University Press.

Rix, Martyn. 1981. *The Art of Botanical Illustration*. New York: Arch Cape Press.

Rix, Martyn. 2013. *Rory McEwen: The Colours of Reality*. Richmond, Surrey: Royal Botanic Gardens, Kew.

Robbins, P., and J. Sharp. 2003. "Producing and Consuming Chemicals: The Moral Economy of the American Lawn." *Economic Geography*, 79: 425–51.

Robertson, Eric. 2006. *Arp: Painter, Poet, Sculptor*. New Haven, CT: Yale University Press.

Robinson, William I. 2014. *Global Capitalism and the Crisis of Humanity*. New York: Cambridge University Press.

Robinson, W., and R. Darke. 2009. *The Wild Garden*. Portland, OR: Timber Press.

Rosenberg, C. 1987. *The Care of Strangers: The Rise of America's Hospital System*. New York: Basic Books.

Roser, M., H. Ritchie, and E. Ortiz-Ospina. 2020. "World Population Growth." OurWorldInData.org. Available online: https://ourworldindata.org/world-population-growth (accessed April 8, 2020).

Rothschuh, K. E. 1983. *Naturheilbewegung, Reformbewegung, Alternativbewegung*. Stuttgart: Hippokrates-Verlag.

Royal Horticultural Society. 2019. "Annual Report and Consolidated Financial Statements for the Year ended 31 January 2019." Available online: https://www.rhs.org.uk/about-the-rhs/pdfs/about-the-rhs/mission-and-strategy/past-annual-reports/rhs-annual-report-2018-19.pdf (accessed May 16, 2021).

Royal Institute of British Architects. 2019. "Modernism: Rejecting Ornament and Embracing Minimalism." Available online: https://www.architecture.com/explore-architecture/modernism (accessed May 16, 2021).

Rudel, T. K. 2007. "Changing Agents of Deforestation: From State-initiated to Enterprise Driven Processes 1970–2000." *Land Use Policy*, 24(1): 35–41.

Russell, J., M. Mascher, I. K. Dawson, S. Kyriakidis et al. 2016. "Exome Sequencing of Geographically Diverse Barley Landraces and Wild Relatives Gives Insights into Environmental Adaptation." *Nature Genetics*, 48: 1024–30.

Ryan, John Charles. 2015. "Tolkien's Sonic Trees and Perfumed Herbs: Plant Intelligence in Middle-earth." In Patrícia Vieira, Monica Gagliano, and John Ryan (eds.), *The Green Thread: Dialogues with the Vegetal World*, ch. 3, 37–58. Lanham, Maryland: Lexington.

Sachs, Susan. 2004. "Summer 2004 Games Crowning Emblem: Uneasy Lie the Heads Wearing the Wreaths." *The New York Times*, August 25. Available online: http://www.nytimes.com/2004/08/25/sports/summer-2004-games-crowning-emblem-uneasy-lie-the-heads-wearing-the-wreaths.html (accessed May 16, 2021).

Sackman, Douglas Cazaux. 2007. *Orange Empire: California and the Fruits of Eden*. Berkeley: University of California Press.

Sagan, C. 2013. *Cosmos*. New York: Random House.

Sage, A. 2018. "Campari Drinks Family Embroiled in Row over Sale of 'Louvre of botany.'" *The Times*, January 8, 35.

Sahba, Fariborz. 2010. "Faith and Form: Contemporary Space for Worship and Pilgrimage." In Karla Cavarra (ed.), *Constructing the Ineffable: Contemporary Sacred Architecture*, 170–82. Yale, CT: Yale University Press.

Salick, J., K. Konchar, and M. Nesbit. 2014. "Biocultural Collections: Needs, Ethics and Goals." In J. Salick, K. Konchar and M. Nesbitt (eds.), *Curating Biocultural Collections*.

A Handbook, ch. 1, 1–13. Richmond and St. Louis: Kew Publishing in association with Missouri Botanical Garden.

Salmón, Enrique. 2012. *Eating the Landscape: American Indian Stories of Food, Identity, and Resilience*. Tucson: The University of Arizona Press.

Saltman, T., and B. Cliff. 1997. "Economic and Environmental Benefits of Biodiversity." *BioScience*, 47: 747–57.

Sampson, Neil R., Nadia Bystriakova, Sandra Brown et al. 2005. "Timber, Fuel, and Fiber." In R. Hassan, R. Scholes, and N. Ash (eds.), *Ecosystems and Human Well-being: Current State and Trends, Volume 1* (Findings of the Condition and Trends Working Group of the Millennium Ecosystem Assessment), 243–70. Washington, DC: Island Press.

Sandifer, P. A., A. E. Sutton-Grier, and B. P. Ward. 2015. "Exploring Connections among Nature, Biodiversity, Ecosystem Services, and Human Health and Well-being: Opportunities to Enhance Health and Biodiversity Conservation." *Ecosystem Services*, 12: 1–15.

Santana-Gálvez, J., L. Cisneros-Zevallos, and D. A. Jacobo-Velázquez. 2019. "A Practical Guide for Designing Effective Nutraceutical Combinations in the Form of Foods, Beverages, and Dietary Supplements against Chronic Degenerative Diseases." *Trends in Food Science and Technology*, 88 (June): 179–93.

Sato, K., M. Yamane, N. Yamaji, H. Kanamori, A. Tagiri, J. G. Schwerdt, G. B. Fincher, T. Matsumoto, K. Takeda, and T. Komatsuda. 2016. "Alanine aminotransferase Controls Seed Dormancy in Barley." *Nature Communications*, 7: Article number: 11625. DOI: 10.1038/ncomms11625.

Saunders, Gill. [1995] 2009. *Picturing Plants: An Analytical History of Botanical Illustration*, 2nd edn. Chicago: KWS.

Saunders, N. J. 2013. *The Poppy: A History of Conflict, Loss, Remembrance & Redemption*. London: One World Publications.

Sayer, J. A., C. S. Harcourt, and N. M. Collins (eds.). 1992. *The Conservation Atlas of Tropical Forests: Africa*. IUCN/WCMC. London: Macmillan Publishers Ltd.

Schama, Simon. 1997. *The Embarrassment of Riches: An Interpretation of Dutch Culture in the Golden Age*. New York: Vintage Books.

Scherer, Agnes. 2016. "The Pre-cosmic Squiggle: Tendril Excesses in Early Modern Art and Science Fiction Cinema." In Dawn Keetley and Angela Tenga (eds.), *Plant Horror: Approaches to the Monstrous Vegetal in Fiction and* Film, 31–53. London: Palgrave Macmillan.

Schiebinger, Londa. 2004. *Plants and Empire: Colonial Bioprospecting in the Atlantic World*. Cambridge, MA: Harvard University Press.

Schiff, P. L. 2006. "Ergot and Its Alkaloids." *American Journal of Pharmaceutical Education*, 70 (5): 98. Available online: https://www.ncbi.nlm.nih.gov/pmc/articles/PMC1637017/ (accessed May 16, 2021).

Schippmann, U., D. Leaman, and A. Cunningham. 2002. *Impact of Cultivation and Gathering of Medicinal Plants on Biodiversity: Global Trends and Issues*. FAO, Rome.

Schmitt, Ronald E. 2007. *Sullivanesque: Urban Architecture and Ornamentation*. Urbana: University of Illinois Press.

Schultes, R. E., A. Hofmann, and C. Ratsch. 1979. *Plants of the Gods: Origins of Hallucinogenic Use*. Rochester, Vermont: Healing Arts.

Schulthess, B. 1994. "The Earth Council's Role in Strengthening Indigenous People's Participation in Implementing the Convention on Biological Diversity." In A. F. Krattiger, J. A. McNeely, W. H. Lesser, K. R. Miller, Y. St. Hill, and R. Senanayake (eds.), *Widening Perspectives on Biodiversity*, ch. 4.4, 225–8. IUCN, Switzerland and International Academy of the Environment, Switzerland.

Schwarz, M. W., D. J. Porter, J. M. Randall, and K. E. Lyons. 1996. "Impact of Nonindigenous Plants." In *Sierra Nevada Ecosystem Project: Final Report to Congress*, vol. II: *Assessments and Scientific Basis for Management Options*, 1203–26. Davis: University of California, Centers for Water and Wildland Resources.

Scott, P., J. A. Burton and R. Fitter. 1987. "Red Data Books: The Historical Background." In Fitter R. and M. Fitter (eds.), *The Road to Extinction*, 1–6. Gland, Switzerland and Cambridge, UK: IUCN/UNEP.

Searchinger, T., R. Waite, C. Hanson, J. Ranganathan, P. Dumas, and E. Matthews. 2019. *Creating a Sustainable Food Future: A Menu of Solutions to Feed Nearly 10 Billion People by 2050*. Washington, DC: World Resources Institute.

Secord, J. 2015. "Overcoming Environmental Determinism: Introduced Species, Hybrid Plants and Animals, and Transformed Lands in the Hellenistic and Roman Worlds." In R. F. Kennedy and M. Jones-Lewis (eds.), *The Routledge Handbook of Identity and the Environment in the Classical and Medieval Worlds*, 210–29. London: Routledge.

Secretariat of the Convention on Biological Diversity. 2009. "The Convention on Biological Diversity Plant Conservation Report: A Review of Progress in Implementing the Global Strategy of Plant Conservation (GSPC)." Montreal, Canada: Secretariat of the CBD.

Secretariat of the Convention on Biological Diversity. 2014. *Global Biodiversity Outlook 4*. Montréal, Canada: Secretariat of the CBD.

Secretariat of the Convention on Biological Diversity. 2015. "Joint Statement by the Liaison Group of the Biodiversity-related Conventions on the Occasion of the United Nations Sustainable Development Summit 25–27 September 2015." Available online: https://whc.unesco.org/en/news/1359 (accessed May 16, 2021).

Seddon, George. 1970. *Swan River Landscapes*. Nedlands: University of Western Australia Press.

Sethi, S. C. 1981. "Photoperiodic Response and Accelerated Generation Turnover in Chickpea." *Field Crops Research*, 4: 215–25.

Seto, K. C., and N. Ramankutty. 2016. "Hidden Linkages between Urbanization and Food Systems." *Science*, 352(6288): 943–5.

Settee, Priscilla. 2018. "Indigenous Food Sovereignty in Canada." In Melissa K. Nelson and Dan Shilling (eds.), *Traditional Ecological Knowledge: Learning from Indigenous Practices for Environmental Sustainability*, ch. 10, 175–87. Cambridge: Cambridge University Press.

Sharrock, S., S. Oldfield, and O. Wilson. 2014. *Plant Conservation Report 2014: A Review of Progress in Implementation of the Global Strategy for Plant* Conservation *2011–2020*. Secretariat of the Convention on Biological Diversity, Montréal, Canada and Botanic Gardens Conservation International, Richmond, UK. Technical Series No. 81.

Sherwood, Shirley. 1996. *Contemporary Botanical Artists: The Shirley Sherwood Collection*. New York: Cross River.

Sheva, Arutz. 2011. "First Ever Talmudic Garden Dedicated in Beit El." *Israel National News*, October 19. Available online: http://www.israelnationalnews.com/News/News.aspx/148923 (accessed July 17, 2020).

Shi J. and J. Lai. 2015. "Patterns of genomic changes with crop domestication and breeding." *Curr. Opin. Plant Biol.* 24, 47–53.

Shprintzen, A. 2018. "Food in 20th-Century American Cities." In *Oxford Research Encyclopedia of American History*, 1–16. Available online: https://oxfordre.com/americanhistory/ (accessed May 16, 2021).

Shteir, Ann B. 1996. *Cultivating Women, Cultivating Science: Flora's Daughters and Botany in England, 1760–1860*. Baltimore, MD: Johns Hopkins University Press.

Silbergeld, E. K. 2016. *Chickenizing Farms and Food: How Industrial Meat Production Endangers Workers, Animals, and Consumers*. Baltimore, MD: Johns Hopkins University Press.

Simon, T. 2015. "Paperwork and the Will of Capital 2015." Available online: http://www.tarynsimon.com/works/pwc/#1 (accessed April 12, 2019).

Simons, K. J., J. P. Fellers, H. N. Trick et al. 2006. "Molecular Characterization of the Major Wheat Domestication Gene Q." *Genetics*, 172: 547–55.

Simpson, Inga. 2017. *Where the Trees Were*. Hatchette: Sydney.

Simpson, Niki, and Peter G. Barnes. 2008. "Photography and Contemporary Botanical Illustration." *Curtis's Botanical Magazine*, 25(3): 258–80.

Slakeski, N., D. C. Baulcombe, K. M. Devos, B. Ahluwalia, D. N. Doan, and G. B. Fincher. 1990. "Structure and Tissue-specific Regulation of Genes Encoding Barley (1→ 3, 1→ 4)-β-glucan Endohydrolases." *Molecular Genetics and Genomics*, 224: 437–49.

Slater, E., and M. Peillon. 2009. "The Suburban Front Garden: A Socio-Spatial Analysis." *Nature and Culture*, 4(1): 78–104.

Slow Food Foundation for Biodiversity. 2019. Available online: https://www.fondazioneslowfood.com/en/what-we-do/the-ark-of-taste/ (accessed October 17, 2019).

Smale, M. 1995. "'Maize is life': Malawi's Delayed Green Revolution." *World Development*, 23(5): 819–31.

Smale, M. 1997. "The Green Revolution and Wheat Genetic Diversity: Some Unfounded Assumptions." *World Development*, 25(8): 1257–69.

Smith, Andrew F. 2013a. "Avocado." In Andrew F. Smith (ed.), *The Oxford Encyclopedia of Food and Drink in America*, 2nd edn., vol. 1, 99–100. New York: Oxford University Press.

Smith, Andrew F. 2013b. "Orange Juice." In Andrew F. Smith (ed.), *The Oxford Encyclopedia of Food and Drink in America*, 2nd edn., vol. 1, 747–8. New York: Oxford University Press.

Smith, I. D. 2002. "Misusing canonical intertexts: Jamaica Kincaid, Wordsworth and colonialism's 'absent things'." *Callaloo* 25(3): 801–20.

Smith, Lionel S., Moth E. J. Broyles, Helen K. Larzleer, and Mark D. E. Fellowes. 2015. "Adding Ecological Value to the Urban Lawnscape. Insect Abundance and Diversity in Grass-free Lawns." *Biodiversity and Conservation*, 24(1): 47–62.

Smith, N. J. H., J. T. Williams, D. L. Plucknett, and J. P. Talbot. 1992. *Tropical Forests and their Crops*. Ithaca and London: Comstock Publishing Associates—a division of Cornell University Press.

Smith, P., and V. Pence. 2017. "The Role of Botanic Gardens in ex situ Conservation." In S. Blackmore and S. Oldfield (eds.), *Plant Conservation Science & Practice—The Role of Botanic Gardens*, ch. 5, 102–33. Cambridge: Cambridge University Press.

Smith, P. P. 2008. "*Ex situ* Conservation of Wild Species: Services Provided by Botanic Gardens." In N. Maxted, B. V. Ford-Lloyd, S. P. Kell, J. M. Iriondo, M. E. Dulloo, and J. Turok (eds.), *Crop Wild Relative Conservation and Use*, 407–12. Wallingford, OX: CABI International.

Smyth, Elaine B. 2010. "Elsie Margaret Stones and the Flora of Louisiana." *Botanical Artist*, 16 (3): 15–17.

Snodgrass, K. 1930. *Margarine as a Butter Substitute*. Fats and Oils Studies of the Food Research Institute, No. 4. Palo Alto, California: Stanford University, Food Research Institute.

Snow, Jon. 2006. "Why I Don't Wear a Poppy on Air." *Channel 4 News*, November 8. Available online: https://www.channel4.com/news/by/jon-snow/blogs/wear-poppy-air (accessed May 16, 2021).

Solness, P. 1999. *Tree Stories*. Sydney: Chapter & Verse.

Soluri, John. 2005. *Banana Cultures: Agriculture, Consumption, and Environmental Change in Honduras and the United States*. Austin: University of Texas Press.

Sommerer, Christa, Laurent Mignonneau, and Florian Weil. 2015. "The Art of Human to Plant Interaction." In Patrícia Vieira, Monica Gagliano and John Ryan (eds.),. *The Green Thread: Dialogues with the Vegetal World*, ch. 13, 229–50. Lanham, Maryland: Lexington.

Soyinka, Wole. 2016. "In Praise—and Dread—of Trees." Toni Morrison Lectures, Princeton University. Available online: https://www.youtube.com/watch?v=bGbckX33Wjg (accessed May 16, 2021).

Spencer-Jones, R. 2007. *1001 Gardens You Must See Before You Die*, 1st edn. London: Cassell Illustrated.

Staub, P. O., M. S. Geck, and C. S. Weckerle. 2011. "Incense and Ritual Plant Use in Southwest China: A Case Study among the Bai in Shaxi." *Journal of Ethnobiology and Ethnomedicine*, 7(1): 43. Available online: http://doi.org/10.1186/1746-4269-7-43 (accessed May 16, 2021).

Stearn, William T. 1990. *Flower Artists of Kew: Botanical Paintings by Contemporary Artists*. London: Herbert Press.

Steffen, W., J. Grinevald, P. J. Crutzen, and J. McNeill. 2011. "The Anthropocene: Conceptual and Historical Perspectives." *Philosophical Transactions of the Royal Society A*, 369: 842–67.

Steffenson, B. J., P. Olivera, J. K. Roy, Y. Jin, K. P. Smith, and G. J. Muehlbauer. 2007. "A Walk on the Wild Side: Mining Wild Wheat and Barley Collections for Rust Resistance Genes." *Crop and Pasture Science*, 58: 532–44.

Stehkämper, B. 2020. "United Nations at 75: A Rose for Peace." *Deutsche Welle*, June 25. Available online: https://p.dw.com/p/3eGAb (accessed May 16, 2021).

Sternberg, Adam. 2015. "Avocados Are Toast." *New York Media*, April 20. Published by the California Avocado Commission.

Stone, Daniel. 2018. *The Food Explorer: The True Adventures of the Globe-Trotting Botanist Who Transformed What America Eats*. New York: Dutton.

Stones, Margaret, and Lowell Urbatsch. 1991. *Flora of Louisiana*. Baton Rouge: Louisiana State University Press.

Stones, Margaret, and Winifred Curtis. 1967–78. *The Endemic Flora of Tasmania*. 6 vols. London: Ariel Press.

Strand, Oliver, and Joe DiStefano. 2010. "Chefs Look for Wild Ingredients Nobody Else Has." *The New York Times*, November 23.

Strong, Jeremy. 2011. *Educated Tastes Food, Drink, and Connoisseur Culture*. Lincoln: University of Nebraska Press.

Stuart, R. 2002. "Entheogenic Sects and Psychedelic Religions." *MAPS [Multidisciplinary Association for Psychedelic Studies] Bulletin*, 12(1): 17–24. Available online: https://maps.org/news-letters/v12n1/12117stu.pdf (accessed May 16, 2021).

Styles, Megan A. 2019. *Roses from Kenya: Labor, Environment, and the Global Trade in Cut Flowers*. Culture, Place, and Nature Series. Seattle: University of Washington Press.

Sukhatme, P. 1961. "The World's Hunger and Future Needs in Food Supplies." *Journal of the Royal Statistical Society. Series A (General)*: 463–525.

Sumitomo, K., H. Akutsu, S. Fukuyama et al. 2015. "Conifer-Derived Monoterpenes and Forest Walking." *Mass Spectrometry*. 4(1), A0042–A0042.

Sumner, J. 2019. *Plants Go to War: A Botanical History of World War II*. Jefferson, NC: McFarland & Co.

Sundin, Sarah. 2016. "Make It Do—Tire Rationing in World War II." Available online: http://www.sarahsundin.com/make-it-do-tire-rationing-in-world-war-ii/ (accessed May 16, 2021).

Swagler, R. 1994. "Evolution and Applications of the Term Consumerism: Theme and Variations." *Journal of Consumer Affairs*, 28(2): 347–60.

Swaminathan, M. S. 2009. "Obituary: Norman E. Borlaug (1914–2009) Plant Scientist who Transformed Global Food Production." *Nature*, 461: 894.

Swanson, Diane L. 2011. "'The Real World': Virginia Woolf and Ecofeminism." In Kristin Czarnecki and Carrie Rohman (eds.), *Virginia Woolf and the Natural World: Selected Papers from the Twentieth Annual International Conference on Virginia Woolf*, 24–34. Clemson, South Carolina: Clemson University Digital Press.

Sydney, Clare. 1986. *Flower Painting*. Oxford: Phaidon.

Synge, H., and H. Townsend. (eds.). 1979. *Survival or Extinction: The Practical Role of Botanic Gardens in the Conservation of Rare and Threatened Plants*. Royal Botanic Gardens, Kew: Bentham-Moxon Trust.

Taketa, S., S. Amano, Y. Tsujino, T. Sato, D, Nomura M. Kakeda et al. 2008. "Barley Grain with Adhering Hulls is Controlled by an ERF Family Transcription Factor Gene Regulating a Lipid Biosynthesis Pathway." *Proceedings of the National Academy of Sciences of the United States of America*, 105: 4062–7.

Talhelm, T., X. Zhang, S. Oishi, C. Shimin, D. Duan, X. Lan, and S. Kitayama. 2014. "Large-scale Psychological Differences within China Explained by Rice versus Wheat Agriculture." *Science*, 344(6184): 603–8.

Tankard, J. B. 2018. *Gardens of the Arts and Crafts Movement*. Timber Press, Portland.

Tapsell, L. C., I. Hemphill, L. Cobiac, D. R. Sullivan, M. Fenech, C. S. Patch, S. Roodenrys et al. 2006. "Health Benefits of Herbs and Spices: The Past, The Present, The Future." *Medical Journal of Australia*, 185(4): S1–S24.

Tardieu, F., B. Parent, C. F. Caldeira, and C. Welcker. 2014. "Genetic and Physiological Controls of Growth under Water Deficit." *Plant Physiology*, 164: 1628–35.

Tate Modern. 2015 "Who is Ibrahim El-Salahi?" Available online: https://www.tate.org.uk/art/artists/ibrahim-el-salahi-16988/who-is-ibrahim-el-salahi (accessed January 24, 2020).

Taubes, G. 2016. *The Case Against Sugar*. Portobello Books.

Tegg, L. 2014. "Past Projects—Grasslands." Available online: http://www.lindategg.com/grasslands.html (accessed May 16, 2021).

Tenga, Angela. 2016. "Seeds of Horror: Sacrifice and Supremacy in *Sir Gawain and the Green Knight*, *The Wicker Man*, and *Children of the Corn*." In Dawn Keetley and Angela Tenga (eds.), *Plant Horror: Approaches to the Monstrous Vegetal in Fiction and Film*, 54–72. London: Palgrave Macmillan.

Tester, M. A., and P. Langridge. 2010. "Breeding Technologies to Increase Crop Production in a Changing World." *Science*, 327: 818–22.

Thann, Myint. 2015. "The Tree Associated with the Birth of Buddha." Blog. Available online: http://sylvanstroll.blogspot.com/2015/03/the-tree-associated-with-birth-of-buddha.html (accessed May 16, 2021).

The Day of the Triffids. 1962. [Film] Dir. Steve Sekely, Prod. George Pitcher. Shepperton, Surrey: Security Pictures.

The Fountain. 2006. [Film] Dir. Darren Aronofsky, Prod. Eric Watson, Arnon Milchan, and Iaian Smith. Burbank, CA: Warner Bros.

The Independent. 2015. "Downing Street photoshopped a poppy onto this picture of David Cameron." Twitter @Independent, September 2. Available online: https://twitter.com/Independent/status/661238930994536448 (accessed May 16, 2021).

The Private Life of Plants. 1995. [TV program] Prod. Neil Nightingale and Keith Scholey. London: British Broadcasting Corporation.

The Ruins. 2008. [Film] Dir. Carter Smith, Prod. Chris Bender et al. Hollywood: Paramount.

The Secret Life of Plants. 1978. [Film] Dir. Walon Green, Prod. Michael Braun. Hollywood: Paramount.

The Thing. 1982. [Film] Dir. John Carpenter, Prod. David Foster and Lawrence Turman. Hollywood: Universal Pictures.

The Thing. 2011. [Film] Dir. Matthijs van Heijningen Jr., Prod. Marc Abraham and Eric Newman. Hollywood: Universal Pictures.

The Thing from Another World. 1951. [Film] Dir. Christian Nyby, Prod. Howard Hawks. Hollywood: RKO Radio Pictures.

The Wicker Man. 1973. [Film] Dir. Robin Hardy, Prod. Peter Snell. Beverly Hills: Anchor Bay Entertainment.

Theophrastus. 1976–90. *De Causis Plantarum*, 3 vols. Translated by Benedict Einarson and G. K. K. Link. Cambridge, MA: Harvard University Press.

Thiers, Barbara M., Melissa C. Tulig, and Kimberly A. Watson. 2016. "Digitization of the New York Botanical Garden Herbarium." *Brittonia*, 68(3): 324–33.

Thomas, C. 2006. "Poverty, Development and Hunger." In J. Baylis and S. Smith (eds.), *The Globalization of World Politics. An Introduction to International Relations*, 3rd edn, 559–81. Oxford: Oxford University Press.

Thomas, P., and A. Farjon. 2011. *Taxus wallichiana*. The IUCN Red List of Threatened Species. Available online: https://www.iucnredlist.org/species/46171879/9730085 (accessed May 16, 2021).

Thomas-Van Gundy, Melissa A., and Michael P. Strager. 2012. "European Settlement-era Vegetation of the Monongahela National Forest, West Virginia." General Technical Report NRS-GTR-101. Newtown Square, PA: US Department of Agriculture, Forest Service, Northern Research Station.

Thoms, U., and J. -P. Gaudilliere (eds). 2014. *The Development of Scientific Marketing in the 20th Century*. London: Pickering & Chatto.

Thuiller, W., S. Lavorel, M Araújo, M. Sykes, and I. Prentice. 2005. "Climate Change Threats to Plant Diversity in Europe." *PNAS*, 102(23): 8245–50.

Timmermann, C. 2001. "Rationalizing Folk Medicine in Interwar Germany: Faith, Business, and Science at Dr. Madaus & Co." *Social History of Medicine*, 14: 459–82.

Tolkien, J. R. R. 1937. *The Hobbit, or There and Back Again*. London: Unwin.

Tolkien, J. R. R. 1954a. *The Fellowship of the Ring*. London: Unwin.

Tolkien, J. R. R. 1954b. *The Two Towers*. London: Unwin.

Tolkien, J. R. R. 2013. *The Letters of J.R.R. Tolkien*. Edited by H. Carpenter. Boston: Houghton Mifflin Harcourt.

Tompkins, Peter, and Christopher Bird. 1973. *The Secret Life of Plants*. New York: Harper & Row.

Toutain, J. C. 1971. *Le consommation alimentaire en France de 1789 à 1964*. Genève: Librairie Droz.

Trimingham, J. S. 1952. *Islam in Ethiopia*. Oxford: Oxford University Press.

Tsing, Anna. 2015. *The Mushroom at the End of the World: On the Possibility of Life in Capitalist Ruins*. Princeton, NJ: Princeton University Press.

Tu, Hongtao. 2018. *Tu Hongtao*. Available online: https://tuhongtao.com/en/artwork (accessed May 16, 2021).

Tu, Youyou. 2015. "Nobel Lecture." December 7. Available online: https://www.nobelprize.org/nobel_prizes/medicine/laureates/2015/tu-lecture.html (accessed May 16, 2021).

Turenscape. 2014. "A Resilient Landscape: Yanweizhou Park in Jinhua City." Available online: https://www.turenscape.com/en/project/detail/4629.html (accessed May 16, 2021).

Twin Peaks. 1990–1. [TV program] Seasons 1 and 2, April 8, 1990–June 10, 1991. ABC.

Tyrrell, M. (ed.). 2018. "Soy Agriculture in the Amazon Basin" in "Global Forest Atlas." Yale School of Forestry and Environmental Science. Available online: https://globalforestatlas.yale.edu/amazon/land-use/soy (accessed June 9, 2020).

U.S. Bureau of Labor Statistics. 2006. *100 Years of U.S. Consumer Spending Data for the Nation, New York City, and Boston, Report 991*. Washington, DC: U.S. Department of Labor.

U.S. Census Bureau. 2015. "Millennials Outnumber Baby Boomers and Are Far More Diverse, Census Bureau Reports," June 25. Available online: https://www.census.gov/newsroom/press-releases/2015/cb15-113.html (accessed November 3, 2019).

Udall, Sharyn Rohlfsen. 2000. *Carr, O'Keeffe, Kahlo: Places of Their Own*. New Haven, CT: Yale University Press.

UIA. 2021. "Flora and Fauna International," in UIA Open Yearbook. Available online: https://uia.org/s/or/en/1100017992 (accessed May 19, 2021).

UN. 2015. "Sustainable Development Goals." Available online: https://sustainabledevelopment.un.org/sdgs (accessed May 16, 2021).

UN. 2019. "UN Sustainable Development Goals: 17 Goals to Transform our World." Available online: https://www.un.org/sustainabledevelopment/sustainable-development-goals/ (accessed December 4, 2019).

UN DESA/Population Division. 2018. "68% of the World Population Projected to Live in Urban Areas by 2050, says UN." Press release, May 16. New York: UN. Available online: https://www.un.org/development/desa/en/news/population/2018-revision-of-world-urbanization-prospects.html (accessed May 16, 2021).

UN General Assembly. 2015. "Transforming our World: The 2030 Agenda for Sustainable Development." Resolution A/RES/70/1, October 21. Available online: https://undocs.org/A/RES/70/1(accessed December 4, 2019).

Underwood, D. 1996. "Snake Charmer." *The Architectural Review*, 200(1197): 46–51.

UNEP. 2011. "Oil Palm Plantations: Threats and Opportunities for Tropical Ecosystems." UNEP Global Environmental Alert Service (GEAS).

UNEP. n.d. "Resource Efficiency as Key Issue in the New Urban Agenda. Advancing Sustainable Consumption and Production in Cities." Available online: http://web.unep.org/ietc/sites/unep.org.ietc/files/Key%20messages%20RE%20Habitat%20III_en.pdf (accessed May 16, 2021).

United Nations Office on Drugs and Crime (UNODC). 2016. *World Wildlife Crime Report: Trafficking in Protected Species*. United Nations, New York.

USDA-ERA. 2020. "Adoption of Genetically Engineered Crops in the United States 1996–2020." Available online: https://www.ers.usda.gov/data-products/adoption-of-genetically-engineered-crops-in-the-us/recent-trends-in-ge-adoption.aspx (accessed June 19, 2020).

Vaidhyanathan, Siva. 2017. *Intellectual Property: A Very Short Introduction*. New York: Oxford University Press.

Van Huylenbroeck, Johan (ed.). 2018. *Ornamental Crops*. Switzerland: Springer.

Van Rheede Tot Drakenstein, H. A. 2003. *Hortus indicus malabaricus* (English trans.). Thiruvananthapuram: University of Kerala.

VanderGoot, Jana. 2018. *Architecture and the Forest Aesthetic: A New Look at Design and Resilient Urbanism*. New York: Routledge.

Vandermeer, John, and Ivette Perfecto. 2005. *Breakfast of Biodiversity: The Political Ecology of Rain Forest Destruction,* 2nd edn. Oakland, CA: Food First Books.

Varghese, J. N., T. P. J. Garrett, P. M. Colman, L. Chen, P. B. Høj, and G. B. Fincher. 1994. "The Three-dimensional Structures of Two Plant β-glucan Endohydrolases with Distinct Substrate Specificities." *Proceedings of the National Academy of Sciences of the United States of America*, 91: 2785–9.

Vavilov, N. I. 1922. "The Law of Homologous Series in Variation." *Journal of Genetics*, 12: 1–45.

Vermeulen, Sonja. 2014. "Climate Change, Food Security and Small-scale Producers: Analysis of Findings of the Fifth Assessment Report (AR5) of the Intergovernmental Panel on Climate Change (IPCC)." Climate Change, Agriculture and Food Security (CCAFS) Info Brief. CGIAR Research Program on CCAFS. Copenhagen, Denmark. Available online: https://cgspace.cgiar.org/bitstream/handle/10568/35215/IPCC_info_note-3April14.pdf (accessed December 4, 2019).

Vernon, C. 1995. "Frank Lloyd Wright, Walter Burley Griffin, Jens Jensen and the Jugendstil Garden in America." *Die Gartenkunst*, 7(2): 232–46.

Vernon, C. 1997. "'A legitimate art distinctive of Australia and Australia alone': The Griffins' Contribution to the Formation of an Australian Landscape Design Ethos." *Landscape Review*, 3(1): 2–27.

Vernon, C. 2002. "Introduction to Wilhelm Miller (1915) The Prairie Spirit in Landscape Gardening." In Wilhelm Miller, *The Prairie Spirit in Landscape Gardening*. American Society of Landscape Architects Centennial Reprint. Amherst: University of Massachusetts Press.

Veuthey, S. and J. -F. Gerber. 2011. "Valuation Contests over the Commoditisation of the Moabi Tree in South-eastern Cameroon. Environmental Values." 20(2): 239–64.

Vieira, Patricia. 2019. "*Phytographia*: Literature as Plant Writing." In Monica Gagliano, John C. Ryan, and Patricia Vieira (eds.), *The Language of Plants: Science, Philosophy,* Literature, 215–33. Minneapolis: University of Minnesota Press.

Vincent, H., J. Wiersema, S. Kell, H. Fielder, S. L. Dobbie, N. Castaneda Alvarez, L. Guarino, R. J. Eastwood, B. Leon, and N. Maxtead. 2013. "A Prioritized Crop Wild Relative Inventory to Help Underpin Global Food Security." *Biological Conservation*, 167: 265–75.

Voeks, Robert A. 2018. *The Ethnobotany of Eden: Rethinking the Jungle Medicine Narrative.* Chicago: The University of Chicago Press.

Wahid, M. B., S. N. Akmar-Abdullah, and I. E. Henson. 2005. "Oil Palm—Achievements and Potential." *Plant Production Sciences*, 8: 288–97.

Wainwright, Joel, and Geoff Mann. 2018. *Climate Leviathan: A Political Theory of Our Planetary Future*. London: Verso.

Walkowiak, S., L. Gao, C. Monat, et al. 2020. "Multiple Wheat Genomes Reveal Global Variation in Modern Breeding." *Nature*, 588: 277–83. DOI: 10.1038/s41586-020-2961-x/.

Walter, K. S., and H. J. Gillett (eds.). 1998. *1997 IUCN Red List of Threatened Plants*. Gland, Switzerland and Cambridge: IUCN.

Walter, Kerry. 2019. Co-owner BG-BASE & former head Threatened Plants Unit, World Conservation Monitor Center, Cambridge, England. Pers. comm. Email, November 15.

Wampole, Christy. 2016. *Rootedness: The Ramifications of a Metaphor.* Chicago: University of Chicago Press.

Wandersee, J. H. and E. E. Schussler. 2001. "Toward a Theory of Plant Blindness." *Plant Science Bulletin*, 47(1): 2–9.

Wang, Catherine. 2016. "Soyinka Discusses the Literal and Metaphorical Role of Trees in Human History." *The Daily Princetonian*, April 18. Available online: http://www.dailyprincetonian.com/article/2016/04/soyinka-discusses-the-literal-and-metaphorical-role-of-trees-in-human-history (accessed May 16, 2021).

Ward, Marilyn, and Martyn Rix. 2006. "Stella Ross-Craig (1906–2006) and Curtis's Botanical Magazine." *Curtis's Botanical Magazine*, 23(3): 256–8.

Ward, W. 1952. "The Lotus Symbol: Its Meaning in Buddhist Art and Philosophy." *Journal of Aesthetics and Art Criticism*, 11(2): 135–46.

Warner, Melanie. 2006. "A Sweetener with a Bad Rap." *The New York Times*, July 2. Available online: https://www.nytimes.com/2006/07/02/business/yourmoney/02syrup.html (accessed April 8, 2020).

Wasson, G. R., A. Hofmann and C. A. P. Ruck. 1978. *The Road to Eleusis: Unveiling the Secret of the Mysteries.* New York: Harcourt Brace Jovanovich.

Waters, M. T., A. M. M. Tiley, E. M. Kramer, A. W. Meerow, J. A. Langdale, and R. W. Scotland. 2013. "The corona of the Daffodil Narcissus bulbocodium shares stamen-like identity and is distinct from the orthodox floral whorls." *The Plant Journal*, 74(4): 615–25.

Watts, M. T. 1957. "The Stylish House or Fashions as an Ecological Factor." In *Reading the Landscape: An Adventure in Ecology*. New York: Macmillan.

Watts, P. 1981. *The Gardens of Edna Walling*. Melbourne: National Trust of Australia (Victoria).

Way, Thaïsa. 2020. Personal communication.

Wearn, J. A. 2016. "Seeds of Change—Polemobotany in the Study of War and Culture." *Journal of War & Culture Studies*, 9(3): 271–84.

Weingarden, Lauren S. (ed.). 1990. *Louis H. Sullivan: A System of Architectural Ornament*. New York: Rizzoli.

Whelan, R. 2016. "'The poor, as well as the rich, need something more than meat and drink': The Vision and Work of the Kyrle Society." In E. Baigent and B. Cowell (eds.), *Octavia Hill, Social Activism and the Remaking of British Society*, 91–117. London: School of Advanced Study, University of London, Institute of Historical Research.

White Jr, L. 1967. "The Historical Roots of our Ecologic Crisis." *Science*, 155(3767): 1203–17.

White Lies. 2013. [Film] Dir. Dana Rotberg, Prod. John Barnett and Chris Hampson. Melbourne: Madman Entertainment.

Whitson T. D. (ed.). 1991. *Weeds of the West*. University of Wyoming, Laramie, Wyoming: Western Society of Weed Science.

Wiggins, S., and S. Keats. 2014. "Rural Wages in Asia." Overseas Development Institute, London.

Wilderness Society, The. 2019. "Our History." Available online: https://www.wilderness.org/about-us/our-team/our-history(accessed May 16, 2021).

Willett, W., J. Rockström, B. Loken, M. Springmann, T. Lang, S. Vermeulen, T. Garnett, D. Tilman, F. DeClerck, A. Wood, and M. Jonell. 2019. "Food in the Anthropocene: The EAT–Lancet Commission on Healthy Diets from Sustainable Food Systems." *The Lancet*, 393(10170): 447–92.

Williams, Jericho. 2016. "An Inscrutable Malice: The Silencing of Humanity in *The Ruins* and *The Happening*." In Dawn Keetley and Angela Tenga (eds.), *Plant Horror: Approaches to the Monstrous Vegetal in Fiction and Film*, 227–41. London: Palgrave Macmillan.

Williams, M. 2003. *Deforesting the Earth: From Prehistory to Global Crisis*. Chicago: University of Chicago Press.

Wilson, E. O. 1984. *Biophilia*. Cambridge: Harvard University Press.

Wimmer, W. 1994. "*Wir haben fast immer was Neues*": *Gesundheitswesen und Innovationen der Pharma-Industrie in Deutschland, 1880–1935*. Berlin: Duncker & Humblot.

Winson, Anthony. 2013. *The Industrial Diet: The Degradation of Food and the Struggle for Healthy Eating*. New York: New York University Press.

Wohlleben, Peter. 2015. *The Hidden Life of Trees: What They Feel, How They Communicate—Discoveries from a Secret World*. Vancouver: David Suzuki Institute, Greystone Books.

Woolf, Virginia. 1921. *Monday or Tuesday*. London: Hogarth.

World Bank. 2007. *World Development Report 2008: Agriculture for Development*. Washington, DC: World Bank.

World Bank. 2010. *World Development Report 2010: Development and Climate Change*. Washington, DC: World Bank.

World Commission on Environment and Development. 1987. *Our Common Future* (Brundtland Report). Oxford: Oxford University Press. Available online: https://archive.org/details/ourcommonfuture00worl (accessed May 16, 2021).

World Economic Forum. 2020. "A Platform for the Trillion Tree Community." Available online: https://www.1t.org/ (accessed March 11, 2020).

WFO 2020. "World Flora Online." Available online: http://www.worldfloraonline.org (accessed January 24, 2020).

World Intellectual Property Organization (WIPO). 2016. "Traditional Knowledge Background Brief No. 10—Intellectual Property and Genetic Resources." WIPO. Available online: https://www.wipo.int/publications/en/series/index.jsp?id=144 (accessed May 16, 2021).

Wright, S. J. 2010. "The Future of Tropical Forests." *Annals of the New York Academy of Sciences*, 1195: 1–27.

Wright, W. P. 1913. *The New Gardening: A Guide to the Most Recent Developments in the Culture of Flowers, Fruits, and Vegetables*. New York: Doubleday, Page & Co.

Wyse Jackson, P. S., and K. Kennedy. 2009. "The Global Strategy for Plant Conservation: A Challenge and Opportunity for the International Community." *Trends in Plant Science*, 14: 578–80.

Wyse Jackson, P. S., and L. Sutherland. 2000. *International Agenda for Botanic Gardens in Conservation*. Botanic Gardens Conservation International, London, UK.

Xia, Y., X. Deng, P. Zhou, K. Shima, and J. A. Teixeira Da Silva. 2006. "The World Floriculture Industry: Dynamics of Production and Markets." In Jaime A. Teixeira Da Silva (ed.), *Floriculture, Ornamental and Plant Biotechnology*, vol. IV, ch. 35, 336–47. London: Global Science Books.

Yamagishi, N., R. Kishigami, and N. Yoshikawa. 2014. "Reduced Generation Time of Apple Seedlings to Within a Year by Means of a Plant Virus Vector: A New Plant-Breeding Technique with no Transmission of Genetic Modification to the Next Generation." *Plant Biotech Journal*, 12: 60–8.

Yan, W. 2017. "A Makeover for the World's Most Hated Crop." *Nature*, 543: 306–8.

Yates, P. 2009. "The Bush Foods Industry and Poverty Alleviation in Central Australia." *Dialogue (Academy of the Social Sciences in Australia)*, 28(2): 47–56.

Yokohama Nursery Co. Ltd. 2019. "The Company History." Available online: https://www.yokohamaueki.co.jp/en/company/history.html (accessed April 14, 2020).

Zakharov, I. A. 2005. "Nikolai I Vavilov (1887–1943)." *Journal of Biosciences*, 30: 299–301.

Zhan, G., and Z. Jin. 2015. "Hani Rice Terraces of Honghe—The Harmonious Landscape of Nature and Humans." *Landscape Research*, 40(6, SI): 655–67.

Zhang, Rui, Penga Fangren, and Li Yongrong. 2015. "Pecan Production in China." *Scientia Horticulturae*, 197: 719–27.

Zhao, L., C. Ma, C. Leipe, T. Long, K. Liu, H. Lu, L. Tang, Y. Zhang, M. Wagner, and P.E. Tarasov. 2017. "Holocene Vegetation Dynamics in Response to Climate Change and Human Activities Derived from Pollen and Charcoal Records from Southeastern China." *Palaeogeography, Palaeoclimatology, Palaeoecology*, 485: 644–60.

Zheng, Y., J. Liu, X. Feng, and X. Gong. 2017. "The Distribution, Diversity, and Conservation Status of Cycas in China." *Ecology and Evolution*, 7: 3212–24.

Zhu, Z., S. Piao, R. B. Myneni, M. Huang, Z. Zeng et al. 2016. "Greening of the Earth and its Drivers." *Nature Climate Change*, 6: 791–5. DOI: 10.1038/NCLIMATE3004.

CONTRIBUTORS

Geoff Bil is a historian of science and Visiting Assistant Professor in the Department of History at the University of Delaware, USA, specializing in nineteenth- and twentieth-century botany, anthropology, empire, Indigenous history, and environmental history in Southeast Asia and the South Pacific.

Frederica Bowcutt is a faculty member at the Evergreen State College in Washington state, USA, where she also serves as Director of the Evergreen Natural History Museum. Her teaching interests and publications are in the fields of floristics, historical ecology, and plant-centric history.

Derek Byerlee AO is a distinguished agricultural economist Member, Board of Trustees, International Food Policy Research Institute, since 2018 and Adjunct Professor, Global Human Development, Georgetown University in Washington, DC, USA.

Kate Cullity is a founding director of TCL (Taylor Cullity Lethlean), an internationally recognized and awarded landscape architect, and environmental artist, with particular skills in the design of public and private gardens, and the integration of public art with landscape and urban design.

Geoff Fincher AO is Emeritus Professor at the University of Adelaide's School of Agriculture, Food and Wine, Australia, renowned for his work in cereal chemistry and the applications of this work in human and animal nutrition, brewing, and in biofuels and biopolymers.

Stephen Forbes is an independent botanist, thinker, and writer who has held leadership roles in Australian botanic gardens in Melbourne, Perth, Sydney, and Adelaide and published widely on botany, gardens, and the relationship between plants and people.

Jean-Paul Gaudillière is senior researcher at the French Institut National de la Santé et de la Recherche Médicale (Inserm) and Professor at École des Hautes Études en Sciences Sociales, Paris (EHESS), France. His research explores the history of the life sciences and medicine during the twentieth century.

Peter Langridge is Emeritus Professor at the University of Adelaide's School of Agriculture, Food and Wine, Australia, and contributes globally to deploying genetic technologies and new strategies to improve the breeding of cereal crops. He is Chair of the Wheat Initiative's Scientific Board.

Jen Lynch is a senior lecturer (industry fellow) and PhD candidate in the landscape architecture program at RMIT University in Melbourne, Australia, and an AILA-registered landscape architect. Her research explores landscape-driven forms of housing development.

Mike Maunder is the Executive Director of the Cambridge Conservation Initiative, based in the University of Cambridge Judge Business School, UK, and has worked in leadership roles in botanical gardens and conservation programs in the US, UK, East Africa, and UAE.

Sara Oldfield OBE is Chair of the IUCN/SSC Global Tree Specialist Group and served as Secretary General of Botanic Gardens Conservation International from 2005–15. She has published numerous books and articles on biodiversity conservation and the role of botanic gardens.

Mark Tredinnick OAM is a celebrated poet, essayist, and teacher based in Sydney, Australia. Mark's work has been widely recognized and has won two of the world's foremost poetry prizes, the Montreal and the Cardiff.

INDEX